Materials in Space—Science, Technology and Exploration

MATERIALS RESEARCH SOCIETY
SYMPOSIUM PROCEEDINGS VOLUME 551

Materials in Space— Science, Technology and Exploration

Symposium held November 29-December 2, 1998, Boston, Massachusetts, U.S.A.

EDITORS:

Aloysius F. Hepp
NASA Glenn Research Center
Cleveland, Ohio, U.S.A.

Joseph M. Prahl
Case Western Reserve University
Cleveland, Ohio, U.S.A.

Theo G. Keith
Ohio Aerospace Institute
Brook Park, Ohio, U.S.A.

Sheila G. Bailey
NASA Glenn Research Center
Cleveland, Ohio, U.S.A.

J. Robert Fowler
Federal Data Corporation
Brook Park, Ohio, U.S.A.

Materials Research Society
Warrendale, Pennsylvania

CAMBRIDGE
UNIVERSITY PRESS

University Printing House, Cambridge CB2 8BS, United Kingdom

One Liberty Plaza, 20th Floor, New York, NY 10006, USA

477 Williamstown Road, Port Melbourne, VIC 3207, Australia

314-321, 3rd Floor, Plot 3, Splendor Forum, Jasola District Centre, New Delhi - 110025, India

79 Anson Road, #06-04/06, Singapore 079906

Cambridge University Press is part of the University of Cambridge.

It furthers the University's mission by disseminating knowledge in the pursuit of education, learning and research at the highest international levels of excellence.

www.cambridge.org
Information on this title: www.cambridge.org/9781558994577

Materials Research Society
506 Keystone Drive, Warrendale, PA 15086
http://www.mrs.org

© Materials Research Society 1999

First published 1999
First paperback edition 2013

Single article reprints from this publication are available through University Microfilms Inc., 300 North Zeeb Road, Ann Arbor, MI 48106

CODEN: MRSPDH

A catalogue record for this publication is available from the British Library

ISBN 978-1-558-99457-7 Hardback
ISBN 978-1-107-41387-0 Paperback

CONTENTS

*Invited Paper

*Invited Paper

*Invited Paper

PART VII: <u>MATERIALS FOR SPACE AND OTHER HOSTILE ENVIRONMENTS</u>

PREFACE

This volume contains 36 papers from the symposium entitled "Materials in Space—Science, Technology and Exploration," held over three days at the 1998 MRS Fall Meeting in Boston, Massachusetts. This symposium was held to commemorate the 40th anniversary of the National Aeronautics and Space Administration; it was also the 25th anniversary of the Materials Research Society. This volume is organized into seven topical areas that follow the organization of the symposium when feasible, but deviate to include contributions from a number of excellent poster presentations (the poster presented by Dr. S. Kishimoto entitled "Development of Metallic Closed Cellular Materials Containing Polymers" captured the spirit of the symposium and a poster award) and to reflect the main themes of the symposium. The symposium began with an excellent plenary session that explored important issues of materials in space. Papers from this session included international efforts related to microgravity materials science and materials for protection of astronauts from space radiation, and are the lead articles in Parts I, V, and VII. Monday afternoon focused on Mars Pathfinder mission results, and materials and technologies for space exploration; these papers round out Part I on space exploration.

Part II focuses on space photovoltaics and presents several contributions from a session on space photovoltaic materials technology, as well as closely related poster presentations. Part III on materials for energy conversion and storage represents contributed and invited presentations from three different symposium sessions. Fundamental studies occupied Tuesday morning, including a session on quantum effects on materials and devices and fundamental studies of microgravity materials science. These two sessions and related poster presentations are included in Parts IV and V, respectively. Wednesday morning was an all-invited session that focused on results of space shuttle microgravity materials science experiments, and included excellent presentations from across the nation; Part VI includes five of the six papers presented in that session. The final section of the proceedings examined materials for space and other hostile environments and represents a cross-section of symposium sessions. The symposium ended with two special events that are not captured in the proceedings. The first event was a panel discussion with many of the participants from the plenary session on Monday morning. The discussion focused on the user microgravity community and the NASA/International Microgravity Materials Science interface. The second event was a keynote session featuring Payload Specialist Albert Sacco (STS-73, USML-2). The focus of the keynote address, involving lively feedback from the audience, was the relationship between payload and mission specialists and performing and analyzing results of materials research in space. In conclusion, the topic of this symposium, like the format of the final sessions, was very positively received, and should be revisited at future meetings.

Aloysius F. Hepp
Joseph M. Prahl
Theo G. Keith
Sheila G. Bailey
J. Robert Fowler

October 1999

ACKNOWLEDGMENTS

The success of this symposium could have not been achieved without the support and contributions of the sponsors (National Aeronautics and Space Administration, Elsevier Science Ltd.) and key personnel in the sponsoring organizations. In addition, the symposium organizers would like to thank the session chairs, invited speakers, and participants for preparing and reviewing the manuscripts. The organizers would also like to express appreciation to MRS officials and staffers for their support and assistance in handling all of the details before, during, and after the Meeting, and helping to produce this proceedings volume. Finally, we would like to salute Col. John H. Glenn, astronaut, retired U.S. Marine pilot, businessman, and senator, on his return to space. Those of us who work at NASA Glenn (formerly Lewis) Research Center must always do our best to live up to the high standards set by this son of Ohio who is truly an American legend.

Invited Speakers, Panel Members, Sponsors, and Session Chairs

J. Barry Andrews, University of Alabama, Birmingham
Joseph H. Armstrong, Global Solar Energy LLC
Bruce A. Banks, NASA Glenn Research Center
Arnon Chait, NASA Glenn Research Center
Donald L. Chubb, NASA Glenn Research Center
Navid S. Fatemi, Essential Research, Inc.
Dale Ferguson, NASA Glenn Research Center
Dennis J. Flood, NASA Glenn Research Center
Donald O. Frazier, NASA Marshall Space Flight Center
M.E. Glicksman, Rensselaer Polytechnic Institute
Brian S. Good, NASA Glenn Research Center
Donald Henderson, Fisk University
Rodney Herring, Canadian Space Agency
Richard W. Hoffman, Ohio Aerospace Institute
William L. Johnson, California Institute of Technology
David Kaplan, NASA Johnson Space Center
Fred J. Kohl, NASA Glenn Research Center
Clifford P. Kubiak, University of California, San Diego
Prashant Kumta, Carnegie Mellon University
Geoffrey A. Landis, Ohio Aerospace Institute
Alex Lehoczky, NASA Marshall Space Flight Center
Douglas M. Matson, Massachusetts Institute of Technology
Philip Mestecky, Elsevier Science Ltd.
Steve Ringel, Ohio State University
Albert Sacco, Jr., Northeastern University
P.W. Vorhees, Northwestern University
Michael J. Wargo, NASA Headquarters
John W. Wilson, NASA Langley Research Center

MATERIALS RESEARCH SOCIETY SYMPOSIUM PROCEEDINGS

MATERIALS RESEARCH SOCIETY SYMPOSIUM PROCEEDINGS

Prior Materials Research Society Symposium Proceedings available by contacting Materials Research Society

Part I

Space Exploration

MATERIALS FOR SHIELDING ASTRONAUTS FROM THE HAZARDS OF SPACE RADIATIONS

J. W. Wilson*, F. A. Cucinotta**, J. Miller***, J. L. Shinn*, S. A. Thibeault*, R. C. Singleterry*, L. C. Simonsen*, and M. H. Kim****
*NASA Langley Research Center, Hampton, VA 23682, john.w.wilson@larc.nasa.gov
**NASA Johnson Space Center, Houston, TX 77058
***DOE Lawrence Berkeley National Laboratory, Berkeley, CA 94720
****NRC/NAS Fellow NASA Langley Research Center, Hampton, VA 23682

ABSTRACT

One major obstacle to human space exploration is the possible limitations imposed by the adverse effects of long-term exposure to the space environment. Even before human spaceflight began, the potentially brief exposure of astronauts to the very intense random solar energetic particle (SEP) events was of great concern. A new challenge appears in deep space exploration from exposure to the low-intensity heavy-ion flux of the galactic cosmic rays (GCR) since the missions are of long duration and the accumulated exposures can be high. Because cancer induction rates increase behind low to rather large thickness of aluminum shielding according to available biological data on mammalian exposures to GCR like ions, the shield requirements for a Mars mission are prohibitively expensive in terms of mission launch costs. Preliminary studies indicate that materials with high hydrogen content and low atomic number constituents are most efficient in protecting the astronauts. This occurs for two reasons: the hydrogen is efficient in breaking up the heavy GCR ions into smaller less damaging fragments and the light constituents produce few secondary radiations (especially few biologically damaging neutrons). An overview of the materials related issues and their impact on human space exploration will be given.

INTRODUCTION

The ionizing radiations in space affecting human operations are of three distinct sources and consist of every known particle including energetic ions formed from stripping the electrons from all of the natural elements. The radiations are described by field functions for each particle type over some spatial domain as a function of time. The three sources of radiations are associated with different origins identified as those of galactic origin (galactic cosmic rays, GCR), particles produced by the acceleration of solar plasma by strong electromotive forces in the solar surface and acceleration across the transition shock boundary of propagating coronal mass ejecta (solar energetic particles, SEP), and particles trapped within the confines of the geomagnetic field. The GCR constitutes a low level background which is time invariant outside the solar system but is modulated over the solar cycle according to changes in the interplanetary plasma which excludes the lower energy galactic ions from the region within several AU of the sun [1]. The SEP are associated with some solar flares which produce intense burst of high energy plasma propagating into the solar system along the confines of the sectored interplanetary magnetic field [2] producing a transition region in which the SEP are accelerated. SEP have always been a primary concern for operations outside the Earth's protective magnetic field and could deliver potentially lethal exposures over the course of several hours [3]. The trapped radiations consist mainly of protons and electrons within two bands centered on the geomagnetic equator reaching maximum intensity at an altitude of 3,600 km followed by a minimum at 7,000 km and a second very broad maximum at 10,000 km [4]. The trapped radiations have limited human operations to altitudes below several hundred kilometers and potentially lethal exposures are obtained over tens of hours in the most intense regions. Low inclination orbits are shielded from extraterrestrial radiations by the geomagnetic field and are mainly exposed to the trapped environment. Inclinations above 45° are sufficiently near the geomagnetic poles for which GCR

3

and SEP exposures can be significant. Indeed, about half of the expected exposures of the International Space Station (ISS) in its inclined orbit of 51.6° will be from GCR [5].

In the usual context, shielding implies an alteration of the radiations through interactions with intervening materials by which the intensity is decreased. This understanding is to some degree correct in the case of the relatively low energy particles of the SEP and the trapped radiations wherein the energy deposited in astronaut tissues can be easily reduced by adding shield material. As one would expect, some materials are more effective than others as the physics of the interactions differ for various materials. The high energies associated with the GCR are distinct in that the energy absorbed in astronaut tissues is at best unchanged by typical spacecraft shielding configurations and use of some materials in spacecraft construction will even increase the energy absorption by the astronaut. For GCR, one must abandon the concept of "absorbing" the radiation by use of shielding. The protection of the astronaut in this case is not directly related to energy absorption within their body tissues but rather depends on the mechanism by which each particle type transmitted through the shield results in biological injury. Even though the energy absorption by the astronaut can be little affected, the mixture of particle types is strongly affected by the choice of the intervening shield material. Knowledge of the specific biological action of the specific mixture of particles behind a given shield material and the modification of that mixture by choice of shield materials is then a critical issue in protecting the astronaut in future human exploration and has important implications on the design and operation of ISS.

Understanding the biological effects of GCR behind intervening material is then key to protection in future NASA activity in either ISS or deep space. As yet no standards on protection against GCR exposures have been promulgated since insufficient information exists on biological effects of such radiations [6,7]. The most important biological effect from GCR exposure of which we are currently aware is cancer induction which relates to mutation and transformation (a specific mutation) events in astronaut tissues. Our knowledge of radiation carcinogenesis in humans is for gamma ray exposures for which excess career risk is proportional to tissue dose (energy absorbed per unit mass) accumulated at low dose rates. Although insufficient data exists to estimate astronaut cancer risks from the GCR high charge and energy (HZE) ions, there exist relatively detailed data on the biological response of several systems including survival, neoplastic transformation, and mutation in mammalian cells and Harderian gland tumor induction in mice. Other biological effects may come to light as exposure of living systems to high energy heavy ion beams continues to be studied. We will discuss the available response models in light of the design criteria used for ISS and the implications for materials research. For further discussion of these issues see "Shielding Strategies for Human Space Exploration" [8]. In the present paper, we review the GCR environment and discuss the issues of shield design in the context of developing a strategy for reducing the health risks of astronauts in future missions. In particular we will examine the role of materials research and development in controlling astronaut health risks from exposure to ionizing radiation in space.

GCR AND BIOLOGICAL RESPONSES

The galactic cosmic rays consist mainly of nuclei (ions) of the elements of hydrogen thru nickel. The energy spectra are broad and extend from tens to millions of MeV (figure 1). The most important energies for protection lies near maximum intensities from a few hundred to several thousand MeV/nucleon (a nucleon is the name given to neutrons and protons of which the ions are composed). The salient feature of these radiations is that a significant number of these particles have high charge which affects the means by which energy is transferred to tissues. Their ion tracks seen in nuclear emulsion are shown in figure 2. The optical density (related to energy deposited) of the track increases as the ion charge squared and the intensity and the lateral extent of the track depend on the ion velocity. The tracks in the figure are for approximately 400 MeV/nucleon ions. Considering that the mammalian cell nucleus size is

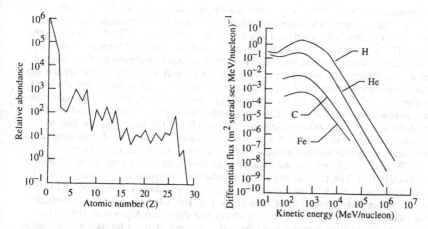

Figure 1. Relative abundances from Mewaldt [9] and selected energy spectra from Simpson [10] for galactic cosmic ray nuclei.

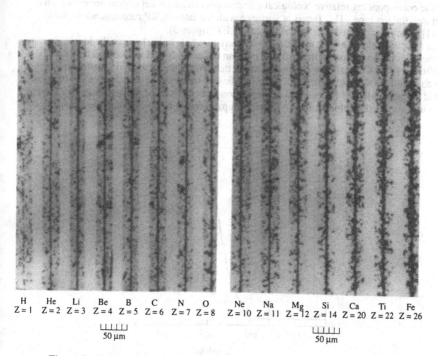

H	He	Li	Be	B	C	N	O	Ne	Na	Mg	Si	Ca	Ti	Fe
Z = 1	Z = 2	Z = 3	Z = 4	Z = 5	Z = 6	Z = 7	Z = 8	Z = 10	Z = 11	Z = 12	Z = 14	Z = 20	Z = 22	Z = 26

50 μm   50 μm

Figure 2. Cosmic-ray ion tracks in nuclear emulsion. (McDonald 1964).

5

several micrometers, it is clear that the passage of a single iron ion through the cell nucleus is a potentially devastating event. The protection standards applied to ISS are those recommended for Space Station Freedom scheduled for a low inclination orbit in which GCR exposures were minimal [6]. These standards were adapted, in part, from those used in the nuclear industry for mainly low energy radiations where ion tracks have very limited lateral extents and biological responses are characterized by mainly the energy lost per ion path length (linear energy transfer, LET). The enhanced effectiveness of high LET radiation to cause cancer for a given absorbed energy is given by an LET dependent quality factor [11] as shown in figure 3. The excess cancer risk is then assumed proportional to the dose equivalent which is the product of quality factor and dose. Although little data exists on human exposures from HZE radiations, the limited studies in mice and mammalian cell cultures allow evaluation of the effects of track structure on shield attenuation properties and evaluation of the implications for dosimetry. The most complete HZE exposure data sets for mammalian cells have been modeled including the mouse embryo cells C3H10T1/2 for the survival and neoplastic transformation data of Yang et al.[12, 13], the hybrid hamster cells V79 for the survival and mutation data of various groups [14], and the mouse Harderian gland tumor data of Alpen et al. [15, 16]. Model results for the Harderian gland tumor data are shown in figure 4 in comparison with data from Alpen et al. [16]. The Harderian target cell initiation cross section (the initiating event in tumor formation is thought to be neoplastic transformation) is shown in figure 5 and compares closely with the transformation cross section found for the C3H10T1/2 cell transformation data of Yang et al. [13]. The most notable feature of the cross sections in figure 5 are the multiple values for a given LET which implies the corresponding relative biological effectiveness (RBE) is dependent not only on the LET but also the ion type. This fact is at variance with the latest ICRP recommended quality factor [11] which is a defined function of only the LET (figure 3).

Track structure related events are difficult to study in whole animals since the local environment within an animal varies across the organ under study and is modified by the surrounding tissues. Cell cultures can be used to better control the local environment and provide an improved system for track structure studies. Among the best studied cell systems is the hybrid hamster cell V79 for survival and mutation end points. The model of the V79 system is shown in comparison with data from various groups in figure 6. As we shall see, these track

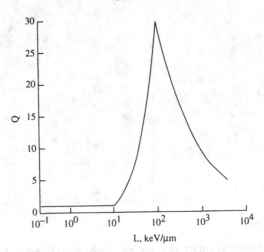

Figure 3. ICRP-60 Recommended Quality Factor.

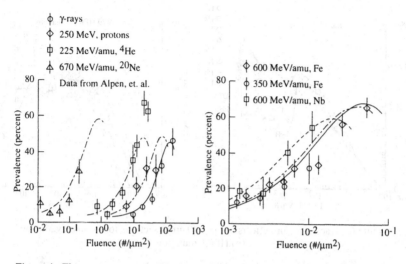

Figure 4. Fluence response for Harderian gland tumors for several radiation types.

structure related features have important implications for attenuation of biological effects within spacecraft shielding materials.

SHIELDING METHODOLOGY

The specification of the interior environment within a spacecraft and evaluation of the effects on the astronaut are at the heart of the radiation protection problem. The Langley Research

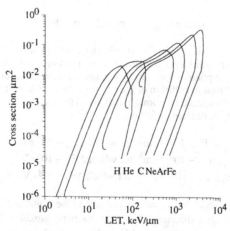

Figure 5. Harderian gland cell initiation cross section obtained from fits to the Alpen *et al.* data.

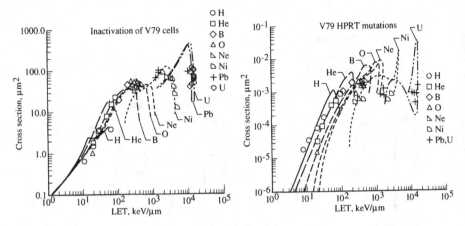

Figure 6. Track structure effects in the V79 cross sections for (a) inactivation and
(b) HPRT mutation.

Center has been developing such techniques. The relevant transport equations are the linear
Boltzmann equations for the flux density ϕ_j (x,Ω,E) of type j particles as

$$\Omega \cdot \nabla \phi_j \ (x,\Omega,E) = \sum \int \sigma_{jk}(\Omega,\Omega',E,E') \ \phi_k(x,\Omega',E') \ d\Omega' \ dE' - \sigma_j(E) \ \phi_j(x,\Omega,E) \qquad (1)$$

where $\sigma_j(E)$ and $\sigma_{jk}(\Omega,\Omega',E,E')$ are the media macroscopic cross sections. The $\sigma_{jk}(\Omega,\Omega',E,E')$
represent all those processes by which type k particles moving in direction Ω' with energy E'
produce a type j particle in direction Ω with energy E. Note that there may be several reactions
which produce a particular product, and the appropriate cross sections for equation (1) are the
inclusive ones. The total cross section $\sigma_j(E)$ with the medium for each particle type of energy E
may be expanded as

$$\sigma_j(E) = \sigma_{j,at} \ (E) + \sigma_{j,el} \ (E) + \sigma_{j,r}(E) \qquad (2)$$

where the first term refers to collision with atomic electrons, the second term is for elastic
nuclear scattering, and the third term describes nuclear reactions. Any realistic calculation must
include not only the description of the primary ion and fragment fields but also the resultant
secondary radiations formed in collision of the nuclei of the shield and target material as well as
the secondary electrons produced in atomic collisions. The microscopic cross sections and
average energy transfer are ordered as follows:

$$\sigma_{j,at} \ (E) \sim 10^{-16} \ cm^2 \ \text{for which} \ \Delta E_{at} \sim 10^2 \ eV \qquad (3)$$

$$\sigma_{j,el} \ (E) \sim 10^{-19} \ cm^2 \ \text{for which} \ \Delta E_{el} \sim 10^6 \ eV \qquad (4)$$

$$\sigma_{j,r} \ (E) \sim 10^{-24} \ cm^2 \ \text{for which} \ \Delta E_r \sim 10^8 \ eV \qquad (5)$$

This ordering allows flexibility in expanding solutions to the Boltzmann equation as a sequence
of physical perturbative approximations. It is clear that many atomic collisions ($\sim 10^6$) occur in
a centimeter of ordinary matter, whereas $\sim 10^3$ nuclear coulomb elastic collisions occur per
centimeter. In contrast, nuclear reactions are separated by a fraction to many centimeters
depending on energy and particle type. Special problems arise in the perturbation approach for

neutrons for which $\sigma_{j,at}$ (E) ~ 0, and the nuclear elastic process appears as the first-order perturbation.

As noted in the development of equation (1), the cross sections appearing in the Boltzmann equation are the inclusive ones so that the time-independent fields contain no spatial (or time) correlations. However, space- and time-correlated events are functions of the fields themselves and may be evaluated once the fields are known. Such correlations are important to the biological injury of living tissues. For example, the correlated release of target fragments in biological systems due to ion or neutron collisions have high probabilities of cell injury with low probability of repair resulting in potentially large relative biological effectiveness (RBE) and associated quality factor. Similarly, the passage of a single ion releases an abundance of low energy electrons from the media resulting in intense fields of correlated electrons near the ion path.

The solution of equation (1) involves hundreds of multi-dimensional integro-differential equations which are coupled together by thousands of energy dependent cross terms and must be solved self-consistently subject to boundary conditions ultimately related to the external space environment and the geometry of the astronaut's body and/or a complex vehicle. In order to implement a solution one must have available the atomic and nuclear cross section data. The development of an atomic/nuclear database is a major task in code development.

Transport Coefficients

The transport coefficients relate to the atomic/molecular and nuclear processes by which the particle fields are modified by the presence of a material medium. As such, basic atomic and nuclear theories provide the input to the transport code data base [17, 18] and requires laboratory validation [19]. It is through the nuclear processes that the particle fields of different radiation types are transformed from one type to another. The atomic/molecular interactions are the principal means by which the physical insult is delivered to biological systems in producing the chemical precursors to biological change within the cells. The temporal and spatial distributions of such precursors within the cell system governs the rates of diffusive and reactive processes leading to the ultimate biological effects. The transport coefficients and their evaluation are described elsewhere [17,18,19].

Transport Solution Methods

The solution to equation (1) can be written in operational form as $\phi = G \phi_B$ where ϕ_B is the inbound flux at the boundary, and G is the Green's function which reduces to a unit operator on the boundary. A guiding principle in radiation-protection practice is that if errors are committed in risk estimates, they should be overestimates (conservative). The presence of scattering terms in equation (1) provides lateral diffusion along a given ray. Such diffusive processes result in leakage near boundaries. If ϕ_Γ is the solution of the Boltzmann equation for a source of particles on the boundary surface Γ, then the solution for the same source on Γ within a region enclosed by Γ_o denoted by $\phi_{\Gamma_o}(\Gamma)$ has the property

$$\phi_{\Gamma_o}(\Gamma) = \phi_\Gamma + \varepsilon_\Gamma \tag{6}$$

where ε_Γ is positive provided Γ_o completely encloses Γ. The most strongly scattered component is the neutron field for which an 0.2 percent error results for semi-infinite media in most practical problems [20,21]. Standard practice in space radiation protection replaces Γ as required at some point on the boundary and along a given ray by the corresponding Γ_N evaluated for normal incidence on a semi-infinite slab. The errors in this approximation are second order in the ratio of beam divergence and radius of curvature of the object [20], rarely exceeds a few percent for space radiations, and are always conservative. The replacement of Γ by Γ_N as a highly accurate approximation [21] for space shielding applications has the added advantages that Γ_N is the natural quantity for comparison with laboratory simulations and has the following properties: If Γ_N is known at a plane a distance x_0 from the boundary (assumed at the origin), then the value of Γ_N at any plane $x \geq x_0$ is

$$G_N(x) = G_N (x - x_0) \, G_N(x_0) \tag{7}$$

Setting $x = x_0 + h$, where h is small and of fixed-step size gives rise to the marching procedures of the HZETRN code [22] used in the present analysis.

IMPACT ON SHIELDING

As noted in previous sections, the GCR are of high energy and the energy absorbed in tissues behind typical shields used in space are nearly independent of shield thickness. This is not to imply that the environment has not been changed by the material, indeed the number of particles is normally increased and the mixture of types changed dramatically. Thus, the composition of the transmitted radiations can be greatly altered by choice of materials and amount but with little affect on energy absorption rates. Evaluation of the effects of shielding requires solution of the transport equation (1) describing the alterations to the GCR resulting from individual atomic and nuclear reactions. We have used the biological response models and the LET dependent quality factor to investigate the attenuation of biologically damaging radiation within shield materials in the space environment. In terms of dose equivalent for a shield of thickness x, H(x), we find that aluminum structures attenuate radiation effects over most of the range of depths used in human rated vehicles (2-10 g/cm^2) as shown in figure 7. Thus, dose equivalent reduction may be a misleading indicator of astronaut satfety [23]. In contrast, track structure models show markedly different attenuation characteristics and, in fact, show that a transformed cell is more likely to result by increasing the aluminum shielding in spite of the decreasing dose equivalent as seen in figure 7.

As a further example of the issues we face, the dose equivalent behind three shield materials is shown in Table I. The first shield is aluminum which is typical of many constructions including ISS. The 1.5148 g/cm^2 thickness is that of the JSC TransHab wall design for a combination of polymers and fillers. The value in parenthesis is the performance advantage of

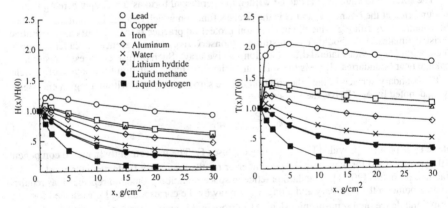

Figure 7. Attenuation of dose equivalent and cell transformation for a
one-year GCR exposure at solar minimum behind various shield materials.

Table I. Annual GCR exposure behind various shields at 1997 solar minimum. Numbers in parenthesis are shield performances relative to aluminum values at the same thickness.

Shield	Dose Equivalent, cSv*	Excess Harderian Gland Tumor Risk, percent*
1.5148 g/cm^2		
Aluminum	130.9(1.00)	3.57(1.00)
TransHab	121.6(1.08)	3.07(1.16)
Polyethylene	113.1(1.16)	2.64(1.35)
5 g/cm^2		
Aluminum	113.9(1.00)	3.37(1.00)
TransHab	99.4(1.15)	2.74(1.24)
Polyethylene	86.4(1.32)	2.20(1.54)

*Unshielded, dose equivalent is 120 cSv and excess tumor risk is 2.23 percent

the given material compared to aluminum. For example, polyethylene is 16 percent more effective than aluminum in controlling dose equivalent with only 1.5148 g/cm^2 of material. Much larger gains are achieved at 5 g/cm^2 thickness. The 5 g/cm^2 thickness is typical for an area within a human-occupied vehicle loaded with equipment. Modest reductions in dose equivalent are found for all three materials at the 5 g/cm^2 thickness. A mouse carried on the same mission of one year duration will have an excess risk of Harderian gland tumors shown in the last column in Table I. A substantial increase in Harderian tumor risk for shield thickness x, HG(x), is found for both thicknesses of aluminum and great amounts of aluminum are required to reduce the risk. Polyethylene shows substantial improvements in performance for controlling Harderian gland tumor risk at both thicknesses. These findings have important implications for deep space exploration but also for ISS which receives half of its exposure from GCR ions.

MATERIALS RELATED ISSUES

The biologically based models show complex dependence on radiation quality which is expressed in terms related to the details of the particle track as distinct from the simple LET dependence of the quality factor used in conventional radiation protection practice including ISS. Even the cancer risk attenuation characteristics of spacecraft shield materials are found to be different for the track structure dependent and LET dependent models leading one to conclude that any useful dosimetric technique must take into account these differences as well. Most important in this respect is that LET dependent quality factors overestimate the effectiveness of most shielding materials (see figure 7) and would falsely indicate reduced cancer risk in many applications. Note also that the relative importance of material choices is clearer in the track structure models. This in part results from the fact that aluminum is inherently poor for protecting against tumors in mice induced by the GCR environment. Clearly the relative importance of design alternatives depends critically on our understanding of the biological action of specific radiation components and our ability to evaluate the transmitted radiations through specific shield materials [7,23].

In this regard, the attenuation properties depend on the atomic/nuclear database which has undergone change in recent years. For example, the dose equivalent relative to the unshielded value H(x)/H(0) is shown in figure 8 where H(x) is the dose equivalent within an aluminum shell of thickness x. We normalize the dose equivalent to the unshielded value H(0) to minimize the effects of changing environmental models in the succession of results shown in the figure. The remaining differences are mainly in nuclear models and transport procedures. The curve labeled Letaw et al. [24] is for the database developed by the Naval Research Laboratory in common use

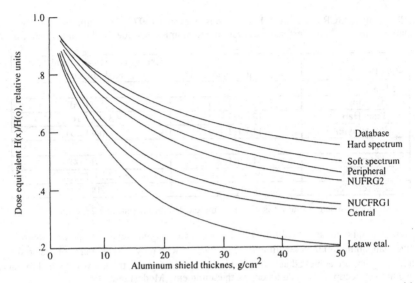

Figure 8. Shield attenuation for solar minimum galactic cosmic ray dose equilvalent resulting from nuclear fragmentation models.

until a few years ago and still used extensively in electronic hardening. NUCFRG1 model is the first database developed by Langley Research Center [25] and already gave higher exposure estimates than the Letaw et al. model for the same environmental model. The NUCFRG2 database is the result of revisions from a set of experiments with iron beams at the Lawrence Berkeley Laboratory Bevalac facility [26]. The hard and soft spectrum results are from direct knockout processes where there is some uncertainty in the knockout spectrum at high energies [27] and are in close agreement with shuttle experiments. Additional revisions in dose equivalent result from environmental model improvements [1]. Although large changes in evaluation of the transmitted spectrum have occurred, it remains an open question how the attenuation will change as new physical processes are added to the transport codes, the nuclear models are improved, and new knowledge of the biological risk comes available [7]. Laboratory validation of the developing database will play an important role in database development [19].

ROLE OF MATERIALS RESEARCH

In table I we included a performance index as the ratio of dose equivalent in aluminum (Al) to dose equivalent in another material (M) as

$$P_H(x) = H_{Al}(x)/ H_M(x) \tag{8}$$

and a similar performance metric for Harderian gland tumor induction, $HG_M(x)$. As noted in the table, the predicted performance of a given material relative to aluminum depends on the biological model used and the specific material. For each shield thickness, we can look at the range of variation over the materials in figure 7 and plot the maximal performance values for the two models related to cancer induction (dose equivalent $H(x)$ and cell transformation, $T(x)$). The maximal performance for each biological model is liquid hydrogen (LH_2) and the result is shown in figure 9. We have added to figure 9 the relative performance factor based on the Harderian gland tumor risk, $HG(x)$. Clearly, large performance factors are possible through

12

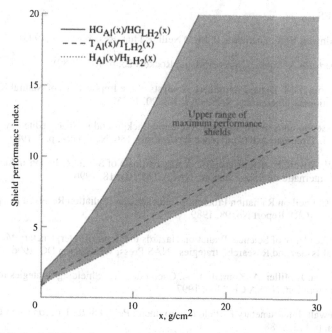

Figure 9. Maximal shield performance factors relative to aluminum with various biological models.

materials development and one would hope to approach the range of maximal performance in figure 9 as closely as possible. Although the actual performance achievable must await an improved understanding of biological response to GCR ions [7], it is clear that high performance shield materials will greatly reduce the mass requirements to protect astronauts in future missions and thereby greatly reduce the associated launch costs.

CONCLUDING REMARKS

The estimation of shield attenuation characteristics of various materials depends on the details of the biological response model. In that the experimental biological evidence displays clear dependence on other parameters in addition to LET, the adequate shield design for future deep space missions needs to reflect this dependence on these other factors (such as track width). Although accumulated data on biological response to heavy ion exposures allows studies of the relative advantage of material choices, the final design of future mission shielding must await a clearer understanding of the human response to GCR radiations. In the meantime, validation of the transport procedures and identification of high performance shield materials should be the focus of current materials research. In that materials with high hydrogen content yield good shield performance, polymeric materials are expected to play an important role in protecting the astronaut on future missions. As part of the consideration of new materials is the practical engineering design process in which high performance shield materials are incorporated into cost effective designs. Many of these engineering issues for use of polymeric composites are already being addressed in the development of light weight aircraft designs. Practical engineering experience gained in the use of polymeric composites in aircraft design is expected to have an important impact on the future of spacecraft design.

REFERENCES

1. G. D. Badhwar, F. A. Cucinotta, P. M. O'Neill, Radiat. Res. 134, P. 9 (1993).

2. D. F. Smart, M. A. Shea, J. Spacecraft and Rockets 26, p. 403 (1989).

3. J. W. Wilson, F. M. Denn, Preliminary Analysts of the Implications of Natural Radiations on Geostationary Operations. NASA TN D-8290, 1976.

4. E. G. Stassenopolous in High-energy Radiation Background in Space, edited by A. C. Rester, J. I. Trombka, AIP Conference Proceedings 186, New York, p. 3, 1986.

5. H. Wu, W. Atwell, F. A. Cucinotta, C. Yang, Estimate of Space Radiation-Induced Cancer Risks for International Space Station. NASA TM-104818, 1996.

6. National Council on Radiation Protection, Guidance on Radiation Received in Space Activities. NCRP Report No. 98, 1989.

7. National Academy of Science, Radiation Hazards to Crews of Interplanetary Missions: Biological Issues and Research Strategies. NAS Press, Washington DC, 1996.

8. J. W. Wilson, J. Miller, A. Konradi, F. A. Cucinotta, eds., Shielding Strategies for Human Space Exploration, NASA CP-3360, 1997.

9. R. A. Mewalt, Interplanetary Particle Environment, Publ. 88-28, Jet Propulsion Laboratory, Pasadena, p. 112, 1988.

10. J. A. Simpson, Ann. Rev. Nucl. Part. Sci. 33 p. 323 (1983).

11. International Commission on Radiological Protection, 1990 Recommendations of the ICRP, ICRP Publication No. 60, Pergamon Press, 1991.

12. J. W. Wilson, F. A. Cucinotta, J. L. Shinn, Cell Kinetics and Track Structure, Biological Effects and Physics of Solar and Galactic Cosmic Radiation, Part A, eds. C. E. Swenberg et al., Press, pp. 295-338, 1993.

13. T. C. Yang, L. M. Craise, Biological Response to Heavy Ion Exposures. In Shielding Strategies for Human Space Exploration. Eds. J. W. Wilson et al. Chpt. 6, pp. 91-109, NASA CP 3360, 1997.

14. F. A. Cucinotta, J. W. Wilson, R. Katz, Int. J. Radiat. Biol. 69, p. 593 (1995).

15. F. A. Cucinotta, J. W. Wilson, Phys. Med. Biol. 39, p. 1811 (1994).

16. E. L. Alpen et al. Radiat. Res. 136, p. 382 (1993), Adv. Sp. Res. 14, p. 573 (1994).

17. F. A. Cucinotta, et al. Computational Procedures and Database Development. In Shielding Strategies for Human Space Exploration. Eds. J. W. Wilson et al. Chpt. 8, pp. 91-109, NASA CP 3360, 1997.

18. H. Tai et al. Comparison of stopping power and range databases for radiation transport study. NASA TP 3644, 1997.

19. J. Miller et al. Acta Astronautica 42, p. 389 (1998).

20. J. W. Wilson et al. Adv. Space Res. 14(10), p. 841 (1994).

21. R. G. Alsmiller, D. C. Irving, W. E. Kinney, and H. S. Moran, "The validity of the straightahead approximation in space vehicle shielding studies," In Second Symposium on Protection Against Radiations in Space. A. Reetz, ed. NASA SP-71, pp. 177-181; 1965.

22. J. W. Wilson et al., HZETRN: Description of a free-space ion and nucleon transport and shielding computer program. NASA TP 3495, 1995.

23. J. W. Wilson et al. Health Phys. 68, p. 50 (1995).

24. J. R. Letaw, R. Silberberg, and C. H. Tsao, "Radiation haxards on space missions outside the magnetosphere." Adv. Space Res. 9(10), p. 285 (1989).

25. J. W. Wilson, L. W. Townsend, and F. F. Badavi, "A semiempirical nuclear fragmentation model." Nucl. Inst. Methods Phys Res. B18, p. 225 (1987).

26. J. W. Wilson, J. L. Shinn, L. W. Townsend, R. K. Tripathi, F. F. Badavi, and S. Y. Chun, "NUCFRG2: A semiempirical nuclear fragmentation model." Nucl. Inst. Methods Phys Res. B94, p. 95 (1995).

27. F. A. Cucinotta, L. W. Townsend, J. W. Wilson, J. L. Shinn, G. D. Badhwar, and R. R. Dubey, "Light ion components of the galactic cosmic rays: Nuclear interactions and transport theory." Adv. Space Res. 17(2), p. 77 (1996).

THE ABRASION OF ALUMINUM, PLATINUM, AND NICKEL BY MARTIAN DUST AS DETERMINED BY THE MARS PATHFINDER WHEEL ABRASION EXPERIMENT

DALE FERGUSON, MARK SIEBERT, DAVID WILT, JOSEPH KOLECKI
NASA Glenn Research Center, Cleveland, OH 44135, ferguson@grc.nasa.gov

ABSTRACT

The Mars Pathfinder Wheel Abrasion Experiment (WAE) spun a specially prepared wheel with strips of aluminum, platinum, and nickel, in the Martian soil. These materials were chosen because of their differing hardnesses, their ability to stick to anodized aluminum, and their comparative chemical inertness under Earth launch and Mars landing conditions. Abrasion of those samples was detected by the change in their specular reflectances of sunlight as measured by a photovoltaic sensor mounted above the wheel. The degree of abrasion occurring on the samples is discussed, along with comparisons to the abrasion seen in Earth-based laboratory experiments using Martian soil analogs. Conclusions are reached about the hardness, grain size, and angularity of the Martian soil particles, and the precautions which must be undertaken to avoid abrasion on moving parts exposed to the Martian dust.

INTRODUCTION AND EXPERIMENTS

The Mars Pathfinder spacecraft landed on Mars on July 4, 1997. Among the experiments on the Sojourner Rover was the Wheel Abrasion Experiment (WAE), designed to determine the abrasion of metals by movement in the Martian soil. Other papers have given details of the experiment [1-5], so only a brief outline will be given here. A photovoltaic sensor, mounted above a wheel with metal samples attached, monitored sunlight reflected by the wheel as it alone spun in the Martian dust. Fifty five wheel spins were accomplished during the mission, some of which with the WAE wheel held above the surface (a so-called "WAE wheelie") to prevent dust caking from negating the measurements. Each data-taking spin consisted of two complete revolutions, of 125 reflectance data points each. True specular reflection was achieved on only a few sols (Martian days).

In addition, ground tests and calibrations were performed in a Mars simulation chamber at the NASA Lewis Research Center. In particular, a wheel nearly identical to that flown was rotated in a tray of Martian soil simulant under Martian atmospheric conditions (see figure 1). Slippage of the wheel simulated spinning the wheel. Comparison of the flight sensor data with data obtained from an identical sensor mounted above the laboratory wheel, under simulated Martian sun illumination, allows us to compare the properties of the Martian soil with the soil simulant being tested. Figure 2 shows the three metal strips containing the metal samples at the end of one laboratory test. These strips were mounted sequentially on the wheel periphery. In this figure, the abrasion of the metal samples is clearly seen, as are the black anodized aluminum baseline samples.

17

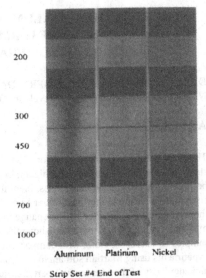

200

300

450

700

1000

Aluminum   Platinum   Nickel

Strip Set #4 End of Test

Figure 1. The laboratory Mars simulation tank, with horizontal simulant tray, and WAE wheel mounted edge-on to the viewer. In operation, the chamber was closed by a door mounted to the round outside flange, and simulated Martian sunlight was admitted through a port at the left.

Figure 2. A photo of the metal strips of Strip Set #4 at the end of testing. Metal film sample thicknesses in Ångstroms are indicated at the left. The black samples are uncoated black anodized aluminum, which was used for a baseline.

In a previous paper [5], we have shown that the flight data imply abrasion and polishing on the aluminum and platinum samples. In order to reach that conclusion, the effects of dust on the reflected sunlight had to be removed. It has been shown [5] that the ratio of the normalized flight sensor signal to the calibration signal may be written as

$$\text{Ratio} = X \, (1 - f \, (1 - (R_d/R_s))),$$

where X is a constant, f is the fraction of the surface area covered by dust, and R_d and R_s are the specular reflectivities of the dust and sample, respectively. Since the metal sample reflectivities are so large, compared to the dust, trends in theses ratios between different samples may be used as indicators of abrasion or polishing. For the sol 53 flight data, values of $f=0.66$ and $R_d = 0.17$ were found [5]. In figure 3, we see the sol 53 data and the preflight calibration data, normalized to the 1000 Å thick samples of platinum and nickel, which showed little or no evidence of wear. Using this normalization, we take ratios of the peaks for each of the metals to the calibration signals to get figure 4. Here, we see the evidence of polishing (ratios significantly higher than one) and abrasion (lower ratio for thin samples than for thick).

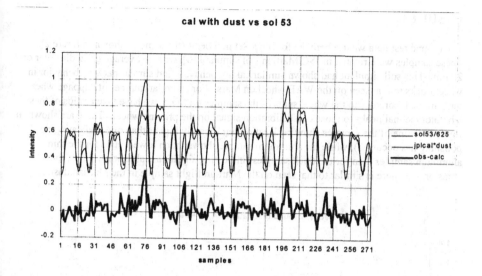

Figure 3. Sol 53 data and calibration data normalized by the peak heights of the 1000 Ångstrom nickel and platinum peaks and the black anodized aluminum background. Sol 53 data and calibration signals are at the top, and the difference is at the bottom.

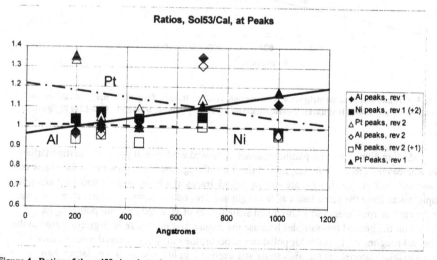

Figure 4. Ratios of the sol53 signals to the calibration signals at the metal reflectance peaks. Lines of least squares linear regression are included. Nickel is squares, aluminum diamonds, and platinum triangles.

RESULTS

The ground test data we use here are for Strip Set number 4 (the same as shown in figure 2). These samples were rotated in JSC Martian soil simulant [6], with an average grain diameter of 25 μm. This soil simulant had shown similar caking behavior and similar packing behavior in wheel tracks to that seen on the WAE wheel on Mars. For a similar number of slippage wheel turns in the laboratory to the wheel spins of the samples on Mars, we have taken ratios of these laboratory signal peaks to those of a calibration signal on the pristine wheel. These are shown in figure 5. None of the samples show peak ratios significantly higher than one (indicating no polishing has occurred), while the aluminum sample shows much lower ratios for the thin aluminum samples than for the thick samples (indicating significant abrasion). Thus, the abrasion seen here is similar to that seen on the Martian flight sample in the Martian dust.

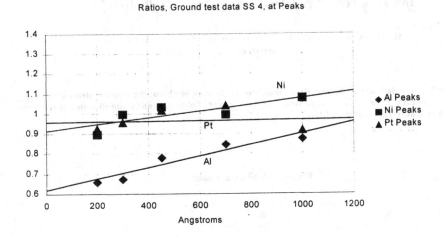

Figure 5. Ground test data on strip set 4, intermediate results. Ratios of signals at peaks to calibration signals. As before, nickel is squares, aluminum diamonds, and platinum triangles. Compare to figure 4 of flight data.

So the Martian aluminum and platinum samples showed evidence of polishing in the Martian dust, whereas the ground test samples did not, even though the abrasion was similar. How can we account for this? In figure 6 are shown Dektak traces of a spare pristine anodized aluminum sample, taken from the same batch as the flight and ground test samples. Here, it can be seen that the surface roughness has a horizontal scale length of 10 to 20 μm. No polishing has occurred on the ground test samples because the ground test grain size is large compared to the horizontal roughness. In order for polishing to occur, the polishing material must be small enough to take material off the high spots and deposit it in low spots. Grains larger than the horizontal roughness scale can only produce scratches in the bulk material. We can infer then that the dust on Mars is fine enough to polish the samples, and thus has a grain diameter or 10 to 20 μm or less.

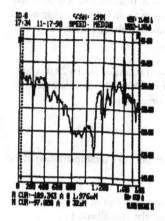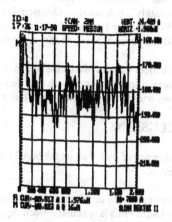

Figure 6. Dektak traces of anodized aluminum with metal coating, similar to flight samples. First trace is from bare anodize to aluminum coating, showing the step at the coating edge. Second trace is on bare anodized aluminum in crosswards direction. Each horizontal division is 200 microns.

In terms of hardness, polishing can only be achieved [7] if the polishing material is 1.2 times as hard as the material being polished. On the Brinell scale [8] aluminum is 16, platinum 64, and nickel about 100. Thus, since the flight samples show polishing on the aluminum and platinum samples, the Martian dust must have a hardness of at least 77 on the Brinell scale. In order for the nickel samples to show no abrasion, the hardness of the Martian dust must be less than 100.

Other Data

It has been shown elsewhere [5] that the electrostatic charging of dust adhering to the WAE wheel, along with the caking and packing of the Martian dust, point to small grain diameters for the Martian dust adhering to the wheel (< 32 μm). In this paper, we have seen that the Martian dust contacted by the wheel and producing polishing on its surface has a small grain diameter. Other data [1] have shown that the Martian dust may be wind-borne and drifting, indicating very small grain sizes.

The Amount Of Abrasion

How much abrasion has occurred on the thinnest aluminum sample of Sol 53? We can find out by determining how much its reflectance has been diminished. From figure 4 we can see that the reflectance of the thinnest sample is now been reduced to only 0.86 that of the thickest samples. Fitting models to the specular reflectance of the flight samples and other samples produced with identical apparatus but different aluminum thicknesses, we have found that the final aluminum thickness is between 40 and 80 Å. Thus, we can estimate the thickness of aluminum lost to be about 140 Å. The total number of wheel revolutions up to and including sol 53 was 55. Of those, 16 were performed on a hard or rocky surface, or in WAE wheelies, where the WAE strips were probably not abraded. Thus, the loss of aluminum may amount to about 140 Å in 39 wheel revs, or roughly 3.6 Å per rev. One wheel revolution corresponds to about 0.4 meter of spin, and with a 5% wheel slip (typical of our ground tests with no tray braking), about 8 meters of travel. For a 100 km traverse, about 45,000 Å would be lost, or 4.5 μm.

It would behoove future Mars missions to take measures to prevent access of the 10-20 μm dust to soft metal or plastic parts which may be subject to wear. Hard metals will see little wear. Aluminum will not be protected by its thin surface layer of aluminum oxide.

CONCLUSIONS

1. The presence of dark dust on the WAE metal samples amounts only to a renormalization of the signal levels.

2. The dust measured on Mars has a specular reflectivity much less than that of the metal samples. (Note - this is specular reflectivity, and does not indicate the dust albedo.)

3. Significant wear (>80% confidence) occurred on Mars on the WAE aluminum samples in 39 abrasive wheel rotations.

4. Evidence for polishing was seen in the Martian platinum and aluminum samples.

5. No evidence is seen for polishing in ground tests with Martian soil simulant of ~25 μm average diameter.

6. Surface roughness on unabraded WAE samples has a horizontal scale length of 10-20 μm.

7. Polishing on the Mars WAE samples implies a dust grain diameter of 10-20 μm or less.

8. Other WAE evidence (electrostatic charging, caking) also lead to grain sizes of <32 μm.

9. Based on the abrasion and polishing data, the Martian dust is harder than aluminum, comparable in hardness to platinum, and softer than nickel.

10. WAE depth-of-dig data indicate the Martian dust may have been wind borne and drifting - this also leads to small dust size estimates.

REFERENCES

1. J.R. Matijevic, J. Crisp, D.B. Bickler, R.S. Banes, B.K. Cooper, H.J. Eisen, J. Gensler, A. Haldemann, F. Hartman, K.A. Jewett, L.H. Matthies, S.L. Laubach, A.H. Mishkin, J.C. Morrison, T.T. Nguyen, A.R. Sirota, H.W. Stone, S. Stride, L.F. Sword, J.A. Tarsala, A.D. Thompson, M.T. Wallace, R. Welch, E. Wellman, B.H. Wilcox, D. Ferguson, P. Jenkins, J. Kolecki, G.A. Landis, D. Wilt, H.J. Moore, and F. Pavlics, "Characterization of the Martian Surface Deposits by the Mars Pathfinder Rover, Sojourner," Nature **278**, p. 1765-1768 (1997).

2. M. Siebert, "Wheel Abrasion Experiment Ground Tests: Initial Results," in Proc. Intersoc. Energy Conversion Conf., Honolulu, Hawaii, July 27-Aug. 1, pp. 743-748 (1997).

3. A.F. Hepp, N.S. Fatemi, D.M. Wilt, D.C. Ferguson, R.W. Hoffman, M.M. Hill, and A.E. Kaloyeros, "Wheel Abrasion Experiment Metals Selection for Mars Pathfinder Mission,' at the 1996 Fall Meeting of the Materials Research Society, Boston, MA, Dec. 2-6, NASA TM 107378 (1996).

4. D.M. Wilt, P.P. Jenkins, and D.A. Scheimann, "Photodetector Development for the Wheel Abrasion Experiment on the Sojourner Miicrorover of the Mars Pathfinder Mission," in Proc. Intersoc. Energy Conversion Conf., Honolulu, Hawaii, July 27-Aug. 1, pp. 738-742 (1997).

5. D.C. Ferguson, J.C. Kolecki, M.W. Siebert, D.M. Wilt, and J.R. Matijevic, "Evidence for Martian Electrostatic Charging and Abrasive Wheel Wear from the Wheel Abrasion Experiment on the Pathfinder Sojourner Rover," J. Geophys. Res. (Planets), in press (1999).

6. J.R. Gaier and M.E. Perez-Davis, "Effect of Particle Size of Martian Dust on the Degradation of Photovoltaic Cell Performance," NASA TM 105232 (1991).

7. ASM Handbook, "Friction, Lubrication, and Wear Technology," **18**, pg. 188 (1992).

8. CRC Handbook of Chemistry and Physics, 43rd ed., Chemical Rubber Company, Cleveland, OH (1961).

MATERIALS FOR THERMAL CONTROL
FOR MARS SURFACE OPERATIONS

Gregory S. Hickey
Jet Propulsion Laboratory, California Institute of Technology, Pasadena, CA 91109

ABSTRACT

The thermal environment for Mars surface exploration provides unique challenges for materials for use in structure and thermal control. The Sojourner Mars Rover has a lightweight integrated structure/insulation that has been environmentally tested and qualified for the Pathfinder mission to Mars. The basic structure with insulation, called the Warm Electronics Box (WEB), accounts for only 10% of the total Rover mass. The WEB is a thermal isolating composite structure with co-cured thermal control surfaces and an ultra-lightweight hydrophobic solid silica aerogel which minimizes conduction and radiation. This design provides excellent thermal insulation at low gas pressures and meets the structural requirements for spacecraft launch loads and for a 60 g impact landing at Mars without damage to the insulation or structure. Since the Pathfinder mission, this basic design concept has been developed and improved for future Mars surface robotic missions.

INTRODUCTION

Thermal control for Mars surface missions is highly dependent on the nature of the mission, the desired lifetime on the surface and the available power. The three types of missions that are either in progress or planned for future opportunities can be described as stationary landers, mobile rovers and surface penetrators. Surface landers would include the Pathfinder lander currently in progress, and the series of Surveyor landers that are planned for the '98 and '01 mission sets. Rovers include the Sojourner microrover that was payload for the Pathfinder mission and its twin, Marie Curie for the '01 mission. The 2003 and 2005 mission will have a much larger rovers, on the order of 30-50 Kg. Each type of mission will have the challenge of a Mars environment that is dynamic. The environmental factors that effect thermal control are temperature variation, the atmospheric density and the atmospheric opacity. Of primary concern to thermal control is temperature variation, which for Mars is characterized by large daily and seasonal and annual temperature variations. The atmospheric density affects the efficiency of insulation materials. The low density Martian atmosphere is primarily CO_2 (95.5%) with small amounts of N_2 (2.7%) and Argon (1.6%). The nominal 8 torr CO_2 will vary on an annual basis by up to +/- 15% depending on latitude due to the condensation and sublimation of CO_2.

The evaluation of Martian surface temperatures are based on three types of data. First, there are the direct measurements from the two Viking landers, which provided data for two specific location of the surface. The second type of data are from remote sensing such as from the Viking Orbiter Infrared Thermal Mappers (IRTM), and from Mars Global Surveyor, which provided seasonal and latitudinal data. The third and more recent is the diurnal temperatures measured during the Pathfinder lander site. All of the available science and engineering data from the Pathfinder mission is available from Reference 1. There have been several papers written on modeling and the interpretation of the Martian surface temperatures [2,3] Kondratev and Hunt provide a summary of the work completed prior to 1981 [4]; and Haberle and Jakosky take a recent look at atmospheric effects on temperature and thermal inertia [5]. For equatorial and mid-latitudes there will be significant temperature variation during the day. Temperatures may

Mat. Res. Soc. Symp. Proc. Vol. 551 © 1999 Materials Research Society

range from a high to 300K to a low of 190K. For polar regions, the daily temperature variation is less, with the peak temperatures around 250 K, and the lower limit bounded by the sublimation temperature of CO_2. Figure 1 presents the daily temperature variation from the Pathfinder ASI/MET experiment [1]. It is of interest to note for thermal design that there is a rapid surface temperature increase at dawn, with the peak temperature nominally two hours after noon, and a slow temperature drop off after sunset for Mars.

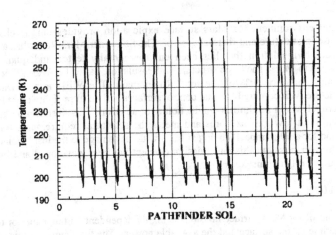

Figure 1: Mars Pathfinder surface temperatures (from Ref. 1)

The primary hardware that requires protection from the thermal environment are the electronics, batteries, and temperature sensitive components of the telecommunication systems. The temperature range for the electronic components on Pathfinder was ± 40 °C. The primary batteries on Pathfinder were the most sensitive of the components. Above 40°C they will autodischarge and below 40 °C will not provide current. Many other electronics can withstand a wider temperature range, but minimizing the temperature swings improves reliability.

Heat transfer is comprised of conductive, convective and radiative components. Thermal insulation for any environment desires to minimize heat transfer. Conduction will be present in any structure due to the need to accommodate mechanical loads. For a Mars environment, convection is low but still significant. It is for this reason that multilayer insulation is not the best choice for thermal insulation. The low pressure environment for Mars provides a challenging thermal environments that is in a region of thermal transport that has not been well studied. The nominal 8 torr CO_2 environment falls in a transition regime between the continuum regime (50 torr and above) and the Knudsen free molecular conduction regime (1 torr and below). Due to the low temperatures on the Mars surface, radiative heat transfer is not large but it is still significant.

There has been a limited amount of evaluation of insulation materials for Mars environments. Wilbert, et.al. conducted a study for Viking that evaluated foam insulations, fibrous, powders and multilayer insulation (MLI) for thermal control [6]. They did an initial selection of materials that were expected to meet the terminal sterilization process that was required to meet planetary protection requirements The result of that study was the primary use of several inches of foam insulation and some MLI on the Viking Landers. Other studies of thermal control for Mars in that time period assumed a foam insulation of 3 to 4 inches thick for potential Mars missions [7,8]. A more recent study revisited the issue of Mars thermal control and evaluated additional porous foam insulations [9]. There has not been a significant amount of work for thermal control for Mars surface exploration until the recent Pathfinder mission to Mars. Planetary protection is a significant restriction on all hardware for Mars missions. The typical process for the Pathfinder mission was to heat subsystems for 5 hours at 125°C. The process for Viking was to sterilize the entire spacecraft. Due to stricter planetary protection for the 2001 and subsequent missions, some compromise between the two extremes is likely. In addition to sterilization, all exterior surfaces will have to be cleanable to leave now biological residue for the 2003 mission and beyond that contain life detection experiments. This paper will survey the current work that has been completed for the Pathfinder mission and for part of the planned '98 and '01 launch opportunities.

INSULATION MATERIALS

Foam insulations are advantageous for many applications. They are inexpensive, easy to handle and machine, and provide moderate insulative properties in environments with atmospheres. They disadvantage is that they tend to be bulky and lose their integrity at very low densities. They can be utilized in structures with minimal difficulty. One restriction on the use of foam insulation for space applications is that they must be able to withstand the depressuration rate to vacuum for the launch vehicle. This limits selection to mostly open cell foams, though a few closed cell foams with strong cohesive mechanical integrity will pass. For a foam insulation, there is no internal convection or radiation. The primary mode of heat transfer is conduction.

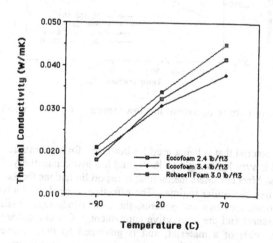

Figure 2: Comparison of foam insulations in a 10 torr CO_2 environment.

Figure 2 shows the thermal conductivity of two different foam insulations. The Eccofoam is an open cell polyurethane foam insulation. The Rohacell is a closed cell polymethacryimide foam The Rohacell has good mechanical properties at low densities. Both are vacuum rated and have acceptable outgassing characteristics. Data is shown for two densities of the Eccofoam. As expected for a foam insulation, as density decreases the thermal conductivity decreases.

Fibrous insulations have advantages over foam insulation in that they can be obtained in lower densities, but they cannot support mechanical loads, so additional structure must be built around them. The modes of heat transfer of fibrous insulation are conduction, convection and radiation. Conduction is significantly less than for foam insulation, but the inclusion of internal radiation and convection does not result in a significant decrease in thermal conductivity for a Mars low pressure environment. Figure 3 shows data for two types of glass fiber insulation. The TG1500 is a 2 lb/ft^3 fibrous insulation that has found several applications for thermal insulation on launch vehicles and the Space Shuttle. The other glass fiber insulation is a 1 lb/ft^3 material that is being considered for use for the '98 Surveyor lander. There is a small effect of density of material, but there is also an effect of pressure on the thermal conductivity of fibrous insulation.

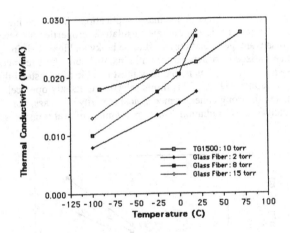

Figure 3: Comparison of fibrous insulations in a low pressure CO_2 environment.

A third type of material that is being used for insulation for Mars missions is aerogel. It has the advantage that it is very lightweight (1 lb/ft^3) and has low conductivity. The use of aerogel for thermal control on Mars considers the thermal transport limitations for the transition between the continuum and free molecular regime. The effective thermal conductance of an aerogel insulation system consists of two components, the solid conductance-radiative conductance (independent of pressure) and the convective component. Gas conductance depends on the mobility inside the voids of a material, and is governed by the relative dimensions and connectivity of the open volume and the gas mean free path. If the interstitial space between material (whether they be voids, particles or fibers), becomes smaller than the mean free path of the gas within the insulation, the mechanism for gas transport shifts from the continuum regime to free molecular conduction in the Knudsen regime. The mean free path of a gas is inversely proportional to the gas pressure. For CO_2 at 10 torr, the mean free path is approximately 3 to 5 microns in the temperature range from - 90°C to 25 °C.

Solid aerogel has pore dimensions 6 to 11 nanometers. This means that the effective gas conduction within the aerogel is a fraction of value in the continuum regime, and is dominated by the conductive-radiative component of the silica aerogel structure. Since low density aerogels approaches 80 to 90% open cells, direct conductance is minimized. Because of the high void fraction and the resulting low mechanical properties of low density aerogels, a lightweight supporting structure has to be designed and integrated into the thermal control. This basic integrated structure and thermal insulation design became the basis for the Sojourner Rover that is the payload for the Pathfinder mission [10,11]. Figure 4 shows the thermal conductivity for two types of aerogel. The first is a silica aerogel that is being used for the Sojourner rover. The second is a Resorcinal-Formaldehye (RF) aerogel that is being considered for future planetary rovers. One basic distinction between the two materials is the silica aerogel is translucent in the infrared and visible spectrum whereas the RF aerogel is opaque.

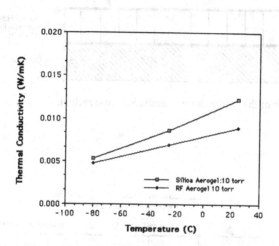

Figure 4: Comparison of aerogel insulations in a 10 torr CO_2 environment.

THERMAL STRUCTURES

Two types of thermal structures were developed for the Pathfinder mission. The first is the Integrated Structural Assembly (ISA) for the Pathfinder lander. The second is the Warm Electronic Box (WEB) for the Sojourner rover. Although each is expected to operate in equivalent thermal regimes, different engineering requirements lead to very different designs. The ISA is a static structure that supports the camera and antenna assembly, nominally 10 kg of mass. It also acts as a barrier for planetary protection. The electronics inside were not held to the same level of sterilization requirements as hardware external to the ISA. The WEB is the primary structure for the rover, but it is designed for mobility. There was a significant operational penalty for mass on the mobility system and strict limitations on volume. The thermal performance of the Sojourner rover is described in detail in Reference 12.

The ISA is basically a conventional composite honeycomb structure. Its cross-section is illustrated in Figure 5. It is comprised of graphite-cyanate facesheets with 2 inch Nomex core. The core is filled with 2 lb/ft³ Eccofoam pressed into the honeycomb cells. On the interior surfaces, there is bonded an additional 2 inch thick piece of Eccofoam, with aluminized kapton as the thermal control surface material. The Sojourner rover WEB used a sheet and spar design with silica aerogel to minimize thermal conduction. Its basic design is illustrated in Figure 6, and described in detail in references [10,11]. Instead of being designed to take loads on any point of the surface, as is typical of a honeycomb design, it was designed to take point loads at the spars by placing them in the direct load paths. This minimized mass and direct thermal conduction through the spars.

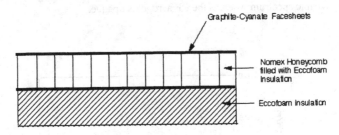

Figure 5: Schematic of the Pathfinder ISA insulation cross section.

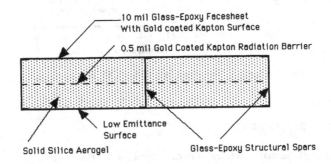

Figure 6: Typical cross section of the sheet and spar design used in the Mars Rover.

Figure 7 shows a comparison of the thermal conductivity of the two structures as a function of temperature. The structural design based on aerogel has almost half the conduction of designs using foam insulation and is approximately one quarter the areal mass for the same effective insulation capability. Also shown in Figure 7 is the effective conductivity of the WEB sheet and spar structural design using the opacified RF aerogel instead of the silica aerogel. This opacified material is being considered for the '03 Rover WEB

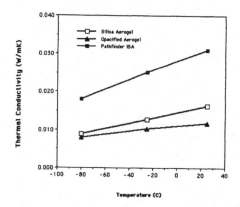

Figure 7: Comparison of thermal insulation structures being used for Mars missions.

ACKNOWLEDGMENTS

This work was conducted at the Jet Propulsion Laboratory, California Institute of Technology under contract from the National Aeronautics and Space Administration.

REFERENCES

1. http://mars.sgi.com/science/science-index.html
2. Morrison D., Sagan C., and Pollock J.B., *Icarus* **11**, 36-45 (1968).
3. Martin T., *Icarus* **45**, 427-446 (1981).
4. Kondratev, K, Hunt, G.E., *Weather and Climate on Planets*, Pergamman Press.
5. Haberle, R.M., Jakosky , B.M., *Icarus* **90**, 187-204 (1991).
6. Wilbers, O.J., Schelden, B.J., Conti J.C., *Progress in Astronautics and Aeronautics* **24**, 630-658, (1970).
7. Nagel R., "Thermal Control Aspects of a Stationary Martian Surface Laboratory," AIAA 3rd Thermophysics Conference, Los Angeles CA, June 1968.
8. Tracey, T., Morey, T.," Mars Lander Thermal Control System Parametric Studies," AIAA 4th Thermophysics Conference, San Francisco CA, June 1969.
9. Fischer, W.P., Ebeling W., Hass B., Keller, K., "Porous Insulation Performance under Mars Environment," SAE 23rd Int. Conference on Environmental Systems, SAE Paper 932116, July 1993.
10. Hickey, G.S., Braun, D., Wen, L.C., Eisen, H.J., "Integrated Lightweight Structure and Thermal Insulation for Mars Rover, SAE 25th Int. Conference on Environmental Systems, SAE Paper 951531, July 1995.
11. Hickey, G.S., Braun, D., Wen, L.C., Eisen, H.J., "Integrated Thermal Control and Qualification for Mars Rover, SAE 26th Int. Conference on Environmental Systems, SAE Paper 961541, July 1996.
12. Eisen, H.J., Hickey, G.S., Braun, D., Wen, L.C., "Sojourner Mars Rover Thermal Performance." SAE 28th Int. Conference on Environmental Systems, SAE Paper 981685, July 1998.

Figure ... Temperature rise ... voltage change used for ... measurements.

ACKNOWLEDGEMENTS

... was conducted at the ... Propulsion Laboratory, ... Institute of Technology, ... under contract to the National Aeronautics and Space Administration.

REFERENCES

1. Thomson, J. Electroanalytical ... Chemistry and ...
2. Morrison, D., ... Gosule J., and Pollock J.B., Nucl. ..., 56, 45 (1969).
3. ... Chemistry, 15, 920-446 (1984).
4. Kennedy, J.H. et al., Electrochemistry of Concrete ... Plenum Press, ...
5. Glasstone, P.S., Laidler, K.M., ... 137-141 (1941).
6. Bockris, J.O., Sockell, A.K.N., and M., Progress in ... and Corrosion, 37, 1950, 2060.
7. Bard, A.J., Electrochemical ..., Academic ... Stationary ... Surface ..., ..., A.J. and ...
   Modern physical chemistry, ..., A.J. ..., ..., CRC ..., ..., Vol. 8, ...
8. ..., ..., "Electrical and Thermal ... ..., Ohio ..., 1992, pp. 80-89.
9. ... and ..., "... Electrical ..., K., ... Power Systems", SAE Paper ..., 1992.
10. ... and ..., "... ..., ..., Environmental Sciences, ..., Paper ..., 1992.
11. ... ..., ..., ..., D.B., Wen, Lee, T., and H.W., "Integrated ... Propulsion ...
    ... ... for ... ..., Mars Rover", SAE 25th ... Conference on Environmental Systems,
    SAE Paper 951531, July 1995.
12. Hickey, G.S., Blaney, D., Wen, L.C., ... H.J., "Integrated Thermal ... for ...
    Exploration of Mars Rover", SAE 24th Int. Conference on Environmental Systems, SAE
    Paper 941311, June 1994.
13. Eagle, B.J., Hickey, G.S., Bhandari, P., ..., "... Sojourner Mars Rover Thermal ...
    ... ...", SAE 28th Int. Conference on Environmental Systems, SAE Paper 951531,
    July 1998.

LARGE SCALE TELEOPERATION ON THE LUNAR SURFACE

Gregory Konesky
ATH Ventures, Inc.
3 Rolling Hill Rd.
Hampton Bays, NY 11946

ABSTRACT

The popular success of the Mars Pathfinder mission, especially in terms of individual access to "live" pictures over the internet, demonstrates substantial mass appeal of space exploration on a personal level. Teleoperation provides an inexpensive approach to remote presence, but is distance-limited due to finite signal propagation times. The proximity of the Lunar surface permits near real-time teleoperation, and through a Virtual Reality approach, presents a sense of "being there" without the difficulty of getting there (and back).

While limited single user teleoperated Lunar Rovers concepts have been proposed (Luna Corp., Inc.), an on-going and economically self supporting venture would require simultaneous teleoperation of many vehicles on a large scale. In addition, each vehicle would carry several simultaneous stereoscopic remote viewing television camera heads, whose position is controlled by the head movements of Earth bound viewers. The overall concept is similar to a small fleet of touring busses, although some smaller scout vehicles may be included, as well as a tow truck. An access fee structure for Earth bound teleoperators provides for return on investment.

Technical issues for an on-going operation in the Lunar environment include temperature extremes, high vacuum, solar radiation, micrometeorites, abrasive nature of Lunar dust and its capacity to electrostatically attach to surfaces are discussed, as are their impact on vehicle design and operation. Optical data links are considered to handle bandwidth requirements of 500 simultaneous users. The economics of scale, on multiple levels, demonstrates the feasibility of large scale teleoperation. Various base-line studies are presented.

INTRODUCTION

Among the innovations of the Mars Pathfinder mission, making newly acquired images available "live" over the internet perhaps had the most surprising result. After the first 30 days of the mission on Mars, over a half billion accesses ("hits") to various NASA images were recorded, with a peak of over 46 million hits on July 8, 1997 alone. This level of internet activity was without precedent. This interest was not limited only to the United States. The French government had to appeal to its citizens to limit internet access as this was compromising their national telephone system. The mass appeal of space exploration on a personal level becomes clear.

Teleoperation involves the remote control of a vehicle or object using feedback cues from the local coordinate system of the vehicle. Vision is a valuable source of such cues, and various techniques used in Virtual Reality can enhance the operators' sense of being on that vehicle. One technique, stereoscopic vision, provides depth cues. By

33

viewing through a head mounted display, the users' head movements are detected and translated into a different view of a given scene. In Virtual Reality, this scene is computed. By mounting a stereoscopic camera head (two CCD cameras spaced an anthropomorphic distance apart) on a teleoperated vehicle, the remote operators' head movements are transmitted to the vehicle and command pan and tilt movements of the stereoscopic camera head, again enhancing the sense of being on that vehicle.

When applied to space exploration, teleoperation suffers from finite signal propagation times, limiting the responsiveness of the perception of remote presence. The Moon, however is sufficiently close that the worst case round trip delay is about two and a quarter seconds, permitting near real-time teleoperation. This has been demonstrated by the Luna 17 robot rover[1] (launched 11/10/70) which operated for 322 days and covered 10.5 kilometers.

Based on the mass appeal demonstrated by Mars Pathfinder, a teleoperated Lunar rover should enjoy wide support. The number of Earth-bound participants can be substantially increased by carrying a large compliment of stereoscopic camera heads on the vehicle as well. A hierarchy of participation is available, from vehicle drivers, to active viewer passengers, to retransmission of current Lunar scenery to a larger base of users without a head mounted display (passive viewers). Multiple vehicles improves overall probability of mission success.

The mission is not proposed simply as entertainment, but permits opportunities for tag-along science, while also satisfying the general publics' desire for personal participation in space exploration. Overall support for space exploration in general will no doubt benefit as well.

VEHICLE MORPHOLOGY

There have been three types of wheeled vehicles operated on the Lunar surface thus far: the Russian Lunokhods, the Apollo 14 Modular Equipment Transporter, and the Apollo Lunar Roving Vehicles[2] (see figure 1). Considerable engineering data is available on these vehicles and represents minimal design challenges.

Since long-term teleoperation is a goal the non-rechargeable batteries employed by the Apollo Lunar Rovers would be replaced by solar panels. An indication of power requirements appears in figure 2.

The inclusion of several vehicle types would improve mission robustness. A touring Bus provides an example of a vehicle with a large ratio of driver(s) to active viewers and permits a large Earth bound user population per vehicle. An intermediate "scout class" vehicle has a somewhat lower ratio, but is more maneuverable, and may scout ahead of the Bus vehicles. Finally, a sort of heavy duty Tow Truck vehicle should be included. Equipped with robot arm manipulators,[3] the Tow Truck serves many functions. If a vehicle gets stuck in deep Lunar dust[3] the Tow Truck may retrieve it. Periodic dusting of camera optics and solar panels may be required and is accomplished with the manipulators.

MISSION CONFIGURATION

Typically, 5 pounds must be launched from Earth to deliver one pound to the Moon[5]

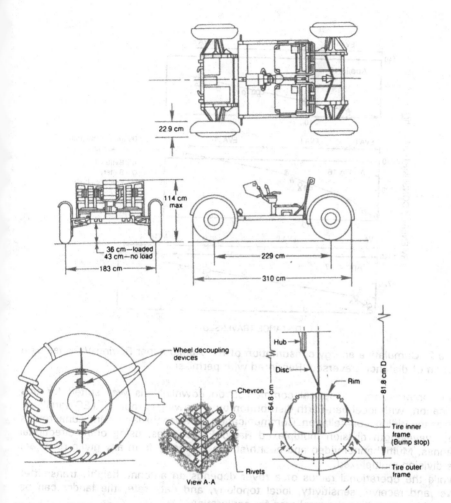

Figure 1. Schematic of the Apollo Lunar Roving Vehicle (Reprinted with permission).

based on Apollo experiences. However, this payload will not return (in the near future), does not require consumables, and can sustain a fairly high Lunar landing shock load. In this respect, the terminal descent air-bag system developed for the Mars Pathfinder should be applicable here as well.

The lander will serve as a communications link between the Lunar rover fleet and Earth, also as was done in Mars Pathfinder. Earth uplink commands consist of rover piloting and individual camera head pan and tilt commands, while downlinks consist of multichannel video and health/status telemetry. Considering a rover bus design carrying 50 stereoscopic camera heads, and a compliment of 10 rovers, the video downlink bandwidth will be on the order of several gigahertz, absent any significant

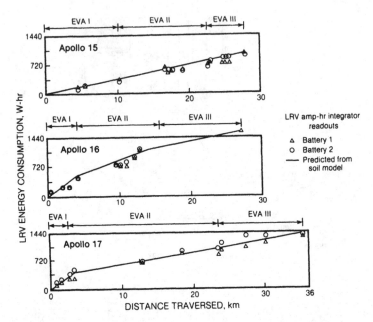

Figure 2. Cumulative energy consumption of the Apollo Lunar Roving Vehicles as a function of distance traversed (Reprinted with permission).

video compression. An optical communication downlink[6] is well suited for this application, with local pan-Earth distribution by a growing number of LEO satellites, such as the Iridium constellation. Communication between the lander and Lunar rovers is by wavelength division multiplexed Radio Frequencies, using omni directional antennas. Multichannel video and health/status telemetry from any given rover are time division multiplexed.

While the operational radius of a rover depends on antenna height, transmitter power and receiver sensitivity, local topology, and data rate, the lander can be relocatable, expanding the total area for exploration.

ACCESS FEE STRUCTURE

If such a project were funded by a not-for-profit organization such as NASA, access to the rovers would be free, although some training procedure would be required for potential rover drivers. The value of increased public participation in space exploration, as well as science tag-along benefits, would be the primary motivation.

However, it can be demonstrated that a fee access structure can provide considerable return on investment, making the project economically self supporting.

Suppose access is divided into 15 minute time slices. Since the mission is solar

Figure 3. Example of a teleoperated vehicle constructed by the author for under $250.

powered, operational availability occurs for at most about 14 Earth days every synodic month. Low sun angle limits this to perhaps 12 working days and since there are 13 synodic months per year, a total of 156 working Earth days per year are available.

Note that a Lunar day is in continuous sunlight 24 hours per Earth day. Therefore, there are 3744 working hours per year, or 14,976 15-minute time slices per year per active channel from the Moon.

Suppose an access fee structure is designed so that drivers (highest value added) are charged at $100/time slice, active viewers at $10/slice, and passive viewers at $1/slice. A 10 vehicle Lunar rover fleet with 500 viewer channels will receive about $15 million from driver revenue and almost $75 million from active viewers per year. It is difficult to assess income from passive viewers since several Earth bound users can share the same look-only stereoscopic video channel. Additional income may derive from cable access television ("the Moon channel") theme park sites, ect.

WHERE TO GO

Wheeled vehicles are necessarily limited to the type of terrain they can accommodate, excluding, for example, some regions of the rugged Lunar highlands. Also, since we are limited to line-of-sight communications with Earth, only Lunar near side operations are possible. Far side operations would require one or several Lunar orbiting relay satellites.

However, the Lunar surface is rich in historical sites, both from the United States and Russia. Care must be exercised not to run over Neil Armstrongs' footprints.

LIFETIME LIMITING FACTORS

Day to day operations on the Lunar surface present several challenges. Surface temperatures can be as high as 123°C and drop to as low as -233°C, requiring some battery reserve to keep warm during the Lunar night. Continuous operation in high vacuum presents challenges in lubrication of moving parts. Lunar dust is of special concern given its abrasive nature and particle distribution size[4] and difficulty to remove in the absence of sunlight. Absent a significant atmosphere, solar VUV wavelengths and micrometeorite hazing will limit solar panel lifetime, as will the occasional Solar flare. Radiation hardening and dust mitigation are essential.

CONCLUSION

An approach is presented to making participation in space exploration available to the general public which relies only on previously demonstrated technology. An access fee structure is considered which may make such a venture economically self supporting. An additional and perhaps greater benefit in personal access to space is the desire to significantly expand human presence there.

ACKNOWLEDGEMENTS

The author is grateful to Dr. David S. McKay, NASA Lyndon B. Johnson Space Center, Dr. Donna L. Shirley, Jet Propulsion Laboratory, and Dr. Grant H. Heiken, Los Alamos National Laboratory for their insights, discussions and assistance.

Permission to reprint figures 1 and 2 was granted by Cambridge University Press. Funding for this project was provided by the author.

REFERENCES

1. Grant H. Heiken, David T. Vaniman and Bevan M. French, *Lunar Sourcebook*, Cambridge University Press, New York, 1991, p. 7.

2. *ibid.*, pp. 522-530.

3. *ibid.*, p. 34.

4. *ibid.*, pp. 477-481.

5. Paul D. Spudis, *The Once and Future Moon*, Smithsonian Institution Press, Washington, D.C., 1996, p. 241.

6. Shlomi Arnon, Stanley R. Rotman and Norman S. Kopeika, Optical Engineering 36(11), pp. 3148-3157, (Nov. 1997).

METAL ALLOY PROPELLANTS PRODUCED FROM LUNAR RESOURCES

Aloysius F. Hepp[‡], Diane L. Linne[‡] and Geoffrey A. Landis[‡‡]
[‡]National Aeronautics and Space Administration, John H. Glenn Research Center, Lewis Field,
Cleveland, OH 44135; [‡‡]Ohio Aerospace Institute, Brook Park, OH 44142

ABSTRACT

The ultimate success of a permanent lunar base will depend upon the use of available resources. Two important resources available on the moon are oxygen and metals. We examine issues, including lunar resources, processing, power, and engine performance, surrounding the use of lunar resources for metal/oxygen engines, including the performance of metal alloys as rocket fuels. Results of this analysis have uncovered several promising candidates including alloys of Al-Ca, Al-Mg, Al-Si, Ca-Mg, and Ca-Si.

INTRODUCTION

A wealth of information about lunar resources was obtained between 1966 and 1976, the era of intensive lunar exploration [1]. The topic of processing of lunar resources has been explored in the scientific and technical literature [2-12]. Two important products obtainable from available lunar resources are oxygen and metals. Oxygen can be used for both life support and as an oxidizer for rocket engines; metals can be used as power materials, as structural materials, and as fuels for rocket engines. There are many benefits to be realized by using indigenous materials for propellants. When launch costs to orbit are counted in thousands of dollars per pound of payload, a reduction in the mass required from Earth can translate to significant cost savings. For lunar missions, if the return propellant can be manufactured at the moon; not only does this mass no longer need to be raised to LEO, but the propellant to transport it to the Moon is also saved [13].

Lunar samples returned from the Apollo and Luna missions indicate that approximately 45 weight percent of the lunar surface material is oxygen [1]. Most of this oxygen is in the form of silicates and other mixed-metal oxides. Oxygen is the clear choice as an *in situ* oxidizer because of its prevalence on the Moon and the accumulated experience in rocket engine combustion. The choice for a fuel, however, is less apparent. The most common elements used in rocket fuels, hydrogen and carbon, are not available in appreciable amounts. Because of this, interest has turned to lunar metals as a potential source of fuel. In an earlier study we summarized available lunar resources, processing methods, and power issues, and calculated engine performance of Al/O, Ti/O, Si/O, and Fe/O as rocket propellants [13]. We have extended our earlier analysis to include the performance of metal alloys as rocket fuels. Results of this analysis have uncovered several promising candidates including alloys of Al-Ca, Al-Mg, Al-Si, Ca-Mg, and Ca-Si.

LUNAR RESOURCES

O_2, Si, Al, Fe, Mg, Ca, and Ti are prevalent on the Moon, but are chemically bound into various minerals. In order to isolate any of these elements, mining, beneficiation, and processing steps are required. Table 1 lists the most important or potentially useful mineral structural types found on the moon. Specific mineral names are included for reference. Lunar rocks and soils are composed of mixtures of silicates and mixed metal oxides (major phases) and metal sulfides and native metals (minor phases). Representative oxide compositions for typical rock and soil

39

samples collected on the moon are listed in table 2. Examples listed are representative of material returned from the Apollo and Luna missions. The geological, mineralogical and chemical data, derived from over 2000 samples, are quite complex. The following discussion of resources focuses on applicability to propellant production.

Highland pristine rocks are of mainly three types [1]. Ferroan anorthosites are mostly plagioclase feldspar with small amounts of pyroxene and olivine. These rocks are quite rich in Ca and Al as expected from the chemical formula of both feldspars. Four other pristine rocks, gabbros, norites, troctolites, and dunites, are described as Mg-rich rocks and contain more pyroxene and olivine. Dunite, for example, is almost pure olivine, accounting for its high concentration of MgO. Troctolites are also composed of relatively higher concentrations of olivine accounting for 20 percent MgO. Finally, KREEP (see table 2) rocks are basaltic lavas with relatively high concentrations of potassium, rare earth elements and phosphorus.

TABLE 1.
MOST COMMON OR POTENTIALLY USEFUL MINERAL TYPES FOUND ON THE MOON (INCLUDING SPECIFIC COMPOUNDS)

NAME	CHEMICAL FORMULA
Silicate Minerals	
1. **Pyroxene**	$(Ca, Fe, Mg)_2Si_2O_6$
Enstatite	$MgSiO_3$
Wollastonite	$CaSiO_3$
Ferrosilite	$FeSiO_3$
2. **Plagioclase Feldspar**	$(Ca, Na)(Al, Si)_4O_8$
Albite	$NaAlSi_3O_8$
Anorthite	$CaAl_2Si_2O_8$
3. **Olivine**	$(Mg, Fe)_2SiO_4$
Fayalite	Fe_2SiO_4
Forsterite	Mg_2SiO_4
Oxide Minerals	
4. **Ilmenite**	$(Fe, Mg)TiO_3$
Geikielite	$MgTiO_3$
Ilmenite	$FeTiO_3$
5. **Spinel**	$(Fe, Mg)(Cr, Al, Ti)_2O_4$
Chromite	$FeCr_2O_4$
Ulvöspinel	Fe_2TiO_4
Hercynite	$FeAl_2O_4$
Spinel	$MgAl_2O_4$
6. **Armalcolite**	$(Fe, Mg)Ti_2O_5$
Ferropseudobrookite	$FeTi_2O_5$
Karrooite	$MgTi_2O_5$
Other Minerals	
7. **Troilite**	FeS
8. **Iron/Nickel Alloys**	(Fe, Ni)
Kamacite	$(Fe, Ni) (Ni < 0.06)$
Taenite	$(Fe, Ni) (0.06 < Ni < 0.5)$
Tetrataenite	$FeNi$

TABLE 2.
APPROXIMATE CHEMICAL COMPOSITION OF SAMPLED LUNAR MATERIALS[a]

MATERIAL[b]	SiO_2	FeO	CaO	TiO_2	Al_2O_3	MgO
Mare Basalts						
High-Ti	40	19	11	11	10	7
Low-Ti	46	21	10	3	9	10
Al low-Ti	46	17	11	3	14	9
Very-low-Ti	46	22	12	1	12	6
Orange Glass	39	22	8	9	6	14
Green Glass	44	21	8	1	8	17
Highland Pristine Rocks						
Ferroan Anorthosites	45	3	17	0	31	3
Gabbros	51	10	12	0	13	13
Norites	51	10	9	0	15	13
Troctolites	43	5	11	0	20	20
Dunites	40	12	1	0	1	45
KREEP[c]	52	10	9	2	16	8
Complex Breccias and Impact Melts						
Fragmental	45	3	17	0	30	3
Glassy Melt	45	5	15	0	27	7
Crystalline	48	8	11	1	18	13
Clast-Poor	47	7	13	1	22	8
Granulitic	45	5	15	0	27	7
Lunar Regolith						
Apollo 12 Site	46	15	11	3	13	9
Apollo 14 Site	48	10	11	2	17	9
Apollo 15 Site	47	14	11	1	15	12
Apollo 16 Site	45	5	16	1	27	6
Apollo 17 Site	43	12	12	4	17	10

[a] - From text reference 1. [b] - Weight fractions listed are composites of several samples from one site or from one mission. [c] - Potassium (K), rare earth element (ree), and phosphorus (P) rock, accounts for approximately 3 to 4 percent, by weight.

Mare basaltic rocks and glasses found on volcanic plains are relatively rich in ilmenite, spinel, and armalcolite. This explains the high concentration of iron oxide. Titanium oxide concentration is variable but generally much higher than found in highland areas [1]. The remaining composition of mare basalts (70 to 90 percent) consists of plagioclase and pyroxene. This accounts for the relatively lower abundance of SiO_2, CaO and Al_2O_3 when compared to highland rocks and breccia. The relatively large amounts of oxides in mare basalts provides a potential source of both iron and titanium.

Breccias and impact melts form a class of materials that range in appearance from homogeneous to composite-like. This is due to the various impact, melting, and cooling processes that result in their formation. The breccias in general consist of clast (fragments) and the matrix that contains them [1]. The majority of the material in various breccias are similar to the pristine rocks, hence the similarities in composition. One potential use for breccias may be as a source of rare platinum-group metals derived from meteoric materials. The lunar regolith, having been disintegrated by mechanical weathering, may be an important source of FeO and Al_2O_3 that requires a minimum of mechanical processing. Finally, lunar regolith (as well as some lunar rocks) is a source of metal powder and alloys. Though a minor component, reduced metals may be an important iron source.

Exploration of the moon to date has sampled only a small fraction of the surface at a superficial level. While cratering has spread samples out over large areas at impact sites, and thus tended to homogenize the lunar regolith, nevertheless, further exploration may reveal mineral types, elements, and concentrations as yet unsuspected. Some authors indicate that differentiation processes may have produced ore deposits significantly enriched in certain components, and that further exploration of the moon may yet reveal new geological entities [2].

PROCESSING LUNAR RESOURCES

Taking advantage of the abundance of metal oxides on the lunar surface as potential sources for *in situ* propellants requires that areas where these raw materials are readily available be identified. The raw material must then be mined and often subjected to a beneficiation process to separate the desired feedstock. Potential propellant elements include O_2, Al, Fe, Si, and Ti. Table 3 lists several of the most reasonable processes proposed for production of oxygen and metals from lunar resources [2-12]. While most of these have terrestrial counterparts, some have evolved to take advantage of unique characteristics of the lunar environment.

Processing methods in table 3 are listed in order of technology readiness [6,13]. Methods that are most developed have terrestrial counterparts. These methods are compatible with the use of solar thermal heating, discussed below, and solar- or nuclear-generated electricity. Unfortunately, these methods often involve use of terrestrial-derived materials such as HF, Na, Li, C, F_2, or Cl_2. Methods that are compatible with space processing involve very high temperatures and relatively large amounts of power.

Mining on the Moon will be necessarily different from terrestrial mining. On Earth, mining relies on an abundant water supply for cooling, lubrication, movement and separation of materials, and collection of metals. Another difference is that the lunar surface has been subjected to a homogenization of the soils by repeated meteor impacts, making the regolith a mixture of many rock and mineral types [1,2,6]. On Earth most ores are recovered below the surface, while on the Moon it is worthwhile to consider surface mining.

TABLE 3.

LUNAR PROCESSING METHODS[a]

PROCESS	FEEDSTOCK	ELECT. POWER (kW / t fuel / year)[b]	THERMAL POWER (kW / t fuel / year)[b]	TEMP. (°C)	PRODUCTS[c] FUEL
Hydrogen Reduction[7]	Ilmenite ($FeTiO_3$)			900	O_2, Fe, FeO, TiO_2
	or pyroclastic glass	0.72	0.18		Fe
Carbothermal[3,5,8,9]	Enstatite ($MgSiO_3$)			1625	O_2, Si, MgO, SiH_4
		0.82	3.28		Si
Carbochlorination[9]	Anorthite ($CaAl_2Si_2O_8$)			675-770	O_2, CaO, Al, Si
		1.33	2.46		Si
		1.38	2.57		Al
HF Acid Leaching[3,5]	Mare Regolith			110	O_2, Al
		8.85	8.85		Al
Reduction by Li or Na[10]	Mare Regolith			900	O_2, Si, Fe, Ti
		2.15	2.15		Si
		3.30	3.30		Fe
		14.55	14.55		Ti
Reduction by Al[6,13]	Anorthite ($CaAl_2Si_2O_8$)			1000	O_2, Si, Al, Ca
		2.56	0.64		Si
		2.64	0.66		Al
Direct Fluorination[2]	Anorthite ($CaAl_2Si_2O_8$)			900	O_2, Al, Si, CaO
		15.52	3.88		Si
		16.16	4.04		Al
Magma Electrolysis[6,13]	Silicate Rock or regolith			1000-1500	O_2, Fe
		0.26	0.26		Fe
Fluxed Electrolysis[3,11]	Silicate Rock or regolith			1000-1500	O_2, Al, Si, Fe
		6.40	6.40		Si
		9.95	9.95		Fe
		19.15	19.15		Al
Vaporization/Fractional Distillation[12]	Regolith			2700	O_2, Al, Si, Suboxides
		1.77	4.13		Si
		5.28	12.32		Al
Selective Ionization[6,13]	Regolith			7700	O_2, Al, Si, Fe, Ti, Mg
		7.90			Si
		12.20			Al
		23.60			Fe
		52.00			Ti

[a] - Normal text indicates terrestrial-derived processes; highlighted text indicates space-derived processes.
[b] - Process power requirements dependent on desired metal product. Some thermal power estimates may not include power needed to reach processing temperature, such as selective ionization. [c] - Products produced by listed method include the major metal-containing species and oxygen.

Surface mining would take advantage of the fact that the surface material is mostly pulverized, helping to reduce mechanical processing of rocks before beneficiation. Other advantages of lunar surface mining include lower gravity relative to Earth (implying easier material transport) and lack of weather or a corrosive atmosphere. One disadvantage is that the moon experiences a 14-day sunlit period followed by 14 days of darkness. This could be a problem if considering solar-derived power for the operation. Additionally, extreme temperature contrasts accompany the day-night cycle, leading to problems with lubrication, friction, and equipment failure [2].

Once raw materials have been mined, feedstocks for various processing techniques may need to be separated from the mined material. This process, called beneficiation, performs the function of concentrating the desired metal oxides. There are two major beneficiation techniques, magnetic and electrostatic [6,13]. Magnetic beneficiation is accomplished by feeding the raw material through the field of one or more magnets, to separate magnetic minerals from non-magnetic materials. The use of magnets with different field strengths further separates the magnetic minerals. Electrostatic separation is more complex, but has the advantage of being able to separate non-magnetic minerals. This process is used to separate materials with respect to their conductive properties: conducting, semi-conducting, or insulating. Most minerals will show some difference in conductive properties. But homogenized regolith may not separate well if many minerals are melted together [1].

There is a variety of propellant processing schemes proposed for potential use on the lunar surface. Although lunar processing methods will model terrestrial modes of operation, there are several concerns that must be considered when processing operations are conducted on the moon [2]. First, there is no air or water, thus depriving the plant of heat sinks provided by these fluids. Traditional energy sources are absent (*i.e.* coal, oil, or gas). Basic processing chemicals are absent (*i.e.* ammonia, salt, chlorine, soda ash, carbon dioxide *etc.*). Finally, since initially there may be no local human operators, the plant will have to be autonomous or teleoperated.

One conclusion that may be drawn from table 3 is that titanium production from lunar materials is quite difficult, requiring large amounts of energy. Production of iron, aluminum, or silicon can be optimized by proper choice of processing method and is dependent upon the feedstock. When anorthite is the feedstock, silicon and aluminum are obtained similarly from this aluminosilicate [9]. Iron may be obtained from reduction of ilmenite [6,7], FeO in glasses, or pyroxenes. Separation of the reduced iron from the process materials may prove challenging.

POWER TECHNOLOGY

Priorities for a power system are reliability and absence of dangerous failure modes. Due to the high cost of transporting cargo to the moon, an additional priority for a surface power system will be low weight. For resource processing, two types of power are needed: thermal and electrical energy. Depending on the processing technology chosen, the relative amount of thermal and electrical process power required can vary considerably. It is much more efficient to use a primary thermal energy source than to produce thermal power from electricity.

There are two main power choices: solar and nuclear. The 354-hour lunar night requires that a solar power system either shut down during the night, or include a large energy storage system [14]. In general, power levels needed for resource processing are so high that energy storage for night operation is not likely to be practical. Thermal power can be produced either from a solar furnace, by the direct use of nuclear core heat, or from electrical power. Concentrator mirrors designed for solar thermal power on Earth have demonstrated the high temperatures needed for most of the thermally-demanding processes proposed. A solar concentrator designed for space solar dynamic power systems produces a working temperature of about 750°C [5]. For mirror concentrator systems, the hot region is at the focus of the mirror, and typically moves as the mirror tracks the sun. Since resource processing equipment is likely to be heavy, a system designed for the moon might require a separate tracking mirror to reflect the sun to the concentrator mirrors, or an active secondary mirror system to redirect light to the processing site.

If a nuclear reactor is used for electrical power, reactor heat could possible be used directly. To date, there has been little discussion of this possibility. The SP-100 nuclear reactor has a working-fluid operating temperature of about 1000°C [13]. Radioactivity from the reactor means it is likely to be located at least a kilometer from human-occupied locations. If the reactor is to supply thermal processing energy, either the processing must be autonomous or remotely operated, hot working fluid must be piped over relatively long distances to a site compatible with man-tended operation, or significant additional shielding must be added to the reactor.

Use of electrical power to produce heat is inefficient. However, a base will require an electrical power system in any case, and it may be more convenient to scale an existing power system up to high powers than to design a separate system. Electrical power may be produced either by a nuclear reactor or by solar panels. A nuclear power system based on the SP-100 reactor would deliver 100 kW of electrical power from a 2.5 MW thermal reactor for a baseline system [13]. Replacing the low-efficiency thermoelectric converters by high efficiency Stirling engines would result in a power level of 825 kW from the same reactor. Less conservative designs incorporating further technological advances may have lower mass.

An alternative source of electrical power is solar panels. Photovoltaics provide power with high reliability and no moving parts. Design considerations for photovoltaic power systems for a lunar base are discussed in [16]; for an advanced system, it may be possible to manufacture the solar cells on the moon [17]. Existing spacecraft use planar photovoltaic arrays. The conversion efficiency of Gallium arsenide (GaAs) cells currently available on the market is over 22%; efficiency over 27% has been achieved in the laboratory with GaAs/GaInP dual-junction cells. An alternative approach is to use thin layers of photovoltaic material on a flexible substrate. Thin-film solar cells have lower conversion efficiencies, but potentially higher specific power [18]. This has not yet been demonstrated in space, although thin-film solar cells have been developed for terrestrial use. A third approach is to concentrate the light onto small, extremely efficient solar cells. This approach has been tested in space only in small-scale experiments.

Of importance to power system analysis is the specific power (power per unit mass). It is possible to measure specific power at the cell, blanket, array level, or power system level. Specific power at the photovoltaic array level (including array structure) for the best arrays developed to date approach 300 W/kg. For currently designed space power systems, the photovoltaic blanket is only about a quarter of the total power generation system mass (excluding batteries used for electrical storage). The array plus structure accounts for half of the power system mass. The power management and distribution (PMAD) system accounts for the remaining half. This provides a powerful incentive to develop new and more efficient PMAD systems and to design new array structures to take advantage of ultra-light blankets.

METAL AND ALLOY PROPELLANTS

A simple analysis was undertaken to compare the theoretical performance or specific impulse (I_{sp}) of six pure metal and eight alloy fuels. These are listed in table 4 along with the melting temperature of the alloys; the alloys were chosen using published phase diagrams and then choosing lowest melting point alloys or eutectics [19].

Specific impulse (I_{sp}) was calculated using equation 1 ($\Delta H/m$ = specific enthalpy change):

$$I_{sp} = (1/g)(2\Delta H/m)^{0.5}. \qquad (1)$$

While simple predictions based on thermochemistry provide adequate comparisons when evaluating potential propellants, a more rigorous theoretical analysis would need to be performed to accurately predict the specific impulse that an actual engine would deliver. Factors that may degrade performance from the ideal values discussed above include incomplete energy release in the chamber due to incomplete mixing of fuel and oxidizer or incomplete burning of the metal particles, finite-rate chemical reactions, growth of a viscous boundary layer in the chamber and nozzle, and thermal or velocity non-equilibrium between the solid and gaseous combustion products. Some losses, such as finite-rate kinetics, cannot be changed or reduced. Other losses, such as incomplete mixing in the chamber and boundary layer growth, can be reduced by proper hardware design. Finally, losses such as incomplete burning of solid metal particles and two phase flow effects can be reduced by proper fuel design. Technology efforts have been initiated to reduce those loss mechanisms that can be affected by hardware or fuel design. A program is underway to establish the technology base needed for the development of engines that use resources indigenous to the moon [13].

TABLE 4.
Melting Temperature and Theoretical I_{sp} of Metal and Alloy Fuels[a]

Alloy or Metal	Melting Temp. (°C)	I_{sp} (sec.)
$Al_{0.3}Mg_{0.7}$	437	565
$Ca_{0.72}Mg_{0.28}$	445	502
$Al_{0.36}Ca_{0.64}$	550	520
Pure Mg	649	557
Pure Al	660	581
$Al_{0.11}Si_{0.89}$	755	563
$Ca_{0.97}Si_{0.03}$	782	488
Pure Ca	839	485
$Fe_{0.28}Ti_{0.72}$	1085	450
$Fe_{0.69}Si_{0.31}$	1215	378
$Si_{0.12}Ti_{0.88}$	1330	502
Pure Si	1412	561
Pure Fe	1536	280
Pure Ti	1670	496

[a]Actual Isp is 50-75% of theoretical due to losses, see text.

In our original study, four metals (Al, Si, Fe, and Ti) were chosen because they are relatively abundant on the Moon, can be obtained by a known terrestrial process, and are candidates for lunar-derived propellants [13]. The particular method(s) and metal(s) or alloys chosen will be a function of the feasibility of the process on the moon (processing materials and power requirements), potential utility of the metal as a propellant (and other applications), and mass trade-offs between the plant requirements and terrestrial-derived substitutes. Theoretical performance of several solid metals burned with oxygen was determined using a one-dimensional chemical equilibrium computer code [13]. This code predicts specific impulse assuming the maximum energy release possible in the combustion chamber less chamber dissociation losses. While lower engine performance can be tolerated from an *in situ* propellant combination because of the benefits of obtaining the propellant at the destination, mission analyses have shown that 280 seconds (see table 4) is too low for iron fuel to be seriously considered as an alternative.

The metal and oxygen propellants can be used as either a monopropellant, with powdered metal suspended in the liquid oxygen, as a bipropellant, with a conventional liquid oxygen feed system and a pneumatic feed system for the powdered fuel, or as a hybrid rocket, with liquid oxygen feed and a solid metal fuel. While most previous discussions of metal propellants have considered only solid or powdered metal [6,12,13], we also considered the possibility of using liquid metals as fuel. This allows well-developed liquid-fuel technology to be used. For this application, a low melting temperature is important, and here we considered a number of low-melting alloys or eutectics. We have identified several Al, Ca and Mg-containing alloys with promising properties that should be considered as alternatives to pure metal propellants.

Research into the ignition and burning of single metal particles in a hot oxygen environment has been started in an effort to reduce potential performance losses. From experience with metal fuels in solid rocket motors and from theoretical calculations, it is known that two keys to reducing performance losses are quick ignition of the metal particles and vapor phase or explosive combustion that minimizes the size of the solid products. To achieve these goals, various aluminum/magnesium alloys are being tested in a shock tube. It is expected that magnesium in an alloy will ignite more quickly than aluminum; differences in boiling temperatures will help promote the vapor phase or explosive combustion. Results from these experiments can be used in future design of rocket engines that use metal/oxygen propellants. Although metals have not been used before as the sole fuel element, the technology work being performed indicates that a metal/oxygen may make a suitable propellant combination for indigenous use on the Moon.

CONCLUSIONS

The case for *in situ* propellant production is a powerful one. Lunar resources are available to provide the necessary metals and oxygen. While our knowledge of the lunar surface and its geology, minerology, and chemistry is extensive, further exploration will be required to fully exploit lunar resources for manned exploration and colonization. Production technology must be developed to deal with, or even take advantage of, the lower gravity, sunlight intensity, and vacuum environment. Lunar production processes must depend as little as possible on nonrenewable Earth-derived chemicals. Power must be obtained from solar or nuclear sources and be compatible with the intended use: thermal or electric energy. The power itself could be derived from local sources.

By obtaining all of the propellants for near-lunar operation on the Moon's surface, significant benefits for future manned lunar missions can be realized. It is also expected that mission architecture will include plans for lunar-derived propellants to fuel further exploration, most likely to Mars. Use of alloy fuels provide several advantages for propulsion and processing: greater flexibility of engine design with low-melting alloys or eutectics, greater control of the combustion process, and simplified processing of raw materials by reduced metals separation.

ACKNOWLEDGEMENTS

We acknowledge the assistance of two NASA Student Researchers, Edward Conroy (Cleveland State University) and James Colvin (University of Arizona), in this study.

REFERENCES

1. *Lunar Sourcebook*, Heiken G., Vaniman, D. and French, B.M. (eds.), Cambridge University Press: Cambridge, UK, 1991.

2. Burt, D.M., *American Scientist*, **77**, 574-579 (1989).

3. Waldron, R.D., Erstfeld, T.E., and Criswell, D.R., in *Space Manufacturing 3*, Grey, J. and Krop, C. (eds.), AIAA: Washington, D.C., 1979, pp. 113-127.

4. Phinney, W.C., Criswell, D.R., Drexler, E. and Garmirian, J., *Prog. Astronaut. Aeronaut.*, **57**, 97-123 (1977).

5. Waldron, R.D. and Criswell, D.R., "Materials Processing in Space," in *Space Industrialization, Vol. 1*, O'Leary, B. (ed.), CRC Press, Inc., pp. 97-130 (1982).

6. Part 1 of *Resources of Near Earth Space*, Lewis, J., Matthews, M.S., and Guerrieri, M.L. eds., University of Arizona Press, Tucson (1993).

7. Williams, R.J., in *Lunar Bases and Space Activities of the 21st Century*, Mendell, W.W. (ed.), Lunar and Planetary Institute: Houston, TX, 1985, pp. 551-558.

8. Cutler, A.H. and Krag, P., in *Lunar Bases and Space Activities of the 21st Century*, Mendell, W.W. (ed.), Lunar and Planetary Institute: Houston, TX, 1985, pp. 559-569.

9. Rosenberg, S.D., Guter, G.A. and Miller, F.E., *Aerospace Chem. Eng.* **62**, 228-234 (1966).

10. Semkow, K.W. and Sammells, A. F., *J. Electrochem. Soc.* **134**, 2088-2089 (1987).

11. Rao, G.M., Elwell, D. and Feigelson, R.S., *J. Electrochem. Soc.* **127**, 1940-1944 (1980).

12. Sparks, D.R., *Journal of Spacecraft and Rockets* **25**, 187-189 (1988).

13. Hepp, A.F., Linne, D.L., Landis, G.A., Wadel, M.F., and Colvin, J.E., *Journal of Propulsion and Power* **10**, 834-840 (1994).

14. Landis, G.A., in *Space Manufacturing 7*, Faughnan, B. and Maryniak, G. (eds), AIAA: Washington, D.C., 1989, pp. 290-296.

15. Landis, G.A., *Journal of Propulsion and Power* **8**, 251-254 (1992).

16. Landis, G.A., Bailey, S.G., Brinker, D.J. and Flood, D.J., *Acta Astron.* **22**, 197-203 (1990).

17. Landis, G.A. and Perino, M.A., in *Space Manufacturing 7*, Faughnan, B. and Maryniak, G. (eds.), AIAA: Washington, D.C., 1989, pp. 144-151.

18. Landis, G.A., and Hepp, A.F., in *Proceedings of the European Space Power Conference*, ESA SP-320, European Space Agency: Noordwijk, Netherlands, 1991, pp. 517-522.

19. *Binary Alloy Phase Diagrams*, 2nd edition, Vols 1-3, Massalski, T.B., et al. (eds.), ASM International: Materials Park, OH, 1990.

Part II

Space Photovoltaics

HIGH EFFICIENCY MONOLITHIC MULTI-JUNCTION SOLAR CELLS USING LATTICE-MISMATCHED GROWTH

R.W. HOFFMAN JR., N.S. FATEMI, M.A. STAN, P. JENKINS, V.G. WEIZER, D.A. SCHEIMAN* AND D.J. BRINKER**
Essential Research, Inc. c/o NASA LeRC, 21000 Brookpark Rd., Cleveland, OH 44135
*Ohio Aerospace Institute, Cleveland, OH
**NASA Lewis Research Center, 21000 Brookpark Rd., Cleveland, OH 44135

ABSTRACT

The demand for spacecraft power has dramatically increased recently. Higher efficiency, multi-junction devices are being developed to satisfy the demand. The multi-junction cells presently being developed and flown do not employ optimized bandgap combinations for ultimate efficiency due to the traditional constraint of maintaining lattice match to available substrates. We are developing a new approach to optimize the bandgap combination and improve the device performance that is based on relaxing the condition of maintaining lattice match to the substrate. We have designed cells based on this approach, fabricated single junction components cells and tested their performance. We will report on our progress toward achieving beginning-of-life AM0 multi-junction device conversion efficiencies above 30%.

INTRODUCTION

Photovoltaic (PV) energy conversion of sunlight has reliably provided power for spacecraft for over three decades. Recently, the demand for space power has increased dramatically due to the large growth of the commercial satellite industry. Higher beginning-of-life (BOL) efficiency photovoltaic devices are desired since the decreased array size provides a weight savings to the satellite. The weight savings can be used to add additional satellite capability or to decrease the launch vehicle size providing further economic benefit. The increased BOL performance may provide enabling technology for certain missions where high end-of life (EOL) performance or survivability is desired.

The theoretical efficiency of a single junction solar cell operated under the space solar spectrum (AM0) has been calculated as a function of the semi-conductor band gap. As the bandgap is increased, the photon absorption and the saturation current density decrease causing a drop in current and an increase in voltage. A maximum theoretical conversion efficiency for a single junction device occurs at a band gap of about 1.6 eV. It has generally been believed that high quality, single crystal semi-conductors are required to maximize minority carrier properties and thus approach the theoretical performance potential.

One method to increase the PV energy conversion efficiency of the solar spectrum is to divide the spectrum in to portions that can be more efficiently converted. Absorbing the higher energy light in a higher band gap material and lower energy light in a lower band gap material allows a wider range of the spectrum to be converted with less energy to be lost as lattice heating. The goal then is to appropriately divide the spectrum, convert each portion efficiently by the use of multiple junctions, each with a unique bandgap, and combine the output of each junction to provide power to a load. Several combinations of junction interconnection are possible in either voltage matched or current matched two, three, or four terminal configurations. It is also possible to mechanically stack individual devices or to fabricate them monolithically.

From the array builder's standpoint, a two-terminal device is the most desirable configuration, thus forcing the individual junctions of a multi-bandgap device to be current matched. A monolithic device consumes only one substrate and avoids problems with mechanical stacking.

Fan performed theoretical calculations of two-terminal, multi-bandgap solar cells under AM0 illumination.[1] His results for two and three junction devices were presented as iso-efficiency contour plots. The plots showed that maximum device performance was achieved using certain optimal combinations of bandgaps. Unfortunately, the optimal bandgap combinations were not lattice-matched to any available substrate. Thus, a trade-off between choosing materials with optimized bandgaps or materials which could be monolithically grown with few lattice defects had to be made. The traditional approach was to place more emphasis on maintaining lattice-match to available substrates than to choose optimal bandgap combinations.

Several monolithic, dual junction devices have been developed based on a lattice-matched approach. The highest performance device was the InGaP/GaAs cell originally developed by Olsen and Kurtz.[2] All layers were grown lattice matched to the GaAs substrate, however, the bandgap combination that resulted was not optimal. To achieve the highest performance for this bandgap combination the top cell was intentionally thinned to allow more light to reach the underlying GaAs junction thus providing current matching between the top and bottom cells. A record efficiency of 25.7% was achieved [2] and recently improved to 26.9% [3] using basically the same InGaP/GaAs approach.

The InGaP/GaAs cell design was chosen for further development under an Air Force/NASA ManTech program. The present commercially available multi-junction solar cells resulting from this program have attained a space solar spectrum (AM0) power conversion efficiency of about 25% on large area cells.[4] As a result, we have observed a relatively rapid transition from Si solar cells to GaAs and more recently to the InGaP/GaAs /Ge multi-junction cells.

We believe that significant efficiency improvements can be attained using another approach to provide monolithic, multi-bandgap solar cells. Our approach is to place more emphasis on using optimized bandgap combinations and relax the traditional requirement of maintaining lattice match to available substrates. In these designs, we maintain lattice match throughout all active device layers in the structure, however, they are not lattice matched to a substrate. We can then design two, three and four junction devices with optimal bandgap combinations and potential for very high conversion efficiencies.

In this paper, we present two of our device designs for two and three junction cells (three and four junction with active Ge substrate). We will present data from individual components developed for implementation in high efficiency cells using a mismatched epitaxial growth approach.

EXPERIMENT

In this section, we describe our approach toward achieving high efficiency multi-junction cells. We present our device designs, including theoretical calculations that predict expected device performance, and describe the experiments used to fabricate and characterize the materials and devices.

Approach

Our approach to increase the multi-junction cell efficiency emphasizes the importance of optimizing the bandgap combinations while lattice matching the active cell layers to each other. The traditional condition that the cells must be lattice matched to available substrates is relaxed, however. This is accomplished through the use of a buffer region grown between the active device layers and the substrate to accommodate the lattice mismatch. This approach is demonstrated in the design of a dual junction cell detailed in the next section.

Figure 1 is a modified version of the familiar crystal grower's chart of semi-conductor bandgap as a function of lattice constant. A lattice-matched condition appears as a vertical line on the figure. The shaded, horizontal rectangles on the figure indicate the range of top and bottom cell bandgap for an optimized dual junction device based on Fan's calculations. The dashed vertical line at 5.65 Å represents the design of the lattice-matched InGaP/GaAs dual junction cell. One can readily see that this combination does not use optimal bandgap pairs. The solid vertical line at a lattice constant of 5.74 Å shows our design for a 1.7 eV bandgap top cell on InGaAs (1.1 eV) dual junction cell. Here one can see that the two cell materials are lattice matched to each other and have an optimal bandgap pair, however, they are mismatched to the GaAs substrate. Indeed this approach allows the optimal bandgaps to be chosen for two, three and four junction devices. The key to high efficiency then is to grow defect-free devices on mismatched substrates. Another advantage of using optimized bandgap combinations is that the current match required for two terminal multi-junction devices becomes independent of individual cell thickness, when equal quality devices are fabricated. In other words, the thickness of cell layers is not adjusted to obtain current matching between cells.

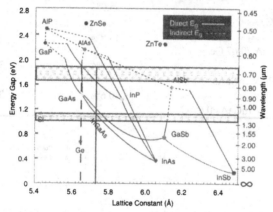

Figure 1. - Crystal growth chart showing traditional designed, lattice matched dual junction cell (dashed line) and our optimal bandgap combination design for two junction cells (solid line).

This approach is different than that taken by Varian researchers where they fabricated dual junction cells using an AlGaAs top cell on an InGaAs bottom cell on a GaAs substrate. [5] A significant lattice mismatch existed between the active cell layers and the top cell was grown under tension which may have limited the device performance. It has been shown that mismatched growth of InGaP under tension on GaAs generates a larger loss of photovoltaic performance then growth in compression.[6]

Theory

We have modeled several bandgap combinations searching for optimized two, and three junction cells based on our approach. These designs are turned into three and four junction designs by activating a junction in the Ge substrate. We further constrained the material selection to the use of compounds readily available by typical organo-metallic vapor phase epitaxy (OMVPE) or molecular beam epitaxy (MBE) III-V growth processes. The model solved the continuity equation in one dimension for up to five layers. We calculated a maximum efficiency for each junction using measured dark current values (J_0) from InGaP and InGaAs cells along with ideal fill factor (FF) and short current density (J_{sc}) values. We also calculated a practical efficiency value taking into account anticipated losses including those due to absorption in the tunnel junction and non-ideal anti-reflection coating.

Our modeling results show that devices exceeding a theoretical AM0 one-sun efficiency of 40% and a practical efficiency of 33.7% can be achieved using our approach. The calculated practical performance from each device in the optimized two, three, and four-junction configuration is shown in Table I. An example of the dual junction cell design is seen in the cross-sectional schematic diagram in Figure 2. In this design, all layers above the buffer are lattice-matched to the top layer of the buffer. The top and bottom cells have surface and interface passivating layers and are interconnected by a tunnel junction.

Table I. - Theoretical modeling results showing maximum theoretical and practical multi-junction device efficiencies using optimized bandgap combinations.

Multi-Junction Device	Maximum Efficiency %	Practical Efficiency %
Dual Junction 1.7 eV/1.1 eV InGaAs	35.0%	29.1%
Triple Junction 1.7 eV/1.1eV InGaAs/0.67 Ge	38.9%	32.3%
Triple Junction 2.1 eV/1.6 eV/1.2 eV InGaAs	37.9%	31.0%
Quad Junction 2.1eV/1.6 eV/1.2 eV InGaAs/0.67 Ge	40.6%	33.7%

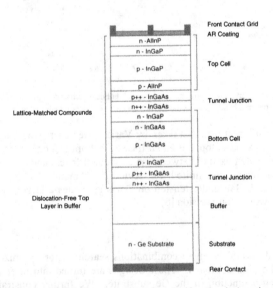

Figure 2. - Schematic diagram showing the cross section of our dual junction device using the optimal bandgap combination approach and lattice-mismatched epitaxial growth.

Procedures

All semiconductor layers were grown by OMVPE using a previously described [7] low pressure reactor and the following precursors: trimethygallium (TMGa), trimethylindium (TMIn), pure arsine, pure phosphine, dimethylzinc (DMZn) and diluted silane in hydrogen. Conductive GaAs substrates having an orientation of (100) cut 2° toward the nearest <110> direction were used. Nomarski interference contract optical microscopy was used to examine the surface morphology of the various compounds. High-resolution x-ray diffraction (HRXRD) was used to determine composition and lattice matching of the InGaAs and InGaP alloys. The bandgap of the device material was determined from the absorption edge of the spectral response data. Transmission electron microscopy (TEM) was used to determine defect structure related to strain relaxation. Dislocation defect densities were estimated from cross-sectional TEM observations. Typically, if one defect is observed in the TEM field of view, then the defect density is above 1×10^7 cm^{-2}.

Individual components (top, middle and bottom cells and interconnecting tunnel junctions) of the multi-junction cell design were fabricated and tested. The InGaAs cell had lattice matched InGaP minority carrier mirror and window layers to reduce the surface or interface recombination velocity. InGaP top and middle cells were fabricated without front or back surface passivating layers, however, they did have an InGaAs contact layer which remained below the front grid metallization. Vacuum evaporated gold-based metallization was used for front and rear contacts. The front grid was fabricated using reverse image photolithography and lift off techniques. Individual cells were fabricated on a single 2" substrate and were isolated by mesa etching into cell areas of either 1 cm^2 having grid shadow of 5% or 0.36 cm^2 having grid shadow of 6%. A MgF$_2$/ZnS dual layer anti-reflection coating was vacuum evaporated to complete the cells.

AM0 conversion efficiencies were measured at 25°C using a single source, Spectrolab X25 solar simulator at the NASA Lewis Research Center in Cleveland, OH. For the InGaAs cell measurements, the light intensity was set using a Si standard cell, which had been well characterized having been flown over 40 times in the NASA Lewis Lear jet calibration facility. InGaP cells were measured using a GaAs standard cell to set-up the simulator.

In addition to testing cells at one sun intensity, we measured cell performance at reduced light intensities (reduced current levels) to obtain accurate fill factor (FF) and open circuit voltage (Voc) values expected in the multi-junction cell stack. Screens were placed in the light path above the test plane to reduce the light intensity. The expected short circuit current density (Jsc) value was obtained by integrating the truncated spectral response of the cell using the bandgap of the overlying cell material as the cutoff. Combining this calculated short circuit current density value with the FF and Voc values measured at the reduced current level allows us to calculate the efficiency of the InGaAs cell below an overlying cell or cell stack.

RESULTS

We will first present results of evaluating the material quality of two InGaAs alloys grown mismatched on GaAs substrates and InGaP alloys grown lattice-matched on the InGaAs alloys. Next, the results of single junction solar cell testing and modeling are presented. Finally, these data are used to predict the expected multi-junction device performance for our designs.

Material Quality

For the InGaAs bottom cell in our multi-junction designs, we are interested in two specific alloys of $In_xGa_{1-x}As$; a 1.1 eV bandgap where x=0.22 (lattice-mismatch of 1.1% to GaAs) and a 1.2 eV where x=0.16 (lattice-mismatch of 0.6% to GaAs). We found that when a single layer of $In_{0.22}Ga_{0.78}As$ was grown above the critical thickness on GaAs, the threading dislocation density in the InGaAs layer was above $3x10^7$ cm^{-2} in agreement with the literature.[8] We have found that we can reduce the threading dislocation density (hereafter referred to as defect) in the active regions of the structure through the use of a proprietary fabrication technique. Devices produced with this technique are effectively defect-free. Measured defect levels are definitely below $1x10^6$ cm^{-2}, and are very probably below $5x10^5$ cm^{-2} since this level must be achieved before high efficiency devices such as those reported here can be realized. [9,10]

The degree of lattice matching for the InGaP on InGaAs alloys was determined by x-ray diffraction. The full width half maximum of x-ray diffraction ([004]) peaks was typically less than 450 arcsec. As such, the layers were lattice matched to within 0.2%. Asymmetric Bragg diffraction from inclined planes indicated that the layers were 95% to 99% relaxed. Peak widths of 150 arcsec were typical for the InGaP alloy lattice matched to GaAs within 0.25%.

The surface morphology of both InGaAs alloys and InGaP alloys were specular to the naked eye, however, the InGaP surfaces tended to demonstrate a slight hazy appearance. Figure 3 shows a comparison of the surface morphology observed for the various alloys. The upper half in each figure is the InGaP alloy surface and the lower half is the surface of the buffer revealed by wet etching down to the top layer of the buffer (either GaAs or InGaAs). Figure 3a shows the surface of a 1.85 eV InGaP alloy lattice-matched to a GaAs buffer, Figure 3b is a 1.65 eV InGaP alloy on a 1.2 eV InGaAs buffer and Figure 3c is a 1.62 eV InGaP alloy on a 1.1 eV InGaAs buffer. The GaAs surface is virtually featureless while both InGaAs alloy layers have a smooth, flat background with superimposed crosshatch pattern along the <110> directions. The crosshatch observed on the surface of the 1.1 eV InGaAs alloy is considerably more defined than the surface of the 1.2 eV InGaAs alloy. The InGaP surfaces exhibited a rough surface texture with two types of defects; small, elongated pyramids aligned with the <110> directions that completely cover the surface and larger faceted mounds at a random distribution. The mound defect distribution was affected by pre-growth exposure to contamination and could be almost completely eliminated by careful substrate handling. The rough surface texture was strongly affected by both growth conditions and the underlying layer surface roughness (crosshatch) or mismatch. The stronger the crosshatch texture in the InGaAs surface, the rougher the InGaP texture.

InGaAs Cell Performance

Both 1.1 eV and 1.2 eV InGaAs cells were fabricated in an n-on-p configuration. These cells were complete with lattice-matched InGaP front window and back surface minority carrier mirror layers. Data from the two best cells is presented.

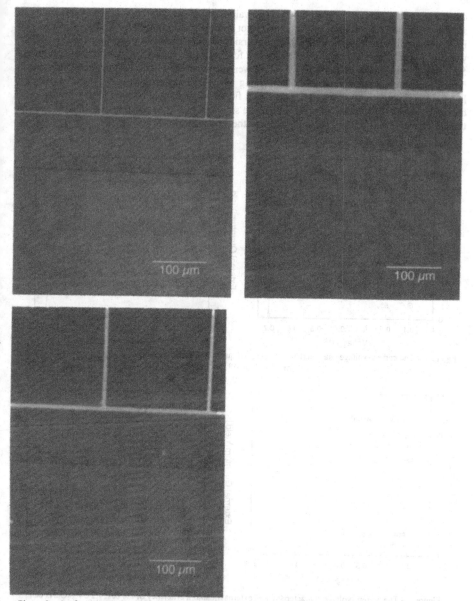

Figure 3. - Surface morphology comparison of three InGaP alloys lattice matched to GaAs or InGaAs buffers on GaAs substrates; a) 1.85 eV InGaP on GaAs (top left); b) 1.65 eV InGaP on 1.2 eV InGaAs on GaAs (top right); c) 1.62 eV InGaP on 1.1 eV InGaAs on GaAs

Figure 4 shows the current-voltage characteristic and the external quantum efficiency of the best 1.1 eV InGaAs cell. An AM0, one-sun conversion efficiency of 16.8% was achieved with outstanding fill factor and open circuit voltage values of 80% and 707 mV respectively. One can also see that the blue response of the cell was not completely optimized for a single junction cell in that a significant loss in blue light conversion occurred due to absorption in the InGaP window. Nearly identical results were obtained for the 1.2 eV InGaAs device as shown in Figure 5. Again, high Voc, FF and efficiency values were demonstrated indicating that the strain due to lattice-mismatch was relieved without forming a significant defect density in the active cell layers in agreement with the TEM results. To further demonstrate how good this result is, we compare the results to those of typical Si space solar cells. Si has a bandgap of 1.1 eV and the typical conversion efficiency is about 16% and the Voc is rarely above 600 mV.

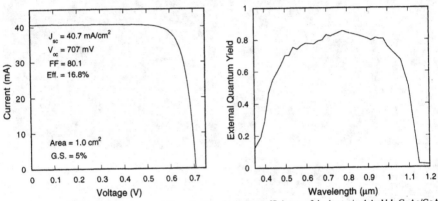

Figure 4 - The current-voltage characteristic and external quantum efficiency of the best n/p, 1.1 eV InGaAs/GaAs cell (ER515-10, AM0, one sun, 25°C).

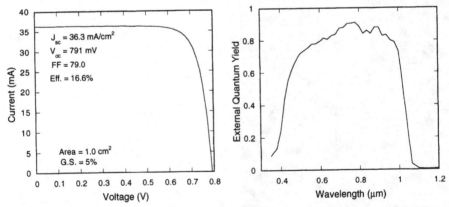

Figure 5. - The current-voltage characteristic and external quantum efficiency of the best 1.2 eV InGaAs cell (ER544-11, AM0, one sun, 25°C).

In general, the mismatched InGaAs cells perform very well indicating that a large number of crystal defects do not propagate into the active cell layers in agreement with the TEM observations.

InGaP Cell Performance

Three specific alloys of InGaP were studied; the InGaP alloy lattice-matched to GaAs (1.85 eV bandgap); the InGaP alloy (1.65 eV bandgap) lattice-matched to the 1.16 eV InGaAs; and the InGaP alloy (1.62 eV bandgap) lattice-matched to the 1.1 eV InGaAs. InGaP cells were fabricated without any surface passivating layers either at the front of the emitter as a window or at the back of the cell base. As a result we expect the current and voltage obtained to date to be below the practical efficiency limit. As previously stated, a strong surface morphology dependence was observed for InGaP alloys. The morphology appears to correlate with cell performance.

The current-voltage characteristic and external quantum efficiency for the InGaP alloy lattice-matched GaAs is shown in Figure 6. Excellent FF ant Voc values of 87.9 and 1.32 V were obtained with the only shortfall being the current collection. Our modeling indicates that an efficiency exceeding 19% could be obtained by application of effective front surface passivation to this cell. Thus, we conclude that we can grow high quality InGaP alloys.

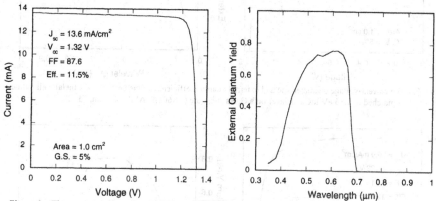

Figure 6. - The current-voltage characteristic and external quantum efficiency of the best 1.85 eV InGaP cell on a GaAs substrate (ER652-4, AM0, one sun, 25°C).

Figure 7 shows the performance of the best 1.65 eV InGaP cell grown lattice-matched to 1.16 eV InGaAs layer on a GaAs substrate. Good performance parameters were obtained, however, the current collection and Voc could be improved through effective surface passivation. This cell is projected to be 16.8 % efficient with the addition of an effective window.

Excellent quality 1.6 eV bandgap InGaP cells were fabricated as indicated by the current-voltage characteristic and quantum efficiency data in Figure 8. The value of FF is lower than expected and we believe it was due to shunt problems associated with the morphological defects at the front surface. The "flat spot" observed in the current-voltage curve, is indicative of a non-ohmic contact directly to the base forming a schottky diode in parallel with the p-n junction. This can occur when the surface is rough and the grid fingers are in contact directly to the base at large surface imperfections. We see a stronger flat spot portion in the I-V curve when we observe an increased surface roughness. None the less, we observed good Voc and Jsc values of 1.12 V and 19.9 mA/cm². The current density is very close to the operating point required for the optimized dual junction cell. As with all the InGaP devices reported here, the performance would benefit from effective front surface passivation. A significant drop in current collection from the blue region of the spectrum occurred, however, modeling indicates that material quality is excellent. Both this cell and the lattice matched 1.85 eV cell had almost identical properties in the base region i.e. diffusion lengths and low surface recombination velocity. Single junction efficiencies of 19% are attainable with the addition of an effective front surface passivation and elimination of the flat spot in the I-V curve. Our data indicates that very high quality InGaP cells can be fabricated on mismatched InGaAs layers on GaAs substrates.

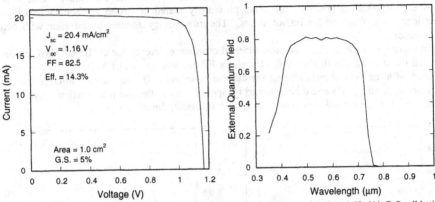

Figure 7. - The current-voltage characteristic and external quantum efficiency of the best 1.65 eV InGaP cell lattice matched to 1.16 eV InGaAs layer on a GaAs substrate (ER608-9, AM0, one sun, 25°C).

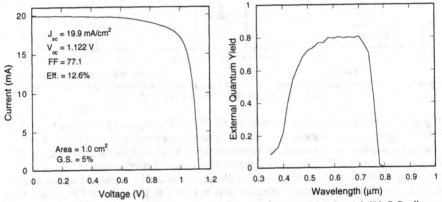

Figure 8. -The current-voltage characteristic and external quantum efficiency of the best 1.6 eV InGaP cell on a GaAs substrate (ER634-5, AM0, one sun, 25°C).

Expected Multi-junction Cell Performance

In the present cells, the n/p configuration is preferred for a bottom cell in a multi-junction device since a higher red response was obtained from n/p configured InGaAs cells compared to p/n configured cells. We calculated the expected performance of the 1.1 eV InGaAs cells in multi-junction configurations by integrating the spectral response with the filtered AM0 spectrum as previously described. The FF and V_{oc} values used in determining the achieved efficiency of the InGaAs cell, however, come from actual measured values using the AM0 spectrum (simulator) at reduced light intensities.

The results of our theoretical modeling of the complete dual junction and triple junction devices are shown in Table II and Table III. As seen by comparing the achieved efficiency to the theoretical modeling results for the individual InGaAs junction, we have in fact demonstrated the required level of performance from the mismatched InGaAs cell to achieve device efficiency of 30% with a practical multi-junction cell. In our approach, the InGaAs cell is the most critical cell since it is closest to the mismatched substrate where the highest dislocation density would be expected. This indicates that our approach toward optimized high efficiency, multi-junction solar cells on mismatched substrates will be successful.

Table II: Modeled practical and achieved efficiency for dual-junction (triple-junction with active Ge substrate) cell and their individual components, under AM0, one sun.

Solar Cell – Eg (eV)	Modeled Practical Efficiency (%)	Achieved Efficiency (%)
InGaAs (bottom) – 1.1	10.4	10.1
InGaP (top) – 1.6	16.6	12.6 (16.6)*
Dual-Junction	27.0	———
Triple-Junction	30.3	———

*Performance for an InGaP top cell projected with front surface passivation window and elimination of flat spot.

Table III. - Modeled practical and achieved efficiencies for triple-junction (quad-junction with active Ge substrate) cells and their individual components, under AM0, one-sun.

Solar Cell – Eg (eV)	Modeled Practical Efficiency (%)	Achieved Efficiency (%)
InGaAs (bottom) – 1.20	7.2	7.3
InGaAsP (middle) – 1.60	9.7	8.5%* (9.9%)**
InGaAlP (top) – 2.1	14.6	———
Triple-Junction	31.5	———
Quad-Junction	33.7	———

*Performance for InGaP middle cell (grown on GaAs) with a bandgap of 1.65 eV.
**The InGaP middle cell with an assumed InAlP window and BSF.

SUMMARY

High quality InGaAs and InGaP cells can be grown on InGaAs layers which accommodate the lattice mismatch to GaAs substrates. The surface morphology of InGaP alloys appears to be strongly dependent on the substrate surface quality with larger mismatch of the InGaAs layer causing a more pronounced cross-hatch feature, increasing the InGaP surface roughness. InGaP cell performance was correlated to surface morphology, however, excellent base properties were obtained. Individual components of optimized bandgap multi-junction cells were demonstrated and projected practical efficiencies show that efficiencies above 30% can be achieved.

ACKNOWLEDGMENTS

This work was supported by two SBIR contracts; NASA contract NAS3-98026 and Air Force contract F29601-98-C-0141. We wish to thank both funding agencies for their support.

REFERENCES

[1] J. C. C. Fan, B-Y. Tsaur, and B. J. Palm, Proc. 16th IEEE Photovoltaic Specialists Conf. (1982) p. 692.

[2] K. A. Bertness, S. R. Kurtz, D. J. Friedman, A. E. Kibbler, C. Kramer, and J. M. Olson, IEEE Proc. First World Conf. on Photovoltaic Energy Conversion (1994), p. 1671.

[3] T. Takamoto, M Yamaguchi, S.J. Taylor, E. Ikeda, T. Agui and H. Kurita, Proc of the 26th IEEE Photovoltaic Specialist Conference (1997) p. 887.

[4] D.N. Keener. Marvin, D.J. Brinker, H.B. Curtis and P.M. Price, Proc. of the 26th IEEE Photovoltaic Specialist Conference (1997) p. 787.

[5] C.R. Lewis, C.W. Ford, G.F. Virshup, B.A. Arau, R.T. Green and J.G. Werthen, Proc. of the 18th IEEE Photovoltaic Specialist Conference (1985) p. 556.

[6] J.M. Olsen, A. Kibbler and S.R. Kurtz, Proc. of the 19th IEEE Photovoltaic Specialists Conference (1987) p. 285.

[7] R.W. Hoffman Jr., N.S. Fatemi, D.M. Wilt, P. Jenkins, D.J. Brinker and D.A. Scheiman, Proc. of the 1st World Conference on Photovoltaic Energy Conversion (1994) p. 1882.

[8] V. Krishnamoorthy, P. Ribas, and R. M. Park, Appl. Phys. Lett. 58, 2000 (1991).

[9] S.M. Vernon, S.P. Tobin, M.M. Al-Jassim, R.K. Ahrenkiel, K.M. Jones and B.M. Keyes Proc. of the 21st IEEE Photovoltaic Specialist Conference (1990) p. 211.

[10] R.K. Jain and D.J. Flood, IEEE Trans.El.Dev., 40 1928 (1993).

HIGH GROWTH RATE METAL-ORGANIC MOLECULAR BEAM EPITAXY FOR THE FABRICATION OF GaAs SPACE SOLAR CELLS

A. FREUNDLICH, F. NEWMAN, L. AGUILAR, M. F. VILELA, C. MONIER
Space Vacuum Epitaxy Center, University of Houston, Houston TX 77204-5507,
alexf@orbit.svec.uh.edu

ABSTRACT

Realization of high quality GaAs photovoltaic materials and devices by Metal-organic Molecular Beam Epitaxy (MOMBE) with growth rates in excess of 3 microns/ hours is demonstrated. Despite high growth rates, the optimization of III/V flux-ratio and growth temperatures leads to a two dimensional layer by layer growth mode characterized by a (2x4) RHEED diagrams and strong intensity oscillations. The not intentionally doped layers exhibit low background impurity concentrations and good luminescence properties. Both n(Si) and p(Be) doping studies in the range of concentrations necessary for photovoltaic device generation are reported. Preliminary GaAs (p/n) tunnel diodes and solar cells fabricated at growth rates in excess of $3\mu m/h$ exhibit performances comparable to state of the art and stress the potential of the high growth rate MOMBE as a reduced toxicity alternative for the production of Space III-V solar cells.

INTRODUCTION

Driven by a demand for satellite with more on-board power, high efficiency GaAs-based single junction and multijunction solar cells have rapidly become one of the industry standards. Major photovoltaic manufacturers are already producing these devices [1,2]and the technology has captured a large share of the space solar cell market [3].

Currently such high efficiency single junction and multijunction III-V space solar cells are produced by metal-organic chemical vapor deposition (MOCVD). MOCVD requires the use of large quantities of group V hydrides such as arsine and phosphine which raises serious safety and environmental concerns, and also affects the final solar cell cost. Production of such solar cells by Molecular Beam Epitaxy (MBE), which uses solid arsenic and phosphorus species and provides access to a large variety of in situ diagnosis tools, has been proposed to minimize toxicity issues and improve yields [4,5]. MBE and related techniques have been shown to be capable of producing very high quality epilayers and devices, including solar cells [5-6]. Despite promising results and development of MBE multi-wafers systems capable of MOCVD production throughputs, MBE growth rates generally in 1 to1.5 microns per hours range (Growth rates in MOCVD systems ranges from 2-10 $\mu m/h$)has generated skepticism about the ability of MBE and related techniques as a production tool for thick (>4 μm) devices such as solar cells.

In this work it is shown that high quality GaAs photovoltaic devices can be produced by MBE with growth rates comparable to MOCVD through the substitution of group III solid sources by metal-organic compounds. The influence the III/V flux-ratio and growth temperatures in maintaining a two dimensional layer by layer growth mode and achieving high growth rates with low residual background impurities is investigated. Finally subsequent to the study of the

Mat. Res. Soc. Symp. Proc. Vol. 551 © 1999 Materials Research Society

optimization of n- and p doping of such high growth rate epilayers, results from a preliminary attempt in the fabrication of GaAs photovoltaic devices such as tunnel diodes and solar cells using the proposed high growth rate approach are reported.

GROWTH AND CHARACTERIZATION

All growth runs are performed in Riber CBE32 reactor. Tri-ethyl-gallium (TEG) and tri-methyl-indium (TMI) are used as group III precursors. The organometallics(OM) are maintained at 40C and introduced mixed with hydrogen through a low temperature(80C) injector. The OM flux is adjusted using mass flow controllers. To allow an optimum operation of the flow controller the down-stream pressure of the OM/H$_2$ mixture is maintained at 40 torr using pressure controllers. The group V element flux was provided by a tetrameric arsenic (As$_4$) source. The Arsenic flux at the substrate was between 1-4 x 10^{-5} torr. Si and Be solid sources are used as n-type and p-type dopants respectively.

GaAs layers are grown with growth rates ranging from 0.5 to 3 monolayers (ML) per second (0.5-3.2 microns per hour) on nominal (100) highly Si doped GaAs wafers. The GaAs surface reconstruction and growth mode is monitored real- time using reflection high energy electron diffraction (RHEED). Substrate temperatures are measured using an IR sensitive pyrometer. The precursor decomposition and species partial pressures (flux) in the reactor were monitored using mass spectrometry.

Influence of growth conditions

The GaAs growth rate as determined through the intensity oscillations of the specular beam in the (2x4) reconstruction RHEED diagram is found to increase linearly with the increasing TEG flux. However the growth rate saturates for fluxes exceeding 16sccm of TEG/H$_2$ mixture to slightly above 3 ML/sec. It is worth noting that beyond this threshold partial and total pressure analysis indicated that the OM pressure in the chamber remained unchanged. The latest suggests that the growth rate is most likely limited by the conductance of our TEG/H$_2$ ¼inch-diameter lines.

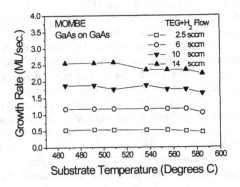

Figure 1: GaAs growth rate as a function of growth temperature for various TEG fluxes 14 sccm (solid up triangle) 10 sccm (solid down triangle) 6 sccm (open circle) and 2.5 sccm (open square). Solid lines are provided only for eye guidance.

The MOMBE growth of GaAs is investigated over a wide range of temperature. A 2x4 RHEED diagram and a nearly constant growth rate are obtained within a range of 460-560 C (Figure 1). A slight drop in the growth rate is noticed for temperatures exceeding 570C suggesting a reduction in the Ga sticking coefficient. Growths performed at substrate temperatures below 500C or/and with high As partial pressures were characterized by a fuzzy RHEED diagram suggesting the epilayer surface roughening. Strong RHEED intensity oscillations demonstrating a layer by layer growth mode are observed in the range of 520-580C. In optimum growth conditions even at growth rates exceeding 3 monolayer/sec over hundred RHEED intensity oscillations are observed (Figure 2).

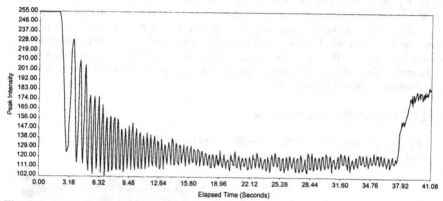

Figure 2: RHEED specular beam intensity oscillations recorded during the growth of GaAs (growth rate 2.62 ML/sec, temperature 570 C)

All GaAs layers grown in optimum conditions (520-590C) exhibit a specular morphology. Background carrier concentrations determined using C-V electrochemical pofilometry indicate the possibility of obtaining high growth rate GaAs layers with residual active impurity levels well below 10^{16}cm^{-3} range. The bound excitonic transitions at 1.512 eV and Carbon neutral acceptor-band (e-C0) at 1.494 eV dominate the 10K photoluminescence spectra of these samples. As for conventional MBE/CBE samples, these photoluminescence analysis suggests carbon as the main residual acceptor impurity in our high growth rate not intentionally doped epilayers.

High growth rate GaAs n and p type doping

n and p type intentional doping of GaAs layers grown at 3 micrometer/hour was performed using Be and Si respectively. Intentional beryllium p-type doping ranging from 5×10^{17} cm^{-3} to 2×10^{20} cm^{-3} and silicon n-type doping ranging from 3×10^{17} cm^{-3} to 8×10^{18} m^{-3} were obtained reproducibly. Doping concentrations are compatible with the development of photovoltaic devices such as solar cells and tunnel diodes. The activation energies associated with dopant incorporation in the solid phase are in agreement with those observed for materials grown with lower growth rates and are in the order of 3.4 eV and 6.2 eV for Be and Si respectively.

PRELIMINARY PHOTOVOLTAIC DEVICE PERFORMANCE

Finally in order to validate the compatibility of the developed high growth rate epilayers for the fabrication of GaAs based advanced solar cells a preliminary attempt is made to fabricate a GaAs p/n solar cell and a tunnel diode.

Device structures are grown at rates of about 3 microns /hour on a highly n-doped 3 inch diameter (100) GaAs wafer at a temperature of 530-560 C. Following the growth process non-alloyed Au metallic layers, deposited by vacuum evaporation, are used as n-type (on substrate) and p-type ohmic contacts. The Au-top contact grid is obtained using a lift off technique.

For the solar cell devices an intermediate highly p-doped GaAs layer is introduced to lower the top contact resistivity. A mesa etching of the structure provides a total area per cell of 5 x 5 mm^2. The top-grid shadowing is estimated to be about 7.5%. The GaAs contact layer was removed from the cell active area using a selective etching and the $Al_{0.8}Ga_{0.2}As$ window layer as an etch stop. Following this step a two layer MgF_2/ZnS anti-reflection coating is implemented to minimize the solar cell reflective losses. Concerning the tunnel diode device processing a mesa is etched directly using the top contact dots as a mask The latest provides a set of devices with mesa diameters of 100 and 200 microns.

GaAs p/n Tunnel diodes characterization

Figure 3 shows the C-V profile of the fabricated tunnel diode stressing the high dopant incorporation and the sharpness of the p/n interface.

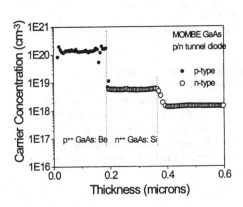

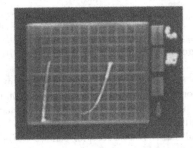

Figure 4: I-V characteristics of a 100 micron diameter MOMBE GaAs tunnel diode.
Vertical: 2mA/div.; horizontal 100mV/div.
Jp= 162 A/cm^{-2}

Figure 3 : Carrier concentration profile of MOMBE grown GaAs tunnel diode.

All diodes exhibit high peak current densities in excess of 150Acm^{-2}, a good peak to valley ratio and a specific resistivity of less than 0.5 mΩcm^2 for a peak current density of 30 A/cm^2. A typical current- voltage characteristic is reported in Figure 4. It is worth nothing that this preliminary tunnel diode device outperforms best MBE-grown GaAs devices[4]

(Jp=46A/cm⁻²) and has a performance comparable to best results achieved by carbon-doped MOCVD tunnel diodes [7]. The associated voltage drop in a tandem device operating using such tunnel diodes even under moderate concentration (x20-x50) is expected to be below few mV and therefore negligible.

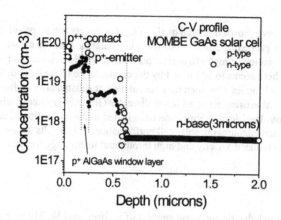

Figure 5: Carrier Concentration profile in GaAs MOMBE solar cell

GaAs p/n Solar cell

All solar devices exhibit good PV characteristics. Typical room temperature spectral response of this preliminary device is provided in Figure 6 and compares well with data reported in the literature for high quality GaAs MOCVD grown solar cells [1,2,8]. The typical device short circuit current exceed 32 mA cm⁻² under one sun AM0 simulation with the best device on the wafer reaching 32.5 mAcm⁻² (comparable to the highest current densities reported for GaAs solar cells). Devices exhibit an open circuit voltage in excess of 950 mV, about 5% below the state of the art GaAs solar cells. Despite a poor fill factor 74-78 % (compared to 86-88% for best GaAs

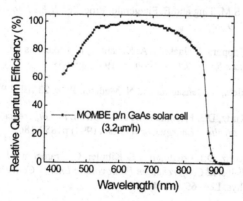

Figure 6: External quantum efficiency of high growth rate MOMBE GaAs p/n solar cell.

solar cell) probably associated with a non optimized GaAs solar cell design/processing all cells exhibit AM0 efficiencies in excess of 16 %. We believe that through the optimization of cell design and processing cells with efficiencies in excess of 20% efficient cell would be easily achieved with high growth rate Metal-organic MBE.

CONCLUSIONS

In this work we have shown that high quality GaAs epilayers can be produced by MOMBE at growth rates exceeding 3 µm/h. The material characteristics are shown to be compatible with the fabrication of high efficiency space photovoltaic devices. As mentioned earlier the growth rates here seems to be limited by the conductance of the TEG delivery line rather than by the growth kinetics. One then may speculate that substantially higher growth rates would technically achievable using standard larger diameter OM delivery line. Preliminary high growth rate GaAs photovoltaic device fabrication has yielded to the realization of record breaking tunnel diodes and encouraging solar cell performance. This results demonstrate the potential of MOMBE as reduced toxicity and high throughput technology for the production III-V space photovoltaics.

AKNOWLEDGEMNET

Authors gratefully acknowledge the technical support of S. Street and W. Thompson. This work was partially supported by the NASA Cooperative agreement NCC8-127.

REFERENCES

1. Y.C.M. Yeh, C.L. Chu, J. Kinger, F.F. Ho, J.M. Olson, M. Timmons, in Proc.25th *IEEE Photovoltaic Specialist Conference*, (1996) p187.

2. E.B. Linder, J.P. Hanley in *Proc. 25th IEEE Photovoltaic Specialist Conference*(1996), p267

3. M. Meyer, R. E. Metzger, Compound Semiconductors (Special Issue on current status of the semiconductor Industry), pp 40-41(1997).

4. S. Ringel, R.M. Sieg, S.M. Ting and E. Fitzgerald, Proc. 26th *IEEE Photovoltaic Specialist Conference*, 1997, p793-799

5. J. Lammasniemi, K. Tappura, R. Jaakola, A. Kazantev, K. Rakennus P.Uusimaa, M. Pessa, in Proc.25th *IEEE Photovoltaic Specialist Conference*, 1996 , p97-100

6. A. Freundlich, M. Vilela, A. Bensaoula and N. Medelci, Proc. 23 IEEE PVSC (1993) p 685

7. K.A. Bertness, S.R. Kurtz, D.J. Friedman and J. M. Olson, *in Conference records of 1st World Conference on Photovoltaic Energy Conversion (*1994) p1859-1862

8. K.A. Bertness, S.R. Kurtz, D.J. Friedman A.E. Kibbler, C. Kramer and J. M. Olson *in Conference records of 1st World Conference on Photovoltaic Energy Conversion* (1994) pp 1671-1675. *ibid* Appl. Phys. Lett. **65**, (1994) p989

INFRARED-SENSITIVE PHOTOVOLTAIC DEVICES FOR THERMOPHOTOVOLTAIC APPLICATIONS

NAVID S. FATEMI[1], DAVID M. WILT[2], RICHARD W. HOFFMAN, JR.[1], AND VICTOR G. WEIZER[1]
[1]Essential Research, Inc., Cleveland, OH
[2]NASA Lewis Research Center, Cleveland, OH

ABSTRACT

One of the critical components in a thermophotovoltaic (TPV) system is the infrared-sensitive photovoltaic (PV) semiconductor device that converts the emitted radiation to electricity. Currently, several semiconductor material systems are under development by various workers in the field. The most common are InGaAs/InP, GaSb, and InGaSbAs/GaSb. These devices normally have electronic energy bandgap values in the range of 0.50-0.74 eV. In addition, the design and structure of these devices fall into two distinct formats: conventional planar and monolithic interconnected module (MIM). The conventional planar devices normally have one semiconductor junction and are high-current/low-voltage devices. In a MIM, small area PV cells are connected in series monolithically, on a semi-insulating substrate. This results in the formation of a single high-voltage/low-current module. Electrical and optical performance results for MIMs with electronic energy bandgaps of 0.60 and 0.74 eV will be presented.

INTRODUCTION

A typical thermophotovoltaic (TPV) system is comprised of a heat source, a radiator (or emitter), an spectrum control element, a heat rejection system, and an infrared-sensitive photovoltaic (PV) device. The heat source provides thermal energy to the radiator. The radiator will in turn emitt radiation (photons) with a characteristic spectral power distribution. The emitted spectrum is shaped by the spectrum control element to better match the spectral response of the PV cell and to recuperate (recycle) the non-convertible radiation. The PV cell converts this radiation to electricity and delivers it to the load. The heat rejection system keeps the temperature of the PV cell to acceptably low levels (i.e., near room temperature) for efficient light to electric energy conversion.

Historically, two types of infrared-sensitive PV devices have been under development for TPV applications: conventional planar one-junction cells and monolithic interconnected module (MIM) devices [1-6]. Most planar cells are composed of InGaAs, GaSb, or InGaSbAs [7-9]. Thus far, however, only InGaAs has been used to fabricate MIM devices.

In a MIM, small area InGaAs photovoltaic (PV) cells are connected in series monolithically, on a semi-insulating (S.I.) InP substrate. This results in the formation of a module with a very desirable power profile, i.e., a high-voltage/low-current configuration. There are other advantages to a MIM device as compared to conventional planar devices. A prominent advantage of MIM is its ability to efficiently recuperate (recycle) the incident non-convertible infrared (IR) radiation.

Optical recuperation is mandatory to achieve high efficiencies in a TPV system. Conventionally, separate front surface filtering elements, such as dielectric stack bandpass, or plasma, or tandem filters are used in front of a PV cell to accomplish optical recuperation. These elements, however, attenuate the useful in-band radiation reaching the PV cell. As a result, the cell's output power density is diminished. They not only add complexity to the design of a TPV system, but they also tend to absorb both in-band and out-of-band radiation.

An alternative approach to optical recuperation is the use of a back surface reflector (BSR). A highly IR-reflective material such as gold can be used on the back side of the semi-insulating InP wafer to accomplish this task. As shown schematically in Figure 1, the fabrication of MIM requires the use of a semi-insulating (S.I.) InP substrate. Unlike doped InP substrates, S.I. InP is transparent to infrared (IR) radiation. Radiation with wavelengths greater than the device bandedge wavelength (i.e., ≥ ~2 μm) can be reflected back to the TPV radiator via the BSR. The reflected radiation is optically recycled by reducing the net heat flux that must be maintained by the heat source.

In fact, spectral utilization (SU) factors greater than 70% have been measured for a MIM with a BSR, and a bandgap of 0.60 eV [10]. The SU factor is defined as the above-bandgap absorbed energy in the active cell layer divided by the total absorbed energy. Therefore, it is desirable for the SU factor to be as large as possible. This SU factor is greater than what has been observed with the most advanced filtering options currently available (i.e., SU of <70%) [11]. The use of a BSR in a MIM configuration, therefore, eliminates the need to use separate front surface filters.

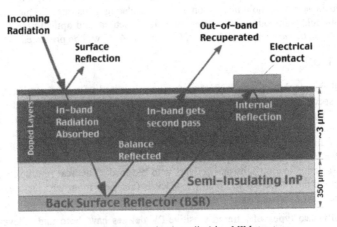

Figure 1.—Optical recuperation (recycling) in a MIM structure.

Another advantage to a MIM design is that both the negative and positive electrical connections are fabricated on the top side of the module, thereby simplifying the array design, interconnection, and thermal management. The completed device may be soldered directly onto the array substrate/heat sink without having to provide electrical isolation. Individual MIMs can then be connected in series or parallel configurations by welding or bonding of metallic interconnect ribbons to adjacent busbars. A photograph of an array, comprised of twelve 1x1-cm conventional MIMs, interconnected by this process is shown in Figure 2.

Figure 2.—Photograph of an array of twelve 1x1-cm MIMs on a heatsink.

In a TPV system, the bandgap of the PV device is ideally selected to match the peak intensity of the radiator (emitter) spectral power density at the operating temperature. At low to moderate radiator temperatures (i.e., 1200-1500 K), the PV cells used normally have bandgap values of 0.74 eV or smaller. The most commonly studied devices are the lattice-matched InGaAs-on-InP and GaSb cells, with energy bandgaps (Eg) of 0.74 and 0.73 eV, respectively. The spectral response of the PV cells with lower bandgaps, however, are better matched to the greybody spectral power density profile for low to moderate radiator temperatures. Examples of cells with lower bandgaps are lattice-mismatched InGaAs-on-InP and lattice-matched InGaAsSb-on-GaSb.

In Figure 3, for example, the spectral response for two high-quality InGaAs/InP MIM devices with bandgaps of 0.74 and 0.60 eV are overlaid on top of the spectral power density curve for a blackbody at a temperature of 1500 K. As shown in the figure, the response of the lattice-mismatched device with Eg=0.60 eV is far better matched to the peak intensity of the blackbody power curve. As a result, in the example shown, the output current of the 0.60-eV device is more than 80% greater than the output current for the 0.74-eV device.

We have fabricated and tested both lattice-matched and lattice-mismatched p/n InGaAs/InP MIMs, grown by organo-metallic vapor phase epitaxy (OMVPE). High performance lattice-mismatched MIMs were developed using a compositionally step-graded InPAs buffer to accommodate a lattice mismatch of 1.1% between the active InGaAs cell structure and the InP substrate. In what follows, we will present data regarding the electrical performance of these devices.

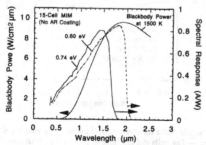

Figure 3.—Measured spectral response data for 15-cell InGaAs/InP MIMs, with Eg = 0.60 eV and 0.74 eV (with no anti-reflection coating), and the blackbody spectral power density curve at 1500 K.

RESULTS

External quantum efficiency (QE) and high-intensity illumination current versus voltage (I-V) measurements were performed to characterize the MIMs. The high illumination tests were carried out using a large-area pulsed solar simulator (LAPSS). We will present data for p/n InGaAs/InP lattice-matched 0.74-eV and lattice-mismatched 0.60-eV MIMs.

External Quantum Efficiency (QE) Data

Initially, conventional planar one-junction cells were fabricated and tested. Excellent current collection over a wide range of wavelengths (i.e., 0.5 to ~2.0 μm) was measured with these devices. This data is shown in Figure 4. Note that the cells did not have an anti-reflection (AR) coating. The estimated internal quantum efficiency was near unity near the bandedge for both 0.60-eV and 0.74-eV devices. The 0.60-eV lattice-mismatched cells had a compositionally step-graded InAsP buffer layer grown between the InP substrate and the active regions of the device. The function of this buffer was to limit the propagation of threading dislocations from the buffer-InP interface to the top active layers of the device. The QE results shown below demonstrate the effectiveness of the buffer.

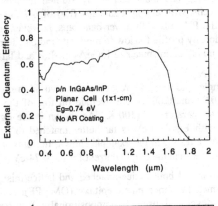

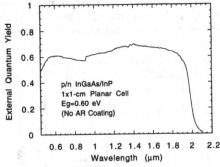

Figure 4.—External quantum efficiency data for conventional planar cells with Eg=0.74 eV (top), and Eg=0.60 eV (bottom) – The cells did not have anti-reflection (AR) coatings.

The QE data for the MIM structures were similar to the data measured for the planar cells. The main difference between the planar cell structure and the MIM structure was that the latter had an extra lateral conduction layer (LCL) layer [1-4]. This layer was composed of highly-doped (1×10^{19} cm^{-3}) n-type InGaAs, and had a thickness of 1 μm. It is positioned underneath the base region of the device and is separated from the base by an InP (InAsP for the 0.60-eV device) stop etch/back surface reflector layer. The QE plots for these MIM structures are illustrated in Figure 5. Note that these devices did not have any AR coating.

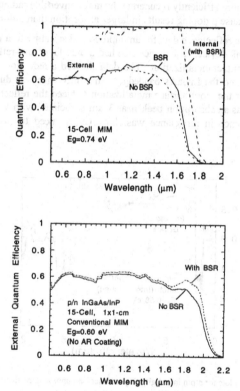

Figure 5.—External quantum efficiency data for MIMs with Eg=0.74 eV (top), and Eg=0.60 eV (bottom). The MIMs did not have anti-reflection (AR) coatings.

The above QE data also show the effect of the BSR in improving the current collection, specially near the bandedge. This is due to the fact that the part of the convertible in-band radiation that was not collected in the first pass through the active regions of the device, was reflected back via the BSR for a second pass through the device, and a second chance for being collected. Also shown in the QE plot for the 0.74-eV device is the calculated internal QE. As seen in the figure, this internal QE was unity near the bandedge.

Infrared (IR) Reflectivity Data

To maximize the optical efficiency, the reflectance of the MIMs (with a BSR), beyond the bandedge wavelength must be as high as possible. This allows for the majority of the non-convertible IR photons to be reflected back to the TPV radiator for recuperation. Therefore, we performed total IR reflectivity measurements of the MIMs, before and after they were processed into complete devices. Both structures had thick (~2 μm) gold BSRs. For these measurements 0.58-eV n/p/n polarity [12] InGaAs/InP interdigitated MIMs [4] were tested. The n/p/n polarity is expected to be able to more efficiently recuperate the non-convertible radiation when compared to a p/n polarity [12], because p doping results in larger absorption than n doping.

Figure 6 compares the reflectance data for an unprocessed structure to a processed MIM for a 0.58-eV n/p/n InGaAs/InP device. The device also had a dual layer anti-reflective (AR) coating (ZnS/MgF$_2$) and Cr/Au contact metallization. The data showed a reduction in the reflectance of the processed device beyond the InGaAs bandedge. This difference was due to several factors. The Cr adhesion layer for the front surface metallization reduced the reflectance in the 2-4 μm region. Overlying that was an absorption peak near 3 μm associated with water absorbed in the AR coating. The difference in reflectance was, however, reduced to insignificant levels by approximately 7 μm.

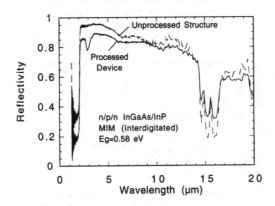

Figure 6.—Reflectivity plots for n/p/n interdigitated processed and unprocessed structures (Eg=0.58 eV)

As shown in the figure, very high non-convertible photon reflectivity values can be measured with the MIM. As a result, a MIM device with a BSR can be optically more efficient than a conventional cell with a separate front surface filtering element.

Current-Voltage (I-V) Data

The MIMs were tested under high-intensity illumination, using a large-area pulsed solar simulator (LAPSS), to assess their performance under simulated operating conditions. The results for the variation in the open-circuit voltage (Voc) and fill factor (FF) with the short-circuit current density (Jsc) are given in Figures 7 and 8, respectively.

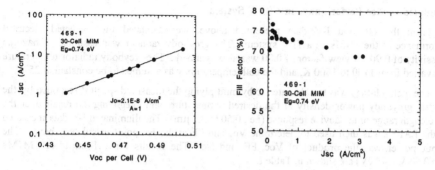

Figure 7.—Variation of Voc (left) and FF (right) with Jsc for a 30-cell InGaAs/InP MIM, with Eg=0.74 eV.

As shown in figure 7 (left), the diode ideality factor (A) for the 0.74-eV MIM was unity. The reverse saturation current density (Jo) was measured to be 2.1×10^{-8} A/cm^2. This value is at or near the best Jo values reported in the literature for a PV device (planar or MIM) with a bandgap of 0.74 eV [13]. We have also observed Jo values, as low as 9.1×10^{-9} A/cm^2 for 1x1-cm planar (non-MIM) p/n InGaAs/InP 0.74-eV cells. Very high FF values were also observed with this MIM. Fill factors remained on or above the 70% mark for Jsc values of up to 2 A/cm^2. The peak FF value of 77% occurred at Jsc of about 0.18 A/cm^2.

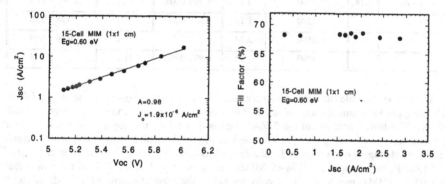

Figure 8.—Variation of Voc (left) and FF (right) with Jsc for a 15-cell InGaAs/InP MIM, with Eg=0.60 eV.

The diode ideality factor (A) for the 0.60-eV MIM was also unity. The reverse saturation current density (Jo) was measured to be 1.9×10^{-6} A/cm^2. This Jo value is more than an order of magnitude lower than the best results reported in the literature for a PV device (planar or MIM) with Eg=0.60 eV [13]. We have also observed Jo values, as low as 4.1×10^{-7} A/cm^2 for 1x1-cm planar (non-MIM) p/n InGaAs/InP 0.60-eV cells.

High fill factor values were also observed with this MIM. The Fill factor values essentially remained flat at about 68-69% for the short-circuit current densities up to about 3 A/cm^2. This result suggests that the total series resistance of the device was negligibly small.

Simulated Device Performance in a TPV System

From the QE and I-V data presented above, we calculated the expected electrical performance of the MIMs in a TPV system. The greybody radiator was assumed to have an emissivity of 0.90. A view factor of 0.90 was also assumed . The greybody radiator temperature was varied from 1100 to 1500 K, and the cell temperature was assumed to be constant at 25°C.

The methodology was to calculate Jsc by multiplying the measured spectral response by the spectral greybody power density at the desired temperature, and integrating the result over the wavelength range of the device response (i.e., 0.50 to 2.1 μm). The illuminated I-V data measured by the LAPSS was then used to determine Voc and FF at the Jsc point calculated above. The output power was the product of Voc, FF, and Isc. The results for 1x1-cm 15-cell MIMs (Eg=0.60 & 0.74 eV) are shown in Table I.

Table I.—Electrical performance of p/n InGaAs/InP MIMs under simulated TPV conditions.

MIM Device Dimensions	Eg (eV)	Radiator Temperature (K)	Jsc (A/cm^2)	Voc (V) / Voc per Cell (V)	Fill Factor (%)	Power (W) / Power Density (W/cm^2)
1 x 1 cm	0.60	1100	0.69	4.69 / 0.313	68.1	0.12 / 0.15
		1300	2.36	5.26 / 0.351	68.3	0.45 / 0.57
		1500	6.00	5.64 / 0.376	64.9	1.15 / 1.46
1 x 1 cm	0.74	1100	0.27	5.70 / 0.380	77.1	0.063 / 0.079
		1300	1.13	6.30 / 0.420	76.1	0.28 / 0.36
		1500	3.30	6.66 / 0.444	75.0	0.86 / 1.10
		1700	7.69	6.95 / 0.463	71.3	2.00 / 2.54

The data in the table show that relatively large electrical output power, in the range of 0.45-1.15 watts, can be obtained with 0.60-eV MIMs at moderate radiator temperature range of 1300-1500 K. The better matching of the 0.60-eV versus 0.74-eV device response to the blackbody peak radiation curve (see Figure 3), and the high quality of the lattice-mismatched devices, result in significantly higher output power for the lower bandgap devices. In the temperature range of 1300-1500 K, for example, the 0.74-eV MIMs produced only about 62-75% of the power that the 0.60-eV MIMs produced. It should be noted that the structure of the MIMs with Eg=0.60 eV was optimized for operation with a radiator temperature of about 1300 K, whereas the structure of the MIMs with Eg=0.74 eV was optimized for operation with a radiator temperature of 1500 K.

We believe that the data presented in Table I are practical values that we can expect to observe under actual TPV conditions. Specifically, in a separate experiment we have demonstrated that our calculated data for the 0.74-eV MIMs are in close agreement with the measured experimental data taken under actual TPV conditions. An array of 12, 0.74-eV, 1x1-cm MIMs (see Figure 2) was coupled to a SiC radiator heated to 1300 K by a combustor. The measured experimental Voc and Isc values were 6% higher and 2% lower than the calculated values, respectively.

SUMMARY

The results presented in this work are summarized below:

1. We have fabricated and tested both lattice-matched and lattice-mismatched p/n InGaAs/InP monolithic interconnected modules (MIM), with bandgaps of 0.74 eV and 0.60 eV, respectively. These devices had a gold back surface reflector (BSR) layer to allow for efficient radiation recuperation, when used in a TPV system.

2. The external quantum efficiency data for the both structures showed very good collection efficiency. The internal QE was estimated to be near unity close to the bandedge.

3. MIM devices were tested under high-intensity illumination. The diode ideality factor (A) and the reverse saturation current density (Jo) for the MIMs with Eg=0.74 eV were near or at unity, and 2.1×10^{-8} A/cm2, respectively. Lower Jo values (i.e., 9.1×10^{-9} A/cm^2) were also observed for the planar (non-MIM) structures. The fill factor values remained at or above 70% for short-circuit current densities up to 2 A/cm2.

4. The diode ideality factor (A) and the reverse saturation current density (Jo) for the MIMs with Eg=0.60 eV were near or at unity, and 1.9×10^{-6} A/cm2, respectively. Lower Jo values (i.e., 4.1×10^{-7} A/cm^2) were also observed for the planar (non-MIM) structures. The fill factor values remained at or above 68% for short-circuit current densities up to 3 A/cm2.

5. From the measured QE and I-V data, the expected output power for the 0.60-eV devices, when coupled to a greybody radiator (emissivity of 0.9 & view-factor of 0.9) was calculated. A relatively large power output, in the range of 0.45-1.15 watts, was calculated at the moderate radiator temperature range of 1300-1500 K.

REFERENCES

[1] N. S. Fatemi, D. M. Wilt, P.P. Jenkins, V.G. Weizer, R.W. Hoffman, C.S. Murray, D. Scheiman, D. Brinker, and D. Riley, "InGaAs Monolithic Interconnected Modules (MIMs)," Twenty Sixth Photovoltaic Specialists Conf. (PVSC), 1997, p. 799.

[2] D. M. Wilt, N. S. Fatemi, P.P. Jenkins, V.G. Weizer, R.W. Hoffman, R.K. Jain, C.S. Murray, and D. Riley, "Electrical and Optical Performance Characteristics of 0.74 eV p/n InGaAs Monolithic Interconnected Modules," Third NREL TPV Conference, 1997.

[3] N. S. Fatemi, D. M. Wilt, P.P. Jenkins, R.W. Hoffman, V.G. Weizer, C.S. Murray, and D. Riley, "Materials and Process Development for the Monolithic Interconnected Module (MIM) InGaAs/InP TPV Devices", Third NREL TPV Conference, 1997.

[4] J. S. Ward, A. Duda, M.W. Wanlass, J.J. Carapella, X. Wu, R.J. Matson, T.J. Coutts, and T. Moriarty, "A Novel Design for Monolithically Interconnected Modules (MIMs) for Thermophotovoltaic Power Conversion," Third NREL TPV Conference, 1997, p. 227.

[5] D. M. Wilt, N. S. Fatemi, P.P. Jenkins, R.W. Hoffman, G.A. Landis, and R.K. Jain, "Monolithically Interconnected InGaAs TPV Module Development," Twenty fifth Photovoltaic Specialists Conf. (PVSC), 1996.

[6] S. Wojtczuk, "Comparison of 0.55 eV InGaAs Single-Junction vs. Multi-Junction TPV Technology," Third NREL TPV Conference, 1997.

[7] D. M. Wilt, N. S. Fatemi, R.W. Hoffman, P. Jenkins, D. Brinker, D. Scheiman, & R. Jain, "High efficiency InGaAs photovoltaic devices for thermophotovoltaic applications," Appl. Phys. Lett., **64**, 2415 (1994).

[8] L.M. Fraas, H.X. Huang, S.Z. Ye, S. Hui, J. Avery, and R. Ballantyne, "Low Cost High Power GaSb Photovoltaic Cells," Third NREL TPV Conference, 1997, p. 33.

[9] C.A. Wang, H.K. Choi, G.W. Turner, D.L. Spears, M.J. Manfra, and G.W. Charache, "Lattice-Matched Epitaxial GaInAsSb/GaSb Thermophotovoltaic Devices," Third NREL TPV Conference, 1997, p. 75.

[10] M.B. Clevenger, C.S. Murray, and D.R. Riley, "Spectral Utilization in Thermophotovoltaic Devices," Materials Research Society (MRS) Fall Meeting, December 1997, Boston, MA.

[11] Data presented by Knolls Atomic Power Labs (KAPL) in a review meeting held in January of 1998 in Schenectady, NY.

[12] D.M. Wilt, C.S. Murray, N.S. Fatemi, and V.G. Weizer, "n/p/n Tunnel Junction InGaAs Monolithic Interconnected Module (MIM)," Fourth NREL TPV Conference, Denver, CO, 1998.

[13] G.W. Charache, J.L. Egley, L.R. Danielson, D.M. Depoy, P.F. Baldasaro, B.C. Campbell, S. Hui, L.M. Fraas, and S.J. Wojtczuk, "Current Status of Low-Temperature Radiator TPV Devices", Twenty-fifth IEEE PVSC, 1996, p.137.

A TESTBED FOR TESTING
MATERIALS PROPERTIES IN SPACE

GEOFFREY A. LANDIS*, SHEILA G. BAILEY**, and ANDREW SEXTON[†]
*Ohio Aerospace Institute, NASA Lewis Research Center mailstop 302-1, Cleveland, OH 44135
**NASA Lewis Research Center, Photovoltaics Branch, mailstop 302-1, Cleveland, OH 44135
†Dynacs, Inc., 2001 Aerospace Parkway, Brook Park, OH 44142

ABSTRACT

The Photovoltaic Engineering Testbed (PET) is a facility to fly on the International Space Station to test advanced solar cell types in the space environment. The purpose is to reduce the cost of validating new technologies and bringing them to spaceflight readiness by measuring them in the in-space environment. The facility is scheduled to be launched in 2002.

PURPOSE

The Photovoltaic Engineering Testbed (PET) is a space-exposure test facility to fly on the International Space Station to test advanced solar cell types in the space environment. The purpose is to reduce the cost of validating new technologies and bringing them to spaceflight readiness by measuring them in the in-space environment. The facility is scheduled to be launched in 2002.

The facility will be used for three primary functions: calibration, measurement, and qualification [1].

Calibration is used to create a space-measured reference cell which can be used to calibrate the performance of cells on the ground, allowing ground measurements to be referenced to an actual space-flown standard of the same type.

Measurement consists of measurements of the performance of cells in space, in particular measuring values of parameters which are sensitive to the space solar spectrum or environment, such as the temperature coefficient.

Qualification consists of verifying that the performance of an interconnected solar-cell coupon does not degrade over time in the space environment.

FACILITY

The PET facility is shown in figures 1 and 2.

The preferred location for the PET facility is on the Exposed Facility of the Japanese Experiment Module (JEM-EF). This location allows simplicity of exchange of samples. A location on the Express pallet, at the end of the truss, is also possible. More details can be found in references 1 and 2.

Mat. Res. Soc. Symp. Proc. Vol. 551 © 1999 Materials Research Society

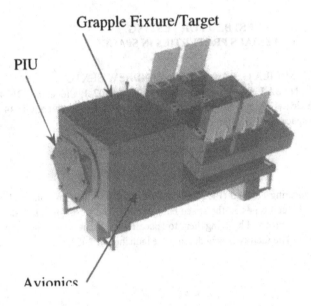

Grapple Fixture/Target

PIU

Avionics

Figure 1:
PET facility (configured for the Japanese Experiment Module Exposed Facility). The hexagonal fixture on the end is the interface to the JEM (labeled PIU, or "Payload Interface Unit"), and the circular disk at the top is the robotic grapple fixture to remove or re-emplace the facility. The beta-axis tracking tray is shown with the four sample holders mounted and the exposure doors open to space. In this view, the zenith (space facing) direction is up, and the nadir (Earth-facing) direction down.

The PET facility is designed for samples to be changed out regularly. Sample change-out is accomplished by the robotic manipulator, by removing the upper tray from the Beta tracking platform. This change-out process is shown in figure 3. The upper tray has four sample holders, which include electronics. New samples can be taken to the station on any resupply flight, and the actual exchange is done inside the pressurized facility, without requirement of extravehicular activity. Once new samples are placed on the upper tray, it is cycled through the airlock and re-emplaced on the PET tracking platform by the robotic manipulator.

The solar spectrum in space is referred to as "Air-Mass Zero" (or "AM0"). This indicates that the mass of air between the solar cell and the sun is zero, or true space conditions. At the surface of the Earth, for light passing perpendicularly through the atmosphere, the mass of air between the cell and the sun is exactly one times the atmospheric thickness; this is thus known as "Air-Mass One" (or AM1). Solar cell performance at air-mass 1 is considerably different from air-mass zero performance, due to the selective absorption of the atmosphere. A purpose of the PET space calibration facility will be to allow routine measurements of performance parameters at true AM0, in actual space sunlight.

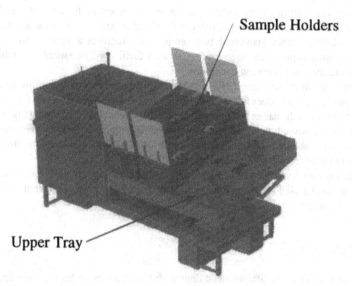

Sample Holders

Upper Tray

Figure 2.
The PET facility viewed from the opposite angle.

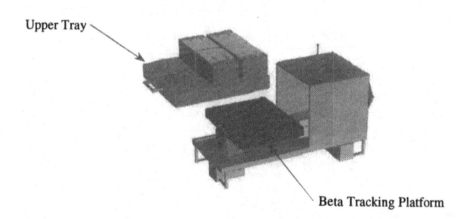

Upper Tray

Beta Tracking Platform

Figure 3
Sample change-out is accomplished by removing the upper tray from the Beta tracking platform. The upper tray has four sample holders, which include electronics.

Since solar cells on-orbit typically operate at temperatures between 45 and 100°C, characteristic of sun-facing surfaces in Earth orbit. The standard measurement temperature for solar cells is 28°C. This standared test temperature requires a temperature coefficient measurement to predict performance in space. The PET facility will measure performance of the cells under actual in-space operating conditions.

Finally, the qualification function of the testbed will be used to verify that the performance of an interconnected coupon does not degrade in the space environment, which is the final critical step leading to flight qualification and acceptance of a new technology. By doing flight exposure on the space station testbed, the qualified samples can be returned for examination after the test. This will give us the ability to diagnose failure mechanisms (if any), allowing a technology to be fine-tuned as required to pass performance specifications.

Although the PET facility was designed for testing photovoltaic cells and coupons, the versatile design can be used for space environment testing of any type of materials or electronic components of interest.

REFERENCES

1. G. Landis and S. Bailey, "Photovoltaic Engineering Testbed on the International Space Station," presented at the 2nd World Conf. on Photovoltaic Energy Conversion, Vienna, Austria, July 1998 (Proceedings in press).

2. G. Landis and A. Sexton, "An Engineering Research Testbed for Photovoltaics," to be presented, STAIF Conference on Space Station Utilization, Albuquerque NM, Jan. 1999.

IMPURITY CHARACTERIZATION OF SOLAR WIND COLLECTORS FOR THE GENESIS DISCOVERY MISSION BY RESONANCE IONIZATION MASS SPECTROMETRY

W. F. Calaway[1], M. P. McCann[1,2] and M. J. Pellin[1]
[1]Materials Science and Chemistry Divisions, Argonne National Laboratory, Argonne, IL 60439
[2]Present address, Dept. of Chemistry, Sam Houston State University, Huntsville, TX 77341

ABSTRACT

NASA's Genesis Discovery Mission is designed to collect solar matter and return it to earth for analysis. The mission consists of launching a spacecraft that carries high purity collector materials, inserting the spacecraft into a halo orbit about the L1 sun-earth libration point, exposing the collectors to the solar wind for two years, and then returning the collectors to earth. The collectors will then be made available for analysis by various methods to determine the elemental and isotopic abundance of the solar wind. In preparation for this mission, potential collector materials are being characterized to determine baseline impurity levels and to assess detection limits for various analysis techniques. As part of the effort, potential solar wind collector materials have been analyzed using resonance ionization mass spectrometry (RIMS). RIMS is a particularly sensitivity variation of secondary neutral mass spectrometry that employs resonantly enhanced multiphoton ionization (REMPI) to selectively postionize an element of interest, and thus discriminates between low levels of that element and the bulk material. The high sensitivity and selectivity of RIMS allow detection of very low concentrations while consuming only small amounts of sample. Thus, RIMS is well suited for detection of many heavy elements in the solar wind, since metals heavier than Fe are expected to range in concentrations from 1 ppm to 0.2 ppt. In addition, RIMS will be able to determine concentration profiles as a function of depth for these implanted solar wind elements effectively separating them from terrestrial contaminants. RIMS analyses to determine Ti concentrations in Si and Ge samples have been measured. Results indicate that the detection limit for RIMS analysis of Ti is below 100 ppt for 10^6 averages. Background analyses of the mass spectra indicate that detection limits for heavier elements will be similar. Furthermore, detection limits near 1 ppt are possible with higher repetition rate lasers where it will be possible to increase signal averaging to 10^8 laser shots.

INTRODUCTION

Last year NASA selected the Genesis mission as its fifth Discovery mission. The intent of the Discovery program is to fund highly focused science projects that meet the "faster, better, cheaper" criteria under which NASA now operates. The purpose of the Genesis Discovery mission is to collect solar matter and return it to earth for analysis.[1] This will be accomplished by equipping a spacecraft with an array of ultra-clean and -pure solar wind collectors. The spacecraft will be inserted into a halo orbit about the L1 libration point where the collectors will be exposed to the solar wind for a period of two years. Afterwards, the collector array will be stowed into a contamination-tight canister within a sample return capsule and returned to earth for mid-air recovery. A key consideration of the Genesis Discovery mission is that the collectors are returned to earth. This allows sophisticated analytical instruments, that are too bulky and complex for onboard spacecraft operation, to be used for the solar wind collector analyses. In addition, some of the collector material will be archived for future analyses when new instruments and new scientific questions arise.

The overall purpose of the Genesis Discovery mission is to determine the isotopic and elemental abundances of the solar wind and thus the sun. These results will be used (1) to expand the present isotopic and elemental abundance data base for the sun, (2) to obtain more precise values for those isotopic and elemental abundances that have already been determined by other methods, and (3) to produce a set of consistent isotopic and elemental abundances for comparison with other objects in the solar system. The goal of the mission would be to use the wealth of new data to improve our understanding of how the highly diverse objects of our solar system originate from a relatively homogeneous solar nebula and to test theories of the origins of the solar system. To accomplish these tasks, solar wind collector samples will be analyzed in Advanced Analytical Instrument Facilities developed as part of the Genesis Discovery mission. A number of state-of-the-art analytical techniques will be brought to bear on the problem including, gas source mass spectrometry (GSMS), secondary ion mass spectrometry (SIMS), resonance ionization mass spectrometry (RIMS), radiochemical neutron activation analysis (RNAA), synchrotron x-ray fluorescence, thermal ionization mass spectrometry (TIMS), and inductively coupled plasma mass spectrometry (ICPMS). This paper presents a discussion of RIMS and how it will contribute to the overall objectives of the Genesis Discovery mission.

Resonance ionization mass spectrometry is a surface analytical method that has a unique combination of capabilities no other single analytical instrument can match, namely unambiguous identification of elemental impurities at trace levels with single monolayer resolution. These features allow a RIMS instrument to produce concentration versus depth profiles of solid materials at levels below 1 ppb. The RIMS instrument capabilities match well the needs of NASA's Genesis Discovery Mission where solar wind constituents will be implanted in collector arrays at levels down to 0.1 ppt and at depths no greater than 100 nm.[2] As part of the Genesis Discovery mission, RIMS instrument development and sample analyzes of collector materials are being conducted at Argonne National Laboratory (ANL). The goal is to improve detection limits for the RIMS technique to the levels required for accurate quantification of the least abundant elements present in the solar wind while contributing to the collector materials selection and verification processes. Recent progress toward this goal is presented in this paper.

EXPERIMENTAL

The RIMS instrument used in this study is referred to as SARISA (surface analysis by resonance ionization of sputtered atoms). It was developed and built at ANL and has been previously describe in detail elsewhere.[3] The basic instrument consists of an ion source for sputtering samples, a laser system for photoionization of elements, a time-of-flight mass spectrometer (TOF MS) housed in an ultra-high vacuum chamber for detection, and a computer system for instrument control and data collection. In a typical measurement, a 500 ns pulse of 5 keV Ar$^+$ ions (300 μm spot diameter) strikes a sample target causing material to be ejected via sputtering. Approximately 300 ns later, light from as many as three tunable dye lasers simultaneously pass through the plume of sputtered material photoionizing neutral ground state atoms of a specific element. (The number of lasers and their frequencies are determined by the ionization scheme for the element of interest.) The photoions are extracted into the TOF MS and detected by a pair of microchannel plates arranged in a chevron configuration.

RIMS is a variation of secondary neutral mass spectrometry (SNMS) that employs resonantly enhanced multiphoton ionization (REMPI) to maximized selectivity and sensitivity. The use of REMPI is a key feature of the RIMS technique. REMPI can photoionize ground state atoms with an efficiency near unity facilitating the detection of a large fraction of the sputtered material. In fact, useful yields (atoms detected/atom sputtered) between 1 and 5% have been measured for SARISA.[4] Particularly important for trace analysis is the fact that REMPI is an efficient way to separate an element of interest from the large number of neutral bulk atoms and to identify unambiguously that specific element.

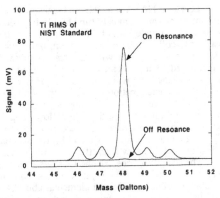

Figure 1. RIMS of Ti in a low alloy steel showing the on resonance signal and the off resonance background.

84

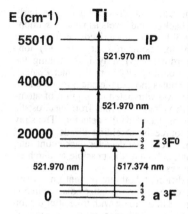

Figure 2. Resonance ionization scheme used for detecting Ti.

Because atomic transitions are very narrow and have large cross sections, tunable lasers can be used to photoionize a specific element without photoionizing other species. Further, a small detuning of the laser allows the background to be measured directly.

Shown in Fig. 1 is a mass spectrum of Ti from a low alloy steel (NIST Standard #1264a; 0.27 at.% Ti) used to calibrate the SARISA instrument. Figure 1 also shows the background level when the ionizing lasers are tuned off the Ti resonance transitions by only 0.1 nm. The resonance ionization scheme used in these measurements was developed at ANL and is rather special.[5] Only two green lasers are needed to ionize Ti from its ground state (a^3F_2) and a multiplet level of the ground state (a^3F_3) 170 cm^{-1} above the ground state. This ionization scheme is shown in Fig. 2.

The two major noise sources in a RIMS measurement are photoionized molecules and secondary ions. Photoionized molecules can be virtually eliminated as a noise source by the proper selection of an ionization scheme and maintaining appropriate laser intensities.[6] Secondary ions are eliminated in the SARISA instrument by imparting different energies to the secondary ions and photoions then using the energy filtering, which is part of the TOF MS, to block the secondary ions.[3] This is accomplished by applying a high voltage pulse to the sample while the primary ion beam strikes the sample. Once the primary ion beam is pulsed off, the target voltage is restored to its normal potential for extraction of photoions (+1100 V) before the lasers are triggered. Employing these techniques, background count rates for SARISA have been previously demonstrated to be near 2 counts in 1000 averages (2 x 10^{-3} counts/average), corresponding to a detection limit of 1 ppb for 10^6 averages.[7]

Two separate studies were conducted and are reported herein. The first study was aimed at improving the detection limit of the SARISA instrument. In this study, measurements of the background levels from a Si target as a function of various instrument parameters were obtained. Measurements were conducted either with the lasers tuned off their resonance frequencies or turned off altogether. Count rates were determined by summing the total counts over a large mass range (typically, 40 to 100 Daltons) and dividing the result by the mass range and number of averages. In the second study, samples from Si and Ge wafers supplied by various vendors have been analyzed to determine baseline levels of Ti. In these measurements, data was collected with lasers tuned on and off the Ti resonance frequencies. Count rates for Ti^{48} were integrated and converted to concentrations using standard calibration procedures.[6] Depth versus Ti concentration profiles were constructed by ion milling into the samples between measurements. Combining background measurements with the Ti analysis results allows a detection limit to be calculated as discussed below.

RESULTS AND DISCUSSION

Noise Suppression

The background for a RIMS analysis is the measured count rate when the laser is detuned from the resonant wavelength of that element. As such, it is the sum of all ions reaching the detector at the appropriate mass that are not resonantly photoionized atoms of the element of interest. To improve the detection limit of SARISA below present value, a detailed study of the sources that cause background counts during RIMS measurements was undertaken. In addition to secondary ions and nonresonant photoions, four other noise sources have been identified. The sum of these four noise sources corresponds to the background count rate of 2 x 10^{-3} counts/average mentioned above. An important feature of all four noise sources is that, for the most part, they do not exhibit a mass signature, but rather are distributed in time fairly uniformly. Each noise source was isolated by monitoring background levels as various sections of the instrument were turned off. From this it was possible to characterized the four noise sources as follows: (1) ions that evolve after primary ion bombardment ends, (2) ions originating from the sample without primary ion bombardment, (3) stray ions striking the detector that have not traveled down the TOF path, and (4) dark counts from the microchannel plate detector. Of the four identified sources, the first contributes most to the background and is the least understood. The second noise source is likely electron stimulated desorption of atoms or molecules. The sample receives a constant flux of high voltage electrons from electrostatic lens elements in the TOF MS that could possibly produce ions that reach the detector. The stray ions striking the detector has been eliminated as a noise source by shielding the detector and TOF path from stray ions and electrons in the vacuum chamber. The detector dark count rate, last noise source, is sufficiently small as not to be important, although it is possible to purchase special "low dark noise" microchannel plates if the need arises.

While the causes of the first two noise sources are different, both are present only when the target potential is set to transmit photoions. The difference is that the latter noise source is constant with time while the former noise source is was found to decrease with time after the ion pulse. Thus, it is possible to fully suppressed the first two noise sources only if the target is held at a nontransmitting potential (e.g., +2500 V) for all times. It is important to note that the background that is observed is from ions that are forming at times after the primary ion pulse in the extraction region of the mass spectrometer. The most likely explanation for the observed phenomenon is delayed formation of ions above the target. The source of these ions while not known could likely be neutral clusters ejected during ion bombardment that fragment to ions or that ionize by some mechanism such as electron bombardment. Neutral clusters are known to sputter from clean metal surfaces and are thought to possess substantial internal energy.[8]

Since ions are forming continuously throughout the time when extraction of photoions must occur, this residual noise source can not be total eliminated by any target pulsing scheme. It can be suppressed, however, by minimizing the length of time that the transmitting target potential is on. In addition, during this study it has been discovered that a negative target potential is better at suppressing noise than a positive potential. A new noise suppression scheme that uses these two observations has been implemented by designing a fast bipolar target pulser with variable width. Using this circuit, the target can remain at a negative nontransmitting potential at all times except for the short time need to extract the photoions. The results of this new design are shown in Fig. 3 where the photoion signal and noise level are plotted as a function of the length of time the target remains in photoion transmitting mode after the primary ion pulse (defined as target pulse width in Fig. 3). As can been seen from Fig. 3, the noise level and signal level are constant for long target pulse widths. As the target pulse width is shortened, the noise is reduced as ions that are formed long after the primary ion pulse are suppressed. The noise rate falls rapidly as the more abundant ions formed at shorter times are suppressed. Finally, as the target pulse is shortened further, photoion extraction efficiency falls off and no further noise suppression can be accomplished without loss of signal. Even so, as can be seen in Fig. 3, noise can be suppressed by a factor of >100 compared the previous method. This background reduction substantially reduces elemental detection limits as discussed below.

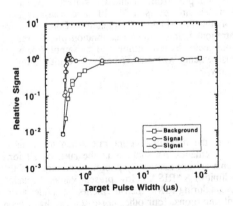

Figure 3. Observed signal and noise for the newly designed target pulsing scheme as a function of target pulse width.

86

Trace Measurements

Pieces of Si and Ge wafers have been analyzed to determine baseline levels of Ti using SARISA. High Ti levels (between 1000 and 100 ppb) were detected on the surface of all samples. Sputter cleaning the sample surface reduced these levels. However, concentration versus depth profiles show a leveling off of the Ti concentration at about 1 ppb at a depth of 1 μm. A series of measurements was conducted that proved memory effects in the instrument were not responsible. Conversely, a correlation between the sputter cleaning spot size and the amount of Ti contaminating the surface suggesting that the wings of the primary ion beam were limiting the dynamic range of the instrument. These problems are presently being addressed with the introduction of a high quality primary ion source. Procedures to minimize contamination during sample preparation are being developed. In addition, a new ion source with substantially shaper wings has been purchased. A new ion optics design for SARISA that incorporates the improved ion gun has begun.

Detection Limits

The background reduction that has been achieved results in a significant reduction of detection limit for the SARISA instrument. For a signal-to-noise ratio of 3, the detection limit, D, (for counting ions when a background is subtracted) is derived from Poisson statistics as

$$D = S \cdot \frac{9 + \sqrt{81 + 72nB}}{2n}, \qquad (1)$$

where S is the instrument sensitivity, n is the number of averages, and B is the background count rate (counts/average).[10] As can be seen from eq. 1, for a large number of averages and a fixed instrument sensitivity, the detection limit goes as the square root of the background count rate. Figure 3 shows that the new noise suppression scheme reduces the background count rate nearly 100 fold leading to a reduction in D of more than an order of magnitude. This is shown in Fig. 4 where the detection limit for Ti is plotted as a function of the number of averages for past and present measured backgrounds. It should be noted that RIMS instrument sensitivities for most metals differ by only a factor of two to three. Therefore, the detection limits plotted in Fig. 4 are approximately valid for any metal where a REMPI scheme can be found that does not cause a measurable photoionization background.

As seen in Fig. 4, the detection limit for Ti has dropped from about 1 ppb to less than 100 ppt for 10^6 averages. For 10^8 averages the 3-σ detection limit is near 5 ppt. With the present low repetition rate (<50 Hz) laser system, it is not possible to achieve 10^8 averages in a reasonable amount of time. (Presently, 10^6 averages require 5.6 hours.) However, a new laser system that can produce the desired tunable light at 5 kHz will allow such low detection limits to be reached. It thus appears possible to achieve the desired detection limits for most heavy metals in the Genesis Discovery mission collectors with only small improvements in instrument sensitivity and background suppression.

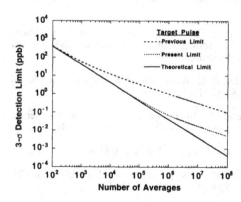

Figure 4. Calculated detection limit for the new noise suppression method compared to the previous detection limit and the theoretical detection limit.

CONCLUSIONS

A systematic study of noise sources in the ANL RIMS instrument (SARISA) has lead to a substantial improvement in the attainable detection limit. Complete elimination of the two dominant sources of background counts, secondary ions and nonresonant photoions, allowed identification of four residual sources. The most substantial of these is ions formed in the mass spectrometer extraction volume after ion bombardment of the sample has ceased. It is proposed that this noise source is derived from neutral clusters that are fragmenting and/or ionizing. Concentration versus depth measurements of Ti in Si and Ge samples were undertaken to test a new noise suppression method and to assess the cleanliness of potential solar wind collector materials. Concentrations near 1 ppb were measured indicating that surface contamination was limiting the dynamic range of the instrument. However, backgrounds were found to be below 5 x 10-5 counts/average. Combining the measured sensitivity for Ti with the measured background, a detection limit for Ti in Si was found to be below 100 ppt for 10^6 averages and near 5 ppt for 10^8 averages.

ACKNOWLEDGMENTS

Work supported by the U.S. Department of Energy, BES-Materials Sciences, under Contract W-31-109-ENG-38 and by NASA under Grant No. 1559.

REFERENCES

1. A detailed discussion of the Genesis Discovery mission can be found on the Internet at www.gps.caltech.edu/genesis/.

2. A. H. Treiman, *Curation of Solar Wind Collector Plates from a Solar Wind Sample Return (SWSR) Spacecraft Mission*, NASA Report No. JSC 26406 (October, 1993).

3. C. E. Young, M. J. Pellin, W. F. Calaway, B. Jörgensen, E. L. Schweitzer, and D. M. Gruen, Nucl. Inst. Method. **B27**, 119-129 (1987).

4. M. J. Pellin, C. E. Young, W. F. Calaway, J. E. Whitten, D. G. Gruen, J. D. Blum, I. D. Hutcheon, and G. J. Wasserburg, Phil Trans. R. Soc. Lond. A **333**, 133-146 (1990).

5. Z. Ma, R. N. Thompson, K. R. Lykke, M. J. Pellin, and A. M. Davis, Rev. Sci. Instrum. **66**, 3168-3176 (1995).

6. W. F. Calaway, R. C. Coon, M. J. Pellin, C. E. Young, J. E. Whitten, R. C. Wiens, D. M. Gruen, G. Stingeder, V. Penka, M. Grasserbauer, and D. S. Burnett, *RIS 92*, Inst. Phys. Conf. Ser. No. 128, pp. 271-274 (IOP Publishing, 1992).

7. C. S. Hansen, W. F. Calaway, M. J. Pellin, R. C. Wiens, and D. S. Burnett, *RIS 96*, AIP Conference Proceeding No. 338, pp. 215-218 (AIP Press, 1997).

8. W. F. Calaway, R. C. Coon, M. J. Pellin, D. M. Gruen, M. Gordon, A. C. Diebold, P. Maillot, J. C. Banks, and J. A. Knapp, Surf Interface Anal. **21**, 131-137 (1994).

9. G. Betz and K. Wien, Int. J. Mass. Spec. and Ion Proc. **140**, 1-110 (1994).

10. W. F. Calaway, D. R. Spiegel, A. H. Marshall, S. W. Downey, and M. J. Pellin, CARRI 96, AIP Conference Proceeding No. 392, pp. 739-742 (AIP Press, 1997).

Part III

Materials for Energy
Conversion and Storage

SYNTHESIS OF THIN FILM BATTERY COMPONENTS BY THE SPRAY DECOMPOSITION TECHNIQUE

K.S. Weil*, P.N. Kumta*, J. D. Harris**, and A. F. Hepp***
*Department of Materials Science and Engineering, Carnegie Mellon University, Pittsburgh, PA 15213; **School of Technology, Kent State University, Kent, OH 44242;
***Photovoltaics and Space Environment Branch, National Aeronautics and Space Administration, John H. Glenn Research Center, Lewis Field, Cleveland, OH 44135

ABSTRACT

Thin film $LiCoO_2$ anodes and SnO_2 cathodes on the order of 1-10 μm thick have been synthesized using a spray decomposition technique, employing a lithium-nickel nitrate/methanol solution and tin ethylhexanoate/methanol solution, respectively. Preliminary electrochemical test results on the films indicate that the $LiCoO_2$ anodes in general display a relatively high open circuit voltage (OCV) of ~4.10-4.25V, good specific capacity on the order of 120 mAh/g, and acceptable cycle-ability, with a 16-25% decay in operating voltage after 50 cycles. The SnO_2 films display excellent performance characteristics, upon an expected initial drop in capacity, with an OCV of 1.00-1.10V, specific capacity of ~600 mAh/g, and virtually no decay in operating voltage after 50 cycles. The solid-state electrolyte, lithium phosphate, was prepared in thin film form by a similar spray processing technique. Additionally, it has been demonstrated that spray decomposition can be used to sequentially deposit the anode, electrolyte, and cathode film layers to form a layered thin film battery (TFB), although electrochemical testing of this thin film device has not yet been conducted.

INTRODUCTION

Over the past several decades the electronics industry has made great strides in reducing the size of components and products, especially in the area of communications. Also, globalization of the world economy and commercialization of space have greatly increased the demand for stand alone power sources for a vast array of items ranging from small personal devices, such as cellular phones, to orbiting satellites. However, miniaturization of power sources has not kept pace with the shrinking size of electronic components. Indeed, many stand-alone products are now designed around their power supply because the battery pack is limiting further size reduction [1]. Therefore, the ability to further develop and manufacture self-contained products depends on the feasibility of developing smaller, longer lasting batteries that retain appropriate power densities without losses. Further size reductions also depend on developing technology for incorporating the power supply into integrated circuits and multi-chip modules. Thin film batteries have several advantages over bulk batteries, particularly for applications in electronics. They could be manufactured using standard techniques familiar to the microelectronics industry. Layers would be thin enough that internal resistance of the solid electrolyte would be reduced to an acceptable level. Batteries could conform to any 2-dimensional shape and deposited onto a variety of substrates.

With miniaturization, the power and current requirements of electronic components has decreased to the point where thin film batteries could adequately meet the energy demands of micro-devices. During the past decade, Bates and co-workers (at Oak Ridge National Laboratory) and Jones and co-workers (at Eveready Battery Company) have independently developed solid-state inorganic thin film batteries that have undergone over 30,000 charge/discharge cycles with little degradation in performance. The batteries also exhibit excellent shelf life [2-4]. Both batteries use a lithium anode; the cells developed at Oak Ridge use Li-intercalated metal oxides for the cathode, whereas the battery fabricated at Eveready uses a TiS_2 cathode. The battery developed at Eveready can be cycled between 1.4 and 2.8V with a current density of 100 mA/cm^2 for a 10 cm^2 battery [3]. Depending on the choice of cathode, the Oak Ridge cells can be tuned to cycle between 1.5 and 4.8V, with current densities approaching 1 mA/cm^2 [5,6]. Both batteries have been cycled in parallel and series configurations. In addition, an Oak Ridge battery performed well after being deposited onto the backside of a multi-chip module [2]. Following the initial success of this thin film battery (TFB) concept, many researchers foresee incorporating TFBs into integrated circuits to serve as standby power sources for CMOS memory or for use as long-term power sources for micro-sensors [3,7,8].

The goal of the research described here was to develop an economical method for synthesizing thin film batteries and to develop a rechargeable battery that could eventually be deposited onto multi-chip modules or incorporated into integrated circuits. Since this study is essentially a proof-of-principle test, it was broken into three phases: (1) the synthesis and characterization of anode, electrolyte, and cathode films separately prepared by spray decomposition or spray pyrolysis, (2) demonstration of the deposition of successive thin film layers to form a TFB, and (3) fabrication and electrochemical evaluation of a TFB prepared by spray decomposition. For ease of preparation and handling, our proof-of-principle work has focused on the synthesis of an all oxide battery. In addition, a Li ion anode was chosen rather than a lithium metal anode to alleviate complications associated with handling and deposition of elemental lithium. Results from the first and second phases of this program are reported in this paper.

EXPERIMENTAL PROCEDURE

1. Synthesis of Films

All of the reagents and solvents employed in this study were used without additional purification. A 0.5M solution of $Sn(OOCC_7H_{15})_2$ (technical grade, Alfa Aesar) dissolved in anhydrous methanol (Aldrich) was used to prepare the tin oxide thin films. The Li_3PO_4 electrolyte films were synthesized from a solution consisting of 2.28 g of $LiOCH_3$ (99%, Alfa Aesar) and 1.42 g of P_2O_5 (98+%, Aldrich) dissolved in 100ml of a pre-mixture containing 98 volume percent ACS grade methanol (Fisher Scientific) and 2 volume percent glacial acetic acid (Fisher Scientific). The cathode films were deposited using a 0.1 M metals solution each of $LiNO_3$ (99%, Alfa Aesar) and $Co(NO_3)_2 \cdot 6H_2O$ (98+%, Aldrich) dissolved in equimolar amounts in ACS grade methanol. All films were prepared on either 0.1 mm thick cobalt or nickel foils measuring 1 cm^2 in area which served not only as the substrate, but also functioned as the current collector in the electrochemical test cells. The foils were stored in methanol after ultrasonically cleaning them sequentially with trichloroethylene, acetone, and methanol. Prior to deposition, the substrate was placed on a stainless steel holder and heated to a set temperature, indicated in Tables 1-3 depending on the system. Deposition was carried out by atomizing the solution through a fine coaxial nozzle using reconditioned air (Matheson Gas) at a pressure of 25 psi, see Figure 1. Atomization was conducted for approximately one to two seconds intermittently at intervals of 30 or more seconds, because of reduction in substrate temperature induced by the evaporating methanolic solution. Once spraying was complete, the film was allowed to anneal on the sample holder for a set period of time, see Tables 1-3. Temperature of the substrate was determined using a thermocouple.

2. Materials Characterization

Crystalline phases in the as-heat treated thin films were characterized by X-ray diffraction (XRD). The samples were fixed in brass holders and analyzed with a Rigaku (Theta-Theta) diffractometer using graphite filtered monochromated CuKα radiation.

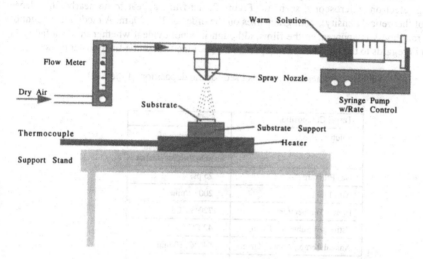

Figure 1. Schematic of spray decomposition apparatus used to prepare thin films.

The XRD data was collected using a scan rate of 1.5°/min and a step size of 0.05°, over a 2Θ range of 10° - 70°. The microstructures of the oxide films were characterized using an AMRAY 1810 scanning electron microscope (SEM) equipped with a Robinson backscatter detector and a windowless EDAX, DX-4 analyzer. To evaluate the electrochemical characteristics of the SnO_2 and $LiCoO_2$ thin films, a three electrode Hockey Puck cell design was used [9] employing lithium as an anode and $LiPF_6$ as the electrolyte after dissolution in a 2:1 solvent mixture of ethylene carbonate and dimethyl carbonate. Half-cell reactions of each thin film electrode were analyzed using an Arbin potentiostat while cycling over the prescribed voltage range described below under a constant current density of 0.25 mA/cm².

RESULTS AND DISCUSSION

1. Synthesis and Characterization of the Thin Film Anode: SnO_2

As mentioned above, the organotin solution was prepared in a 0.5M concentration by dissolving tin 2-ethylhexanoate in anhydrous methanol. The clear solution was heated under stirring for several minutes to a temperature of ~ 60°C. The warmed solution was then forced by a syringe into the coaxial spray nozzle where it was mixed with and atomized by pressurized air and sprayed onto a heated cobalt substrate. Cobalt substrates were weighed prior to and after coating to determine film weight and allow calculation of estimated film thickness. Thickness calculations were made assuming that films were ~95% dense, as estimated by SEM observation. The rate of film growth normal to substrate surface appeared to be a function of several factors: solution concentration, carrier gas flow rate, solution flow rate, and substrate temperature.

While a thorough study of these processing conditions is planned, a series of exploratory experiments demonstrated that an electrochemically active tin oxide film could be prepared using the processing conditions listed in Table 1. XRD analysis of the films prepared under these conditions demonstrated that they were highly crystalline, single phase SnO_2. As observed by scanning electron microscopy, seen in Figure 2, the films appear to be nearly fully dense (~95% of theoretical density), with grain sizes on the order of 10 - 20μm. A moderate amount of surface roughness is apparent in the films, although it is not evident whether this is a factor of concern in regards to the electrochemical performance of the film or the final power device.

Table 1. Synthesis conditions used for the deposition of the SnO_2.

Target Composition	SnO_2
Solution Used	0.5M Mixture of: Tin 2-ethylhexanoate in anhydrous methanol
Gas Pressure	25 psi
Air Flow Rate	200 cc/min
Heater Temperature	720±30°C
Estimated Substrate Temp.	420°C
Anneal Temp., Anneal Time	540°C, 10 min
Weight of Applied Film	1 mg
Estimated Film Thickness	1.2 μm

Figure 2. SEM micrograph showing the surface structure of a tin oxide film deposited on a heated cobalt substrate under the conditions listed in Table 1.

The electrochemical properties of the 1 mg anode film were measured in a lithium anode - liquid electrolyte (EC/DMC/LiPF$_6$) test cell by cycling the cell between 0.01 and 1.00V. The film displayed an open circuit voltage (OCV) typical of bulk SnO$_2$ (1.02V), a very high specific capacity of 587 mAh/g after the first cycle loss, and excellent performance, with virtually no decay (<1%) in capacity after 50 cycles, see Figure 3.

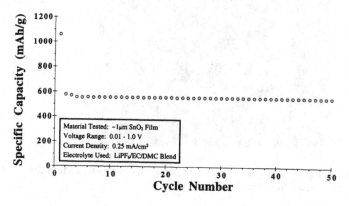

Figure 3. A plot of capacity vs. cycle number for the first 50 cycles of the tin oxide film deposited on a heated cobalt substrate under the conditions listed in Table 1.

2. Synthesis and Characterization of the Thin Film Electrolyte: Li$_3$PO$_4$

The lithium-phosphorus solution was prepared in a glovebox due to hygroscopic nature of P$_2$O$_5$. As with the organotin solution, once the lithium-phosphorus solution was prepared, it was heated under stirring to a temperature of 60°C prior to spray pyrolysis. The heated solution was then sprayed onto heated cobalt substrates using the conditions listed below in Table 2.

Table 2. Synthesis conditions used for deposition of Li$_3$PO$_4$ electrolyte.

Target Composition	Li$_3$PO$_4$
Solution Used	In 100 ml Methanol: 5ml Glacial Acetic Acid 2.28 g LiOCH$_3$ 1.42 g P$_2$O$_5$
Gas Pressure	25 psi
Air Flow Rate	120 cc/min
Heater Temperature	715±30°C
Estimated Substrate Temp.	435°C
Anneal Temp., Anneal Time	530°C, 5 min
Weight of Applied Film	1 mg
Estimated Film Thickness	2.0 μm

Figure 4. SEM micrograph showing surface structure of a lithium phosphate film deposited on a heated cobalt substrate under conditions listed in Table 2.

XRD analysis of the films prepared under these conditions show that they are weakly crystalline, although the two-theta positions of the broad peaks correspond to those of crystalline Li_3PO_4. SEM micrographs of the films, of the type shown in Figure 4, suggest that the lithium phosphate films are quite dense (estimated to be ~90% of theoretical density), but do display a number of obvious pores (~0.2 - 0.5μm in diameter) and cracks.

It is likely that the pores originate from gas bubbles that arise from the evolution of methanol vapor during solvent evaporation and from the gaseous byproducts released due to the decomposition of lithium methoxide. The cracks seen in Figure 4 are probably due to both thermal expansion mismatch between the substrate and the Li_3PO_4 film and thermal shock that occurred upon the removal of the coated substrate from the heater after annealing. The latter problem can easily be corrected by allowing the film to cool more slowly. The electrochemical properties of this film have not yet been examined.

3. Synthesis and Characterization of the Thin Film Cathode: $LiCoO_2$

Three separate solutions, each differing in total metals (Li + Co) concentration, 0.05M, 0.10M, and 0.20M, were prepared by dissolving lithium nitrate, $LiNO_3$, and cobalt nitrate hexahydrate, $Co(NO_3)_2 \cdot 6H_2O$, in reagent grade ethanol at ~60°C under stirring. Once the salts were completely dissolved, each of the warm solutions was sprayed onto heated nickel substrates. As observed with the tin oxide thin films, film growth normal to the substrate surface depended on several processing factors, including solution concentration. As a means of developing $LiCoO_2$ films for electrochemical analysis, the conditions listed above in Table 3 were chosen. XRD analysis verified that the thick black films formed on the nickel substrates were predominantly $LiCoO_2$, although minor peaks corresponding to a second unidentified phase were observed. As evident from SEM micrographs of the type shown in Figure 5, the films display ~50 x 50 µm patches of dense material surrounded by regions which contain low to moderate levels of fine porosity. It is not known how these inhomogeneities in film morphology arise, but it is likely that they are largely due to the reactions accompanying the decomposition of the nitrates. Experiments are currently underway to evaluate use of metal-organic salts of lithium and cobalt as possible precursors for formation of less porous $LiCoO_2$ films.

Table 3. Synthesis conditions used for the deposition of the $LiCoO_2$ cathode.

Target Composition	$LiCoO_2$
Solution Used	0.1M Mixture of: $LiNO_3$ $Co(NO_3)_2 \cdot 6H_2O$ in ACS Methanol
Gas Pressure	25 psi
Air Flow Rate	150 cc/min
Heater Temperature	740±30°C
Estimated Substrate Temp.	470°C
Anneal Temp., Anneal Time	600°C, 30 min
Weight of Applied Film	4 mg
Estimated Film Thickness	5.2 µm

Figure 5. SEM micrograph showing the surface structure of a lithium cobalt oxide film deposited on a heated nickel substrate under the conditions listed in Table 3.

A 4 mg LiCoO₂ film prepared using the processing conditions listed in Table 3 was electrochemically tested in a lithium anode - liquid electrolyte (EC/DMC/LiPF₆) test cell by cycling the cathode between 3.00 and 4.25V. The film displayed a relatively high open circuit voltage (OCV) of 4.19V, good specific capacity of 116 mAh/g, and moderate cyclability performance, characteristic of bulk LiCoO₂, with a 25.8% decay in operating voltage after 50 cycles, as shown in Figure 6. While this performance is acceptable for a proof of principle test in a TFB test cell, both the OCV and cyclability can likely be improved with only minor modifications in processing. These include: (1) the addition of magnesium (using a Mg(NO₃)₂·6H₂O source) to the initial precursor solution in ~ 1 - 1.5 wt % to improve the electrical conductivity of LiCoO₂ films [10], (2) use of cobalt substrates in place of nickel to avoid the potential formation of LiNiO₂ or mixed transition metal oxides at the interface of the LiCoO₂ and metal substrate surfaces, and (3) removal of the thin oxide scale that inevitably builds up on the uncoated surface of the substrate during deposition, possibly by mechanical abrasion, in order to improve the electrical conductivity through the cathode side of the test cell.

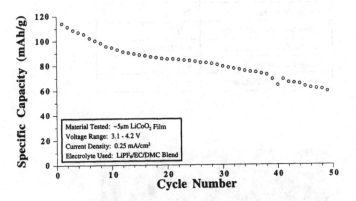

Figure 6. A plot of capacity vs. cycle number for the first 50 cycles of the lithium cobalt oxide film deposited on a heated nickel substrate under the conditions listed in Table 3.

4. Deposition of Successive Layers to Fabricate an Electrochemical Device

A lithium phosphate film was deposited directly on the SnO₂ thin film described above. XRD analysis verified that this film is poorly crystalline Li₃PO₄. As seen from the central micrograph in Figure 7, the surface morphology of this film is nearly identical to the one deposited directly on cobalt (Figure 4). This two-layer film was then used as the substrate on which a lithium cobalt oxide film was deposited. As seen in the micrograph on the right in Figure 7, the surface morphology of this film is quite different from that deposited directly on nickel (compare with Figure 5). The surface of the three layer film is composed of porous spheroidal shaped oxide agglomerates which are approximately 5- 10μm in diameter. These agglomerates appear to be partially sintered to one another, leading to a porous microstructure containing macropores approximately 2μm in size.

SnO₂   SnO₂ + Li₃PO₄   SnO₂ + Li₃PO₄ + LiCoO₂

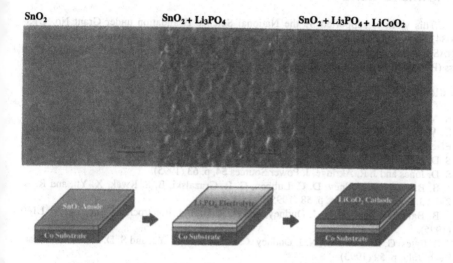

Figure 7. SEM micrographs displaying surface structure of a tin oxide film deposited on a cobalt substrate, a lithium phosphate film deposited on tin oxide film, and a lithium cobalt oxide film deposited on lithium phosphate film, using conditions listed in Tables 1,2, and 3, respectively.

XRD analysis verified that this top layer is predominantly LiCoO₂, although again an unidentified minor phase was observed. Although the microstructure and phase purity of the LiCoO₂ film is not optimal, it appears that a three-layer TFB can be built by sequentially depositing the anode, electrolyte, and cathode layers on a cobalt substrate using the spray decomposition technique. Additional devices, in which the amounts of SnO₂ and LiCoO₂ are balanced to account for the first cycle loss of lithium in SnO₂, are being constructed using this technique for future electrochemical testing.

CONCLUSIONS

We have successfully demonstrated the use of the spray decomposition technique in preparing good quality, adherent films of SnO₂, Li₃PO₄, and LiCoO₂ on cobalt and nickel substrates. Both the SnO₂ and the LiCoO₂ films display good electrochemical characteristics, although further improvements in the phase purity and morphology of the LiCoO₂ films are desired. Additionally, we have demonstrated that the spray decomposition technique can be used to sequentially deposit SnO₂, Li₃PO₄, and LiCoO₂ to fabricate thin film solid state electrochemical cells. The next goal in this feasibility study will be to fabricate good quality layers of electrolyte and electrodes well balanced in the amount of lithium and to analyze the electrochemical characteristics of such a cell.

ACKNOWLEDGMENTS

This work is supported by the National Science Foundation under Grant No. CTS-9700343. One of the authors (KSW) would also like to acknowledge the award of a Fellowship by NASA. Acknowledgement is also made to Changs Ascending, Taiwan and to Pittsburgh Plate Glass (PPG), Pittsburgh for providing partial support.

REFERENCES

1. K. Murata, Electrochem. Acta **40**, p. 2177 (1995).
2. B. Wang, B., J. B. Bates, F. X. Hart, B. C. Sales, R. A. Zuhra, and J. D. Robertson, J. Electrochem. Soc. **143**, p. 3203 (1996).
3. S. D. Jones and J. R. Akridge, Solid State Ionics **86-88**, p. 1291 (1996).
4. S. D. Jones and J. R. Akridge, J. Power Sources **54**, p. 63 (1995).
5. J. B. Bates, N. J. Dudney, D. C. Lubben, G. R. Gruzalski, B. S. Kwak, X. Yu, and R. A. Zuhr, J. Power Sources **54**, p. 58 (1995).
6. J. B. Bates, D. Lubben, N. J. Dudney, and F. X. Hart, J. Electrochem. Soc. **142**, p. L149 (1995).
7. J. B. Bates, G. R. Gruzalski, N. J. Dudney, C. F. Luck, X.-H. Yu, and S. D. Jones, Solid State Tech. July, p. 59 (1993).
8. P. Birke, W. F. Chu, and W. Weppner, Solid State Ionics **93**, p. 1 (1997).
9. P. N. Kumta, D. Gallet, A. Waghray, G. E. Blomgren, and M. P. Setter, J. Power Sources **72**, p.91 (1998).
10. H. Tukamoto and A. R. West, J. Electrochem. Soc. **144**, p. 3164 (1997).

HIGHLY TRANSPARENT ARCPROOF FILMS FOR SPACE APPLICATIONS

A. J. ADORJAN, C. ALEXANDER, T. L. BLANCHARD, E. BRUCKNER,
R. FERRANTE, S. M. GAJDOS, K. MAYER, A. M. PAL, P. A. WALTERS,
P. D. HAMBOURGER*, J. A. DEVER, S. RUTLEDGE**, HENRY FU***
*Department of Physics, Cleveland State University, Cleveland, OH 44115
**NASA Lewis Research Center, Cleveland, OH 44135
***Ohio Aerospace Institute, Cleveland, OH 44142

ABSTRACT

Highly transparent films with tailorable sheet resistivity were prepared by ion-beam sputtering of indium tin oxide (ITO) with MgF_2 or SiO_2 in the presence of high-purity air. Sheet resistivities of 10^3-10^{11} ohms/square ($\Omega/\square$) and visible transmittances as high as 92% (not corrected for substrate absorption) were obtained in films ~30 nm thick. Resistivity increased by as much as two orders of magnitude in the first year after preparation; however, thicker films (e.g. 80 nm) were much more stable but somewhat less transparent. Preliminary data from exposure of film samples to atomic oxygen in a plasma asher indicate minimal degradation in optical properties. Heat-treating pure ITO in air produced transparent, slightly conductive films but with poorer stability of sheet resistivity in air than co-deposited ITO with either SiO_2 or MgF_2. Electrical transport measurements yielded new information on the electronic properties of ITO and related materials. These films show promise as low-absorption static bleedoff coatings for space photovoltaic arrays as well as CRT faceplates and other commercial applications.

INTRODUCTION

Photovoltaic panels and other nonconductive spacecraft surfaces can become electrically charged due to the solar wind. This may cause arcing and resulting interference or damage to computers and communications equipment. Charging can be prevented by applying a conductive coating with sheet resistivity less than ~10^8-10^9 $\Omega/\square$.[1] ITO has been used for this application since it offers high transparency and sheet resistivity typically as low as 100-1,000 $\Omega/\square$.[2-4]

The low sheet resistivity of ITO may actually be detrimental to photovoltaic arrays in low earth orbit since accidental contact between the coating and array wiring could result in high current flow between the power system and the conductive plasma present at low altitudes. In addition, ITO's high carrier concentration and mobility cause substantial free-carrier reflection and absorption at visible wavelengths, reducing transparency. To meet the transparency and sheet resistivity requirements with "pure" ITO of reasonable thickness would require reducing its carrier concentration to roughly 10^{13} cm^{-3}. The electrical properties of such a film would be highly sensitive to oxygen and impurity content.

We are pursuing an alternate approach based on co-deposition of ITO with a transparent, insulating coating known to be successful in previous space applications. In this paper we report electrical and optical properties of co-sputtered ITO-MgF_2 and ITO-SiO_2

101

films. Although some problems remain with stability and reproducibility of sheet resistivity, our results suggest these films may be useful in a variety of space and commercial applications.

EXPERIMENT

Films were deposited on 0.75-mm-thick fused silica by ion beam co-sputtering from a 2-component target in the presence of high purity air which provided oxygen to assure proper stoichiometry during deposition. The ITO target used was a mixture of indium oxide (91 mole %) and tin oxide (9 mole %). Wedges of either MgF_2 or SiO_2 were placed on top of the ITO target to form a 2-component target.[1,5] Content of MgF_2 or SiO_2 in the films, calculated from relative target areas and sputter rates for each constituent, ranged from 4 to 21 volume %.

Each substrate for sheet resistivity measurements was covered with an aluminum mask to produce a bar-shaped sample measuring $0.3x1.9$ cm^2 with contact arms along the edge for Hall-effect and four-lead resistance measurements.

Electrical connections to samples were made with spring-loaded pressure contacts. Resistance and Hall measurements were made with direct currents of 5 pA-100 μA using guarded, shielded cabling and high-impedance electrometers to measure the sample voltages. Magnetic field for Hall measurements was 1.2 Tesla. Measurements below room temperature were made in an exchange-gas cryostat with the sample in helium gas.

The oxygen content of some samples was increased by heat treatment at 250 °C in high-purity air at pressures up to 1 atmosphere. Sample resistance was monitored in-situ to determine when to discontinue heat treatment. We heat-treated some samples as many as 17 times, taking low-temperature electrical measurements after each treatment, to investigate the relationship between carrier concentration and electronic properties.

Some samples were exposed to simulated space environment in a plasma asher with or without a Faraday cage to protect them from ultraviolet and high-energy electrons.

RESULTS

Sheet Resistivity and Optical Transmittance vs Composition

Sheet resistivity and solar transmittance for typical 65 nm thick ITO-MgF_2 films are plotted vs MgF_2 content in Fig. 1. These data illustrate the wide range of sheet resistivities and substantial improvement in transmittance obtainable with the co-deposition approach. Optical transmittances as high as 92% (not corrected for substrate absorption) were obtained in 30 nm thick films. Similar results were obtained in ITO-SiO_2; however, these films showed somewhat higher reflectance than ITO-MgF_2.

Stability of Sheet Resistivity vs Time

Sheet resistivity of several ITO-MgF_2 and ITO-SiO_2 samples is plotted vs time after deposition in Fig. 2. Resistivity increases substantially with time, even one year after deposition. Data for 30 nm thick samples of both materials are similar; however, the thicker ITO-SiO₂ samples are markedly more stable. This suggests that film thickness is an important determinant of stability.

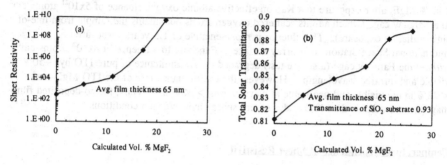

FIG. 1. Sheet resistivity ($\Omega/\square$) (a) and total solar transmittance (b) of 65 nm thick ITO-MgF₂ films vs MgF₂ content.

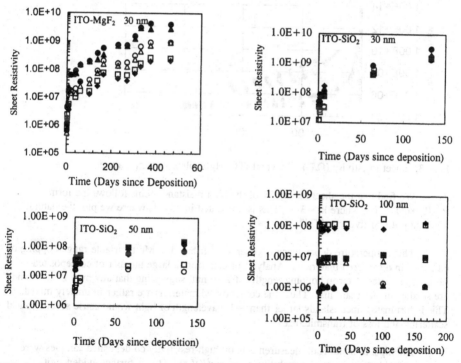

FIG. 2. Sheet resistivity ($\Omega/\square$) of several ITO-MgF₂ and ITO-SiO₂ films vs time since deposition (stored in ambient air).

Plasma Asher Exposures

No degradation of transmittance was detected in samples of pure ITO and ITO-17.2 vol. % MgF_2 after exposure to a Kapton effective atomic oxygen fluence of 5×10^{21} atoms/cm^2 in a Faraday cage (which admits scattered oxygen atoms but shields the sample from line-of-sight radiation exposure). This fluence is representative of 1.5 years' exposure of International Space Station solar array surfaces. Exposure to a fluence of 8×10^{20} atoms/cm^2 *without* the Faraday cage (a severe test) degraded the transmittance of pure ITO by ~50% at visible and infrared wavelengths. However, the transmittance loss of the ITO-MgF_2 was <22% in the visible and negligible at wavelengths >1400 nm. Thus, the co-deposited films may be more durable than ITO under severe space environmental conditions.[1]

Temperature Dependence of Sheet Resistivity

The sheet resistivity of several ITO/MgF$_2$ samples is plotted vs temperature in Fig. 3.

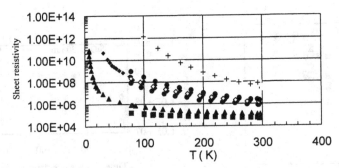

FIG. 3. Sheet resistivity ($\Omega/\square$) of several ITO-MgF_2 films vs temperature.

We find that the sheet resistivity of the high-resistance films follows the form $R = R_0 \exp(T_0/T)^{1/k}$, where k ~3-4. This is illustrated in Fig. 4, where we plot the natural logarithm of resistivity vs $1/T^{1/4}$.

This temperature dependence is generally attributed to Mott variable-range hopping of electrons in disordered materials. Analysis of data from a large number of co-deposited samples indicates electron hopping lengths of 2-60 nm, suggesting that any pure ITO "islands" are smaller in size than this. Thus, the co-deposited materials are rather intimately mixed. This is fortunate, since islands larger than ~1 wavelength of light would cause severe optical scattering and loss of transmittance.

Attempts to make Hall measurements on high-resistivity co-deposited samples were unsuccessful, indicating carrier mobility <0.3 cm^2/(V s). This is further evidence of extremely short electron mean free paths in these materials.

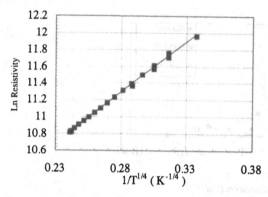

FIG. 4. Natural log of resistivity vs $1/T^{1/4}$ for an ITO-MgF$_2$ film.

"Oxygenated" Samples

The resistivity of films of pure ITO and ITO with a low concentration of insulator was increased many orders of magnitude by heating in pure air, thus filling oxygen vacancies which act as donors.[6] These samples were insufficiently stable for the intended applications. However, they yielded important information about ITO and related materials since they had sufficient mobility to permit Hall measurements.

Heat treatment of samples in air reversibly reduced their room-temperature carrier concentration from $\sim 10^{20}$ cm^{-3} to as low as $\sim 10^{14}$ cm^{-3} while the resistivity developed an increasingly large negative temperature coefficient. Similar phenomena were reported in oxygenated disordered indium oxide.[6] The authors of Ref. 6 attributed their results to the passage of the Fermi energy from extended conduction-band states to a continuum of disorder-induced localized states in the energy (or "mobility") gap.

We suggest a somewhat different explanation for our samples, based on Hall-effect data which allowed us to measure the activation energy for excitation of electrons from the Fermi energy to the bottom of the conduction band. In Fig. 5 we plot this activation energy (E_a) vs room-temperature carrier concentration for several ITO, ITO-MgF$_2$, and ITO-SiO$_2$ films with varying degrees of oxygenation. Unlike the results of Ref. 6 on indium oxide, this plot shows that, as the donor concentration is reduced, E_a eventually stops increasing and levels off at ~ 0.25 eV. This may indicate the presence of a discrete donor level ~ 0.2-0.3 eV below the bottom of the conduction band which should be taken into account in future studies of the electronic properties of indium oxide and related materials. A possible reason for the discrepancy with Ref. 6 is the previous authors' use of resistivity measurements (which are affected by temperature dependence of the mobility) rather than Hall measurements to determine E_a.

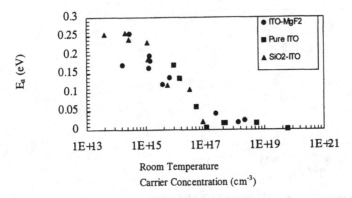

FIG. 5. Activation energy (E_a) vs room temperature carrier concentration for several oxygenated samples.

CONCLUSIONS

Co-sputtering of ITO with MgF_2 or SiO_2 produces thin films with tailorable sheet resistivity up to $\sim 10^{10}$ ohms/square and substantially greater transparency than pure ITO films of similar thickness. The co-deposited films may be more durable than ITO under exposure to the space environment. Stability of sheet resistivity with time is inadequate but might be improved by using thicker films (consistent with adequate transparency). Hall measurements in oxygenated films suggest a discrete donor energy level ~ 0.2-0.3 eV below the bottom of the conduction band. Work continues on development of these materials.

REFERENCES

1. Joyce A. Dever, et al., NASA Technical Memorandum 1998-208499, August 1998

2. C. K. Purvis, H. B Garrett, A. C. Whittlesey, and N. J. Stevens, NASA Technical Paper 2361, September 1984.

3. D. C. Ferguson, AIAA Paper No. 93-0705, January 1993.

4. A. C. Tribble, AIAA Paper No. 93-0614, January 1993.

5. B. A. Banks, et al., 39th annual Technical Conference Proceedings, p. 431 (1996).

6. B. Claflin and H. Fritzsche, J. Electron. Mater. 25, 1772 (1996) and references therein.

ACKNOWLEDGMENTS

The authors at Cleveland State University gratefully acknowledge the support of NASA Lewis Research Center under Cooperative Agreements NCC3-339, 3-383, and 3-522.

RARE EARTH DOPED YTTRIUM ALUMINIUM GARNET (YAG) SELECTIVE EMITTERS

Donald L. Chubb,* AnnaMaria T. Pal,** Martin O. Patton** and Phillip P. Jenkins**
*National Aeronautics and Space Administration, Lewis Research Center, 21000 Brook Park Rd, Cleveland, Ohio 44135, donald.chubb@lerc.nasa.gov
**Essential Research, Inc., 23811 Chagrin Blvd., Cleveland, Ohio 44122

ABSTRACT

As a result of their electron structure, rare earth ions in crystals at high temperature emit radiation in several narrow bands rather than in a continuous blackbody manner. This study presents a spectral emittance model for films and cylinders of rare earth doped yttrium aluminum garnets. Good agreement between experimental and theoretical film spectral emittances was found for erbium and holmium aluminum garnets. Spectral emittances of films are sensitive to temperature differences across the film. For operating conditions of interest, the film emitter experiences a linear temperature variation whereas the cylinder emitter has a more advantageous uniform temperature. Emitter efficiency is also a sensitive function of temperature. For holminum aluminum garnet film the efficiency is 0.35 at 1446K but only 0.27 at 1270 K.

INTRODUCTION

A selective emitter is a material that emits optical radiation in a few emission bands rather than in a continuous spectrum like a blackbody or a gray body (constant emittance). Thermophotovoltaic (TPV) energy conversion is the main application for selective emitters. In TPV energy conversion, the selective emitter converts thermal energy to near infrared radiation at wavelengths where photovoltaic energy conversion is efficient. A TPV system is rather simple, consisting of three main components, a heat source, an emitter and a photovoltaic cell array.

The ideal selective emitter would have a single emission band with an emittance approaching one within the band and negligible emittance outside the emission band. For the photon energy or wavelength region of interest for TPV energy conversion (1000 to 3000 nm), an electronic transition of an atom or molecule is required to produce the desired radiation. However, when atoms are compressed to solid state densities the emission is not characterized by narrow band emission as with an isolated atom, but by a continuous emission spectrum.

Fortunately, there is a group of atoms that even at solid state densities behave nearly like isolated atoms. These are the lanthanides or rare earth atoms. For doubly and triply charged ions of these elements in crystals the orbits of the valence 4f electrons, which account for emission and absorption, lie inside the 5s and 5p electron orbits. The 5s and 5p electrons "shield" the 4f valence electrons from the surrounding ions in the crystal. As a result, the rare earth ions in the solid state emit in narrow bands rather than in a continuous gray body manner. For the temperatures of interest ($1200 \leq T \leq 2000K$) the rare earths of most interest have a strong emission band in the near infrared ($800 \leq \lambda \leq 3000$ nm) resulting from electron transitions from the first excited state manifold to the ground state manifold. The rare earths of most interest for selective emitters are ytterbium (Yb), thulium (Tm), erbium (Er), holmium (Ho) and dysprosium (Dy). The spectra of the rare earth ions in crystals have been extensively studied. Most of this work is summarized in the text of Dieke.[1]

The first selective emitters investigated[2] were made by sintering rare earth oxide powders. These emitters showed the strong emission bands. However, emittance outside the emission bands was also large so that the emitter efficiency was low. In the late 1980's Nelson and Parent[3,4] reported a large improvement in rare earth oxide emitters. Their emitters are

107

constructed of bundles of small diameter (5 to 10 μm) fibers similar to the construction of the Welsbach mantle used in gas lanterns. The very small characteristic dimension of these emitters results in low emittance outside the emission band and thus greatly increased efficiency. The fibrous selective emitters are well suited to a combustion driven TPV system where the fibrous mantle surrounds the flame. However, for coupling to other thermal energy sources likely to be used for a space TPV system, such as nuclear or solar, the fibrous emitter is not so well suited. As just stated, it was the small characteristic dimension that made the fibrous emitters efficient. Another geometry for achieving a small characteristic dimension and also easily coupling to any thermal source is a film. A film containing a rare earth on a low emittance substrate, which blocks radiation from the thermal source, can be easily attached to any thermal source. In addition, a film is more durable than a fibrous geometry. Therefore, we began a theoretical and experimental investigation of rare earth containing film selective emitters.[5-12]

Our first attempt at producing a film selective emitter was by electron beam evaporation of pure rare earth oxides. However, we soon learned that film thicknesses on the order 1 to 10 μm were not sufficient to produce large emittance. Film thicknesses of 0.1 to 1 mm (100 to 1000 μm) are required. Evaporation is not applicable for films of that thickness so we looked for other methods. We knew that yttrium aluminum garnet (YAG) could be doped with rare earths and grown in single crystal form. Therefore, with material in this form we could cut and polish a sample to any thickness. Thus our first successful selective emitters were single crystal rare earth doped YAG. Although there are many other possible high temperature host materials for rare earths, this study is confined to rare earth doped YAG emitters.

The optical properties that characterize a selective emitter are the extinction coefficient, α_λ, which is the sum of the absorption coefficient, a_λ, and the scattering coefficient, σ_λ, and the index of refraction, n. Knowing α_λ and n, the spectral emittance, ε_λ, can be calculated. In the next section the theoretical spectral emittance model for both a film and a cylinder will be presented. Figure 1 shows the extinction coefficient for single crystal erbium aluminum garnet $(Er_3Al_5O_{12})$. Although the results in Fig. 1 apply for erbium aluminum garnet, qualitatively similar results will occur for any of the possible erbium containing materials. The major emission band for Er occurs at wavelength, $\lambda \approx 1500$ nm, with smaller bands occurring at $\lambda \approx 970$ nm, 800 nm and 640 nm. Most all high temperature ceramic materials have large extinction coefficient and thus large absorptance and emittance at long wavelengths. For YAG and the rare earth aluminum garnets this region begins at $\lambda \approx 5000$ nm. Obviously this long wavelength

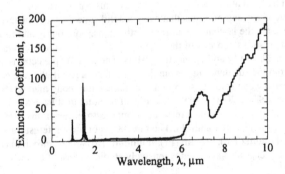

Fig. 1. Extinction coefficient for erbium aluminum garnet $(Er_3Al_5O_{12})$.

region of high emittance is undesirable for a selective emitter. An efficient selective emitter is one that emits most of its energy in the large emission band.

The next section presents the spectral emittance model for both an emitting film and a cylinder, which is an approximation to the fibrous emitter. After that the effect on ε_λ of the temperature drop between the front and back surfaces of an emitting film and rare earth doping level are presented. Finally, we present spectral emittance and emitter efficiency results for a film emitter of erbium aluminum garnet ($Er_3Al_5O_{12}$) and for both a film and cylindrical holmium aluminum garnet ($Ho_3Al_5O_{12}$) emitter.

Theoretical Spectral Emittance

Spectral emittance is usually thought of as a surface property. However, emitted photons originate at various distances below the surface. For metals these distances are very small (<1 μm). But in the case of a dielectric, such as rare earth doped YAG, these distances are greater than 100 μm. Since rare earth doped YAG also has low thermal conductivity (0.01 W/cm K) this means that significant temperature changes can occur even for small distances. As a result, since thermal radiation is a sensitive function of temperature the spectral emittance is also sensitive to the temperature changes.

Temperature changes depend on geometry, as well as, thermal conductivity and radiation transfer. For steady state conditions the conservation of energy is given by the following equation.

$$\nabla\left[k_{th}\nabla T - Q\right] = 0 \tag{1}$$

Where k_{th} is the thermal conductivity, T is the temperature and Q is the total radiation flux.

$$Q = \int_0^\infty q_\lambda d\lambda \quad W/cm^2 \tag{2}$$

and q_λ is the radiation flux at wavelength, λ, and has the units W/cm^2 nm.

Now consider Eq. (1) for a film emitter as shown in Fig. 2(a). In this case the film thickness is small compared to the dimensions perpendicular to the film. As a result, we assume the temperature and radiation depend only on the x-coordinate. Therefore, Eq. (1) becomes the following.

$$-k_{th}\frac{dT}{dx} + Q(x) = Q_{in} = constant \tag{3}$$

Where Q_{in} is the thermal energy entering through the substrate. The radiation flux will always be less than the blackbody flux $\sigma_{sb}T_s^4$, where σ_{sb} is the Stefan-Boltzmann constant (5.67×10^{-12} W/cm^2K^4) and T_s is the substrate temperature. Therefore, defining the following dimensionless variables.

$$\bar{T} = \frac{T}{T_s}, \quad \bar{Q} = \frac{Q}{\sigma_{sb}T_s^4}, \quad \bar{x} = \frac{x}{d} \tag{4}$$

results in the following energy equation.

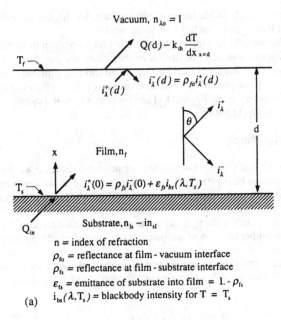

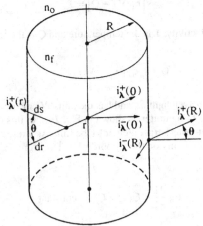

n_f = index of refraction of cylinder
n_0 = index of refraction of surroundings
$\cos\theta \, ds = dr$

(b)

Fig. 2. Schematics of emitting film and cylinder. Schematic of emitting film. (b) Schematic of emitting cyclinder.

$$-\frac{d\overline{T}}{dx} + \gamma\overline{Q} = \gamma\overline{Q}_{in} = \text{constant} \tag{5}$$

Where γ is the ratio of the radiation flux to the thermal conduction flux.

$$\gamma \equiv \frac{\sigma_{sb}T_s^4}{k_{th}\dfrac{T_s}{d}} = \frac{\sigma_{sb}T_s^3 d}{k_{th}} \tag{6}$$

Thus, if $\gamma \ll 1$ the radiation term can be neglected and the solution to Eq. (4) is the following.

$$\overline{T} = 1 - \overline{x}\Delta T \tag{7}$$

Where,

$$\Delta T = \frac{T_s - T_f}{T_s} \tag{8}$$

For most of the materials considered for rare earth selective emitters $k_{th} \geq 0.01$ W/cmK, $T_s \leq 2000$ K and $d \leq 0.05$ cm. Thus, γ will be less than 0.2. The linear temperature gradient given by Eq. (8) will be used in calculating the spectral emittance of a film emitter.

Now consider the energy equation for a cylindrical emitter as shown in Fig. 2(b). Assuming that T and Q depend only on the radial coordinate, r, Eq. (1) becomes the following.

$$r\left[k_{th}\frac{dT}{dr} - Q\right] = \text{constant} \tag{9}$$

In order to avoid the term in brackets being singular at $r = 0$ it must vanish for all r. Thus at all r the conduction and radiation fluxes balance.

$$k_{th}\frac{dT}{dr} = Q \tag{10}$$

In other words, for steady state conditions all the thermal energy being conducted into the cylinder at the outer radius, $r = R$, leaves the cylinder as radiation.

$$\overline{Q} = \frac{Q}{\sigma_{sb}T_s^4} \quad \overline{T} = \frac{T}{T_s} \quad \overline{r} = \frac{r}{R} \tag{11}$$

Where again, T_s is the temperature at $r = R$. In this case Eq. (10) becomes the following.

$$\frac{d\overline{T}}{d\overline{r}} = \gamma\overline{Q} \tag{12}$$

Where again, γ is the ratio of the radiation flux to the thermal conduction flux.

$$\gamma \equiv \frac{\sigma_{sb}T_s^4}{k_{th}\dfrac{T_s}{R}} = \frac{\sigma_{sb}T_s^3 R}{k_{th}} \tag{13}$$

Therefore, for $k_{th} \geq 0.01$ W/cmK, $T_s \leq 2000$ K and $R < 0.1$ cm it is a reasonable approximation to neglect the right hand side of equation (12) and obtain the result $\overline{T} = $ constant ($T = T_s$).

Thus depending on the geometry, there are two different results for temperature in the case of small γ. For a planar film geometry the temperature is a linear function of x. But in the case of a cylinder the temperature is a constant for $\gamma \ll 1$. In calculating the spectral emittance for a cylinder the temperature was assumed constant.

To determine the spectral emittance, ε_λ, of a film emitter the radiation flux leaving the film at $x = d$, $q_\lambda(d)$, must be calculated since ε_λ is defined as follows.

$$\varepsilon_\lambda \equiv \frac{q_\lambda(d)}{e_{bs}(\lambda, T_s)} = \frac{q_\lambda(d)}{\pi i_{bs}(\lambda, T_s)} \tag{14}$$

Similarly, for a cylinder the spectral emittance is the following.

$$\varepsilon_\lambda \equiv \frac{q_\lambda(R)}{e_{bs}(\lambda, T_s)} = \frac{q_\lambda(R)}{\pi i_{bs}(\lambda, T_s)} \tag{15}$$

Where $e_{bs}(\lambda, T_s)$ is the blackbody emissive power and T_s is the substrate temperature for the film or the cylinder temperature.

$$e_{bs}(\lambda, T_s) = \pi i_{bs}(\lambda, T_s) = \frac{2\pi h c_0^2}{\lambda^5 \left[\exp(hc_0 / \lambda k T_s) - 1\right]} \tag{16}$$

The radiation flux, q_λ, is obtained by solving the radiation transfer equation with the appropriate boundary conditions. These boundary conditions depend on the reflectances at the film-vacuum, ρ_{f0}, and film-substrate, ρ_{fs}, interfaces and the emittance at film-substrate interface, $\varepsilon_{fs} = 1 - \rho_{fs}$. In solving the radiation transfer equation the intensity, i_λ, is split into positive going (+x or +r direction) and negative going (–x or –r direction) components [14]. In this case the flux q_λ is given as follows.

$$q_\lambda = q_\lambda^+ - q_\lambda^- = 2\pi \int_0^{\pi/2} i^+ \cos\theta \sin\theta d\theta - 2\pi \int_{\pi/2}^{\pi} i_\lambda^- \cos\theta \sin\theta d\theta \tag{17}$$

The emittance theory neglecting scattering for the film emitter is presented in Ref. 15 and for the cylinder in Ref. 16. Since the rare earth doped YAG emitters are single crystal, scattering will be small compared to absorption and emission. For a film emitter with a linear temperature gradient the spectral emittance is the following

$$\varepsilon_\lambda \equiv \frac{q_\lambda(K_d)}{e_{bs}(\lambda, T_s)} = \frac{2n_f^2(1-\rho_{f0})}{DEN} \left\{ \left[\frac{\varepsilon_{fs}}{n_f^2} + 2\rho_{fs}\overline{\Phi}_-(K_d) \right] h_- + \overline{\Phi}_+(K_d)h_+ - \overline{\Phi}_M\left(\frac{K_d}{\mu_M} \right) h_M \right\}$$

film, no scattering with linear temperature gradient   (18)

Where,

$$DEN = 1 - \rho_{fs}E_3(K_d) \left[\rho_{fo}E_3(K_d) + (1-\rho_{fo})\mu_M^2 E_3\left(\frac{K_d}{\mu_m} \right) \right] \tag{19}$$

$$h_- = E_3(K_d) - \mu_M^2 E_3\left(\frac{K_d}{\mu_M} \right) \tag{20}$$

$$h_+ = 1 - 4\rho_{fs}\mu_M^2 E_3(K_d)E_3\left(\frac{K_d}{\mu_M} \right) \tag{21}$$

$$h_M = 1 - 4\rho_{fs}E_3^2(K_d) \tag{22}$$

$$\overline{\Phi}_+(K_d) = K_d(e^u - 1)\int_0^1 \frac{E_2[K_d(1-v)]}{\exp\left[\dfrac{u}{1-v\Delta T} \right] - 1} dv \tag{23}$$

$$\overline{\Phi}_-(K_d) = K_d(e^u - 1)\int_0^1 \frac{E_2[K_d v]}{\exp\left[\dfrac{u}{1-v\Delta T} \right] - 1} dv \tag{24}$$

$$\overline{\Phi}_M\left(\frac{K_d}{\mu_M} \right) = \mu_M K_d(e^u - 1)\int_0^1 \frac{E_2\left[\dfrac{K_d}{\mu_M}(1-v) \right]}{\exp\left[\dfrac{u}{1-v\Delta T} \right] - 1} dv \tag{25}$$

Where ΔT is the temperature gradient given by Eq. (8) and

$$u = \frac{hc_o}{\lambda k T_s} \tag{26}$$

$$v = \frac{K}{K_d} = \frac{x}{d} \tag{27}$$

$$E_n(x) = \int_0^1 z^{n-2} \exp\left(-\frac{x}{z}\right) dz \tag{28}$$

$$\mu_M^2 = 1 - \left(\frac{n_o}{n_f}\right)^2 \tag{29}$$

As Eq. (18) shows, an important variable for determining ε_λ is the optical depth, K_d, which is the product of the extinction coefficient, α_λ, and the thickness, d.

$$K_d = \alpha_\lambda d \tag{30}$$

Another important parameter for determining ε_λ for a film emitter is the temperature change ΔT. This will be discussed later.

From Eq. (18) we see that the emittance is made up of three parts. The coefficient of the h_- term represents the radiation leaving the substrate plus the radiation reflected back into the film from the film-substrate interface. This part of the emittance decreases with increasing optical depth, K_d. The second part of the emittance, $\overline{\Phi}_+ h_+$, represents the radiation emitted within the film and increases with increasing K_d. The last part of ε_λ is the negative term, $-\overline{\Phi}_M h_M$, which represents the radiation with angle of incidence $\theta > \theta_M$ that is totally reflected back into the film at the film-vacuum interface. This part increases with increasing optical depth.

Spectral emittance results for a cylinder of radius, R, at uniform temperature, T_s, can be obtained by setting $\varepsilon_{fs} = 0$ ($\rho_{fs} = 1$) and $\Delta T = 0$ in Eqs. (19) to (25).[15] When this is done the following result is obtained.

$$\varepsilon_\lambda = \frac{n_o^2(1 - \rho_{fo})\left[1 - 4E_3^2(K_R)\right]}{1 - 4E_3(K_R)\left[\rho_{fo}E_3(K_R) + \mu_M^2(1 - \rho_{fo})E_3\left(\frac{K_R}{\mu_M}\right)\right]} \tag{31}$$

cylinder at uniform temperature and no scattering

In this case the optical depth is the following.

$$K_R = \alpha_\lambda R \tag{32}$$

Effect of Temperature Gradient on Spectral Emittance of Emitting Film

To calculate ε_λ the extinction coefficient, α_λ, and index of refraction, n_f, must be known. These quantities were obtained from measured transmittance and reflectance data.[15] Also, the reflectances at the film-vacuum interface, ρ_{fo} and film-substrate interface, ρ_{fs}, ($\varepsilon_{fs} = 1 - \rho_{fs}$) must be known. For ρ_{fo} the value for normal reflectance was used.

$$\rho_{fo} = \left(\frac{n_f - n_o}{n_f + n_o}\right)^2 \qquad (33)$$

For ρ_{fs} the value for normal reflectance for a metal into a dielectric[14] was used.

$$\rho_{fs} = \left(1 - \varepsilon_{fs}\right) = \frac{\left(n_s - n_f\right)^2 + n_{sI}^2}{\left(n_s + n_f\right)^2 + n_{sI}^2} \qquad (34)$$

Included in Eq. (34) are the real, n_s, and imaginary parts, n_{sI}, of the substrate index of refraction.

Since radiation is strongly dependent on temperature there will be a significant effect of temperature gradient, ΔT, on the spectral emittance of an emitting film. This can be shown by using the spectral emittance model of the previous section. Using the extinction coefficient data for erbium aluminum garnet, $Er_3Al_5O_{12}$, shown in Fig. 1, ε_λ (Eq. (18)) was calculated for an emitter with a platinum substrate, a thickness $d = 0.63$ mm and a substrate temperature, $T_s = 1635$ K. Results for the platinum index of refraction were obtained from Ref. 17. The ε_λ results for $800 \leq \lambda \leq 2000$ nm are shown in Fig. 3 for $\Delta T = 0$ and $\Delta T = 0.08$. As can be seen, in the emission bands centered at $\lambda \approx 1000$ nm and $\lambda \approx 1500$ nm, where the extinction coefficient is large, the spectral emittance is greatly reduced in going from $\Delta T = 0$ to $\Delta T = 0.08$. Outside the emission bands, where the extinction coefficient is much smaller, the spectral emittance is not greatly effected by ΔT.

Obviously ΔT can be reduced by decreasing the thickness, d. However, decreasing d also reduces the optical depth, $K_d = \alpha_\lambda d$, which will result in reduced ε_λ. Thus varying d produces counteracting effects on ε_λ. Decreasing d will reduce ΔT, which will increase ε_λ. But decreasing

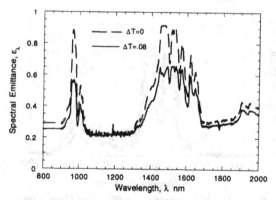

Fig. 3. Effect of temperature gradient on spectral emittance of $Er_3Al_5O_{12}$ with platinum substrate for $T_s = 1635$ K and $d = 0.63$ mm.

d will also reduce K_d, which will decrease ε_λ. As a result, there will be an optimum thickness, d, to obtain a maximum ε_λ.

Effect of Doping Level on Extinction Coefficient of $Ho_xY_{3-x}Al_5O_{12}$

The extinction coefficient, α_λ, is the critical optical property for determining spectral emittance, ε_λ, since the optical depth, $K_d = \alpha_\lambda d$, is directly proportional to α_λ. Large α_λ will result in large ε_λ. Therefore, to evaluate the effect of doping level on the emittance of holmium doped YAG we measured the extinction coefficient for a series of doping levels. The results of these measurements are shown in Fig. 4 for the wavelength range $1700 \leq \lambda \leq 2200$ nm. This shows the main emission band of Ho which results from electronic transitions from the first excited state manifold, $^5I_7$, to the ground state manifold, $^5I_8$.

As Fig. 4 shows the extinction coefficient is a monotonically increasing function of the Ho doping level. The maximum extinction coefficients in the emission bands double in going from $Ho_{1.2}Y_{1.8}Al_5O_{12}$ to pure holmium aluminum garnet, $Ho_3Al_5O_{12}$. However, a doubling of α_λ in an emission band where α_λ is already large does not translate into a doubling of ε_λ. For large α_λ and thus large K_d the emittance is independent of α_λ[15]. The benefit of a large extinction coefficient is that the thickness, d, can be reduced while still maintaining a large optical depth $K_d \geq 1$). Thus the smaller d will result in a smaller temperature gradient which in turn will increase ε_λ.

The same effect of doping level on extinction coefficient has been established for $Er_xY_{3-x}Al_5O_{12}$[15]. Due to their similar atomic structure we expect the same increase in α_λ with increasing doping level for all the rare earths. Panitz[18] has found similar results for Yb doping of YAG.

Spectral Emittance of Emitting Film of $Er_3Al_5O_{12}$ and $Ho_3Al_5O_{12}$

Figure 5 compares the measured and calculated film spectral emittance of erbium aluminum garnet, $Er_3Al_5O_{12}$, for a sample of thickness, d = 0.63 mm, with a Pt foil substrate. The substrate temperature was $T_s = 1635$ K and the measured temperature gradient was $\Delta T = 0.08$. The

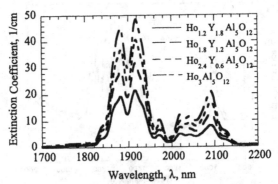

Fig. 4. Effect on extinction coefficient of Ho-doping level in YAG. Extinction coefficient for $Ho_{1.2}Y_{1.8}Al_5O_{12}$, $Ho_{1.8}Y_{1.2}Al_5O_{12}$,

$Ho_{2.4}Y_{0.6}Al_5O_{12}$, and $Ho_3Al_5O_{12}$.

experimental measurement of ε_λ is explained in reference 15. Figure 5(a) shows the wavelength region $600 \leq \lambda \leq 10,000$ nm while Fig. 5(b) shows an expanded view of the emission band region $800 \leq \lambda \leq 2000$ nm. Er has four emission bands associated with electronic transitions from the first 4 excited state manifolds to the ground state. The most intense band centered at $\lambda \approx 1500$ nm results from transitions from the first excited state manifold ($^4I_{13/2} \rightarrow {}^4I_{15/2}$). The other three bands centered at $\lambda \approx 1000$ nm, 800 nm and 640 nm result from transitions of the next three excited state manifolds ($^4I_{11/2} \rightarrow {}^4I_{15/2}$, $^4I_{9/2} \rightarrow {}^4I_{15/2}$ and $^4F_{9/2} \rightarrow {}^4I_{15/2}$). The large emittance in the region $\lambda \geq 5000$ nm results from the vibrational modes of the crystal lattice. As pointed out in the Introduction, this long wavelength cutoff region occurs for most all high temperature ceramic materials.

There is good agreement between the measured and calculated (Eq. (18)) spectral emittance except for the region $1800 \leq \lambda \leq 6000$ nm. In the region $1800 \leq \lambda \leq 4000$ nm where the extinction coefficient is small the main contribution to the emittance comes from the Pt substrate. Error in the measured extinction coefficient is the reason the calculated ε_λ is larger than the measured value in the $1800 \leq \lambda \leq 4000$ nm region[15].

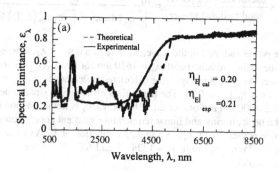

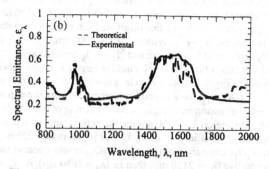

Fig. 5. Comparison of experimental and calculated spectral emittance of Erbium aluminum garnet ($Er_3Al_5O_{12}$) with platinum substrate. Thickness, d = 0.63 mm, substrate temperature, T_s = 1635 K, front face temperature, T_f =1503 K, ΔT = 0.0807. (a) $600 < \lambda < 10,000$ nm. (b) $800 < \lambda < 2000$ nm.

For the region $4000 \leq \lambda \leq 6000$ nm the vibrational modes of the crystal lattice are the primary source of the emission. The extinction coefficient was measured at room temperature. Therefore, the calculated ε_λ for $4000 \leq \lambda \leq 6000$ nm corresponds to room temperature conditions. Since the calculated ε_λ is less than the experimental ε_λ in the $4000 \leq \lambda \leq 6000$ nm region we conclude that the crystal lattice structure is charging with temperature such that higher energy (shorter wavelength) modes exist. These new modes produce the increase in extinction coefficient in the $4000 \leq \lambda \leq 6000$ nm region. In the emission band region, $600 \leq \lambda \leq 1800$ nm electronic transitions of the Er ions account for the radiation. Therefore, widening of the emission bands is the expected effect for increasing temperature rather than increasing extinction coefficient. This conclusion is substantiated since there is good agreement between experimental and calculated ε_λ for $600 \leq \lambda \leq 1800$ nm with the experimental emission bands being somewhat broader. The resolution of the extinction coefficient data was much greater than experimental ε_λ data. As a result, the calculated ε_λ shows more structure in the emission bands.

As a measure of the effectiveness of $Er_3Al_5O_{12}$ as a selective emitter we define the emitter efficiency as follows.

$$\eta_E \equiv \frac{\text{useful radiated power}}{\text{total radiated power}} = \frac{\int_0^{\lambda_\ell} q_\lambda(d)d\lambda}{\int_0^\infty q_\lambda(d)d\lambda} = \frac{\int_0^{\lambda_\ell} \varepsilon_\lambda e_{bs}(\lambda, T_s)d\lambda}{\int_0^\infty \varepsilon_\lambda e_{bs}(\lambda, T_s)d\lambda} \tag{35}$$

The numerator is the power radiated in the wavelength region $0 \leq \lambda \leq \lambda_\ell$, where λ_ℓ is the wavelength at the end of the main emission band ($\lambda_\ell = 1650$ nm for $Er_3Al_5O_{12}$). The denominator is the total radiated power. As shown in Fig. 5 the calculated efficiency is $\eta_E|_{cal} = 0.20$ and the experimental efficiency is $\eta_E|_{exp} = 0.21$. The theoretical efficiency is smaller because $\varepsilon_\lambda|_{cal}$ is too large in the $1800 \leq \lambda \leq 4000$ nm region. As Eq. (18) shows, ε_λ is not a function of T_s but is a function of ΔT for the no scattering and linear temperature gradient approximations. However, the blackbody emissive power, $e_{bs}(\lambda, T_s)$, is a sensitive function of T_s. As a result η_E is also a sensitive function of T_s. For $d = 0.3$ mm, $T_s = 1234$ °K and $\Delta T = 0.094$ the measured efficiency is reduced to $\eta_E|_{exp} = 0.065$.

Figure 6 compares calculated and measured values of ε_λ for $Ho_3Al_5O_{12}$ on a Pt substrate with $d = 0.3$ mm. The substrate temperature was $T_s = 1446$ K and $\Delta T = 0.0692$. Similar to $Er_3Al_5O_{12}$, $Ho_3Al_5O_{12}$ has emission bands that originate from transitions to the ground state from the 4 lowest excited states; $^5I_7 \rightarrow {}^5I_8$ centered at $\lambda \approx 2000$ nm, $^5I_6 \rightarrow {}^5I_8$ centered at $\lambda \approx 1200$ nm, $^5I_5 \rightarrow {}^5I_8$ centered at $\lambda \approx 900$ nm and $^5I_4 \rightarrow {}^5I_8$ centered at $\lambda \approx 650$ nm. The first excited state to ground state transition ($^5I_7 \rightarrow {}^5I_8$) has the largest intensity. Also similar to $Er_3Al_5O_{12}$, but not shown in Fig. 6, the long wavelength region of large emittance begins at $\lambda \approx 5000$ nm. Similar to $Er_3Al_5O_{12}$, there is good agreement between the experimental and theoretical emittance in the emission band regions. Also, because of experimental error in the extinction coefficient for the region $\lambda > 2200$ nm the theoretical emittance is too large. For $Ho_3Al_5O_{12}$ the measured emitter efficiency using $\lambda_\ell = 2150$ nm in Eq. (35) is $\eta_E|_{exp} = 0.34$. The larger η_E for $Ho_3Al_5O_{12}$ compared to η_E for $Er_3Al_5O_{12}$ results because the integration limit, λ_ℓ, in Eq. (35) is larger for Ho ($\lambda_\ell = 2150$ nm) than Er ($\lambda_\ell = 1650$ nm). At $T_s = 1270$ K and $\Delta T = 0.062$ the measured efficiency is $\eta_E|_{exp} = 0.27$ for the sample in Fig. 6.

Spectral Emittance of Emitting Cylinder of $Ho_3Al_5O_{12}$

Using Eq. (31) the spectral emittance of a cylinder of radius, $R = 1.0$ mm, of $Ho_3Al_5O_{12}$ was calculated. These results are shown in Fig. 7. Comparing the film results for $Ho_3Al_5O_{12}$ (Fig. 6)

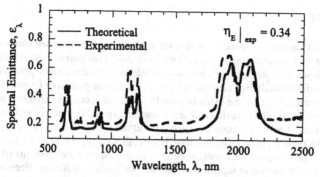

Fig. 6. Comparison of experimental and calculated
spectral emittance of Holmium aluminum garnet
($Ho_3Al_5O_{12}$) with platinum substrate. Thickness,
d = 0.3 mm, substrate temperature, T_s = 1446K,
front face temperature, T_f = 1346 K, ΔT = 0.0692.

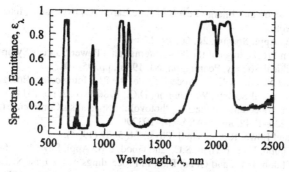

Fig. 7. Calculated emittance for a cylinder of
Holmium aluminum garnet ($Ho_3Al_5O_{12}$)
with a radius, r = 0.1 cm.

to Fig. 7 we see that ε_λ is significantly larger in the emission bands for the cylindrical geometry. This results because the cylinder is at uniform temperature whereas the film has a linear temperature gradient.

Obviously a cylinder is more desirable than a film for obtaining large emittance in an emission band. Fibrous emitters of rare earth oxides[3,4] are used for combustion driven TPV systems.[19] As mentioned in the Introduction, the fibrous emitters are well suited for coupling energy from a flame. However, for other TPV systems where the source of thermal energy is solar or nuclear the film type emitter is more applicable.

CONCLUSION

Rare earth ions in high temperature host materials emit in several narrow bands in the near infrared and the visible spectrum. A spectral emittance model for film and cylindrical emitters, where optical depth is the key variable, has been developed. There are many possible host materials although this study includes only rare earth aluminum garnets of Er and Ho. Experimental measurements of the extinction coefficients and spectral emittances of these garnets have been made. The experimental and theoretical spectral emittance results for the film emitter are in good agreement. Maximum spectral emittances of 0.6 to 0.8 occur in the emission bands. At $\lambda > 5000$ nm all these materials have large emittance ($\varepsilon_\lambda > 0.8$) as a result of emission from the vibrational modes of the crystal structures. Small temperature gradients ($\Delta T < 0.1$) across the film emitters result in significant reductions in spectral emittance. Because spectral emittance increases with optical depth but decreases with temperature gradient there is an optimum film thickness for maximum emittance. For operating conditions of interest, the film emitter has a linear temperature variation whereas the cylindrical emitter has a uniform temperature. Thus, the cylinder has larger spectral emittance in the emission bands. Emitter efficiency is a sensitive function of temperature. For holmium aluminum garnet ($Ho_3Al_5O_{12}$) the efficiency at $T_s = 1446$ K for film thickness, d = 0.3 mm, and temperature gradient, $\Delta T = 0.069$ is $\eta_E|_{exp} = 0.34$ and at $T_s = 1270$ K with $\Delta T = 0.062$ the efficiency is $\eta_E|_{exp} = 0.27$.

REFERENCES

1. Dieke, G.H., Spectra and Energy Levels of Rare Earth Ions in Crystals, Interscience, New York, 1968.
2. Guazzoni, G.E., Appl. Spectra. 26, 60, 1972.
3. Nelson, R.E., in Proceedings of the 32nd International Power Sources Symposium (Electronchemical Society, Pennington, NJ, 1986), pp. 95–101.
4. Parent, C.R. and Nelson, R.E., in Proceedings of the 21st Intersociety Energy Conversion (American Chemical Society, Washington, D.C., 1986), vol. 2, pp. 1314–1317.
5. Chubb, D.L., in Proceedings of the 21st Photovoltaic Specialists Conference (IEEE, New York, 1990), pp. 13226–1342; also NASA TM–103290.
6. Chubb, D.L., and Lowe, R.A., J. Appl. Phys., 74, 5687 (1993).
7. Lowe, R.A., Chubb, D.L., Farmer, S.C., and Good, B.S., Appl. Phys. Lett. 64, 3551, 1994.
8. Lowe, R.A., Chubb, D.L., and Good, B.S., in Proceedings of the First NREL Conference on Thermophotovoltaic Generation of Electricity, (American Institute of Physics Conference Proceedings 321, 1994), pp. 291–297.
9. Lowe, R.A., Chubb, D.L., and Good, B.S, in Proceedings of the First World Conference on Photovoltaic Energy Conversion, (IEEE, 1995).
10. Lowe, R.A., Good, B.S., and Chubb, D.L., in Proceedings of 30th Intersociety Energy Conversion Engineering Conference (IECEC), (American Society of Mechanical Engineers, 1995), vol. 2, pp. 511–515.
11. Chubb, D.L., Lowe, R.A., and Good, B.S., in Proceedings of The First NREL Conference on Thermophotovoltaic Generation of Electricity, (American Institute of Physics Conference Proceedings 321, 1994), pp. 229–244.
12. Good, B.S., Chubb, D.L., and Lowe, R.A., in Proceedings of The First NREL Conference on Thermophotovoltaic Generation of Electricity, (American Institute of Physics Conference Proceedings 321, 1994), pp. 263–275.

13. Chubb, D.L., Good, B.S., Clark, E.B., and Chen, Z., Effect of Temperature Gradient on Thick Film Selective Emitter Emittance, presented at The Third NREL Conference on Thermophotovoltaic Generation of Electricity, AIP Conference Proceedings 401, 1997.
14. Siegel, R. and Howell, J.R., Thermal Radiation Heat Transfer, 2nd edition, Washington, DC, Hemisphere, 1981, Chs. 4, 14.
15. Chubb, D.L., Pal, A.T., Patton, M.O., and Jenkins, P.P., "Rare Earth Doped High Temperature Ceramic Selective Emitters," NASA TM–208491.
16. Chubb, D.L., "Emittance Theory for Cylindrical Fiber Selective Emitter," NASA TM.
17. Lide, D.R., (Ed.), CRC Handbook of Chemistry and Physics, 71st ed., CRC Press, 1990.
18. Panitz, J.-C., Schubnell, M., Durisch, W., and Geiger, F., "Influence of Ytterbium Concentration on the Emissive Properties of Yb:YAG and Yb: Y_2O_3," presented at The Third NREL Conference on Thermophotovoltaic Generation of Electricity, AIP Conference Proceedings 401, 1997.
19. Becker, F.E., Doyle, E.F., and Kailash, S., "Development of a Portable Thermophotovoltaic Power Generator in Proceedings of the Third NREL Conference on Thermophotovoltaic Generation of Electricity, AIP Proceedings 401, 1997, pp. 329–339.

ELECTRODEPOSITED CIS-BASED SOLAR CELL MATERIALS

R.P. RAFFAELLE*, T. POTDEVIN*, J.G. MANTOVANI*, R. FRIEDFELD*, J. GORSE**,
M. BREEN***, S.G. BAILEY***, AND A.F. HEPP***
*Physics and Space Sciences Department, Florida Institute of Technology, Melbourne, FL
32901, rpr@pss.fit.edu
**Baldwin-Wallace University, Berea, OH 44135
***NASA Lewis Research Center, Cleveland, OH 44135

ABSTRACT

We have been investigating the synthesis of novel materials and device structures based
on $CuInSe_2$ or CIS system. We are interested in developing CIS for use in thin film photoltaic
solar cells used in space power systems. Non-stoichiometric native defects allow CIS to be
selectively doped n or p type by controlling the Cu to In ratio. The Cu to In ratio of an
electrodeposited CIS thin film is shown to be directly proportional to the deposition voltage.
This behavior has allowed us to deposit different semiconductor-type films, pn junctions, and
multilayer structures from the same aqueous solution. The affect of annealing in an Argon
atmosphere and the addition of a wetting agent to our deposition solutions on the structural and
electrical behavior of our films was measured. A dramatic improvement in the crystallinity,
which was proportional to the annealing temperature was observed. The addition of a wetting
agent to the solution was shown to improve the surface morphology and provide uniformity in
the grain size. The electrical characterization of Schottky barriers on the CIS showed a decrease
in several orders of magnitude of the carrier density with annealing. The ability to incorporate
our CIS-based films in pn junction devices using CdS window layers was demonstrated by the
rectifying current versus voltage behavior of these junctions.

INTRODUCTION

The inexpensive production of device-quality copper indium diselenide (CIS) thin
films is one of the major goals of the international photovoltaics community. This is due to
the fact that CIS is a nearly ideal candidate for thin film solar cells, considering that it has a
direct bandgap of about 1.1 eV, it is well lattice-matched to the large bandgap (window)
material CdS, it can be made p-type with good electrical characteristics, it has good terrestrial
stability and radiation resistance, and it has an extremely high optical absorption coefficient
[1]. In addition, the substitutional doping of Ga for In in this material can be used to raise the
bandgap to better match the solar spectrum. Conversion efficiencies of 17.8% have already
been achieved for solar cell based on CIS doped with Ga [2]. Its nearly ideal photovoltaic
properties and radiation resistance in its polycrystalline for make is an outstanding candidtate
for large area flexible arrays used in space power systems.

There have been a wide variety of vacuum-based techniques for producing thin films
CIS such as co-evaporation of the elements, r.f. sputtering, ion beam sputtering, magnetron
reactive sputtering, close-space chemical vapor transport, and molecular beam epitaxy.
Although these techniques have been shown to produce high-quality films, their high vacuum
and thermal and/or cryogenic requirements make them expensive and inhibit their scalability.
We have been investigating the use of electrochemical deposition as a simple, cost-effective,
and easily scalable alternative to the more standard vapor deposition techniques for
producing CIS thin films [3].

Electrochemical deposition or electrodeposition provides a technique of growing thin films of variable thickness and stoichiometry. The stoichiometry of the films is potentiostatically controlled, and the thickness is based on the product of the current density and the deposition time. We have previously shown that the Cu to In ratio will demonstrate a linear dependence with deposition potential over a fairly wide range [4]. CIS has been shown to be electrically and structurally stable over a wide range of stoichiometries. Stoichiometric variations in polycrystalline CIS leads to a high concentration of electrically active native defects. Additionally, CIS has been shown to undergo a n- to p- type change as the Cu to In ratio is raised above unity. The stoichimetrically controlled native defects in this system have allowed us to electrodeposit thin films with different electrical, optical properties, and semiconductor types from the same aqueous solution by simply varying the deposition voltage. An example of this was the deposition of a pn junction from a single aqueous solution using a step-function potential [5].

Unfortunately, as-electrodeposited CIS thin films usually have a rather small grain size, poor homogeneity, and extremely high defect densities. These properties have drastically limited the utility of this synthesis method to produce devices. The effect of these large defect densities is less of a problem than was expected due to a recently discovered defect pairing mechanism [6]. In addition, post-deposition techniques involving heat treatment have successfully been used to address these drawbacks [7].

EXPERIMENT

A series of one micron thick films were electrodeposited on mechanically polished molybdenum (Mo) substrates using deposition potentials chosen to give p-type films (-1.1 to -1.3 V vs. SCE) [5]. The deposition solutions consisted of 1mM $CuSO_4$, 10 mM $In_2(SO_4)_3$, 5 mM SeO_2, 25 mM Na-citrate. A second solution was also used which had the addition of 10% ethyl alchohol. Selected samples were subjected to a post-deposition annealing at temperatures ranging at 600 °C for 1 hour in a flowing argon (Ar) atmosphere. The deposition potentials used to deposit our thin films were generated and monitored by a CHI Instruments computer controlled potentiostat.

Following the deposition, the crystallinity of the films was investigated by X-ray diffraction using a Phillips PW 3710 XRD. The composition of these films was also determined by energy dispersive spectroscopy (EDS) in a Hitachi S4700 scanning electron microscope. Images of the surface morphology were also taken in the SEM to determine the effects from the addition of the alcohol to our solution and post-deposition annealing.

The thickness of the films was estimated using an equation based upon Faraday's law. The thickness of a film is proportional to the current delivered and the deposition time and inversely proportional to the deposition area. These parameters combined with the theoretical structure, stoichiometry, and valence of the transferred ions are used to complete the calculation [5]. This equation only provides an estimate of the film thickness, as variations in the composition and crystallinity will effect the result

In order to investigate the electrical characteristics of our films, we e-beam evaporated a set of 1200 Angstroms thick aluminum pads. The barrier height of a metal to p-type semiconductor junction is the difference between the metal work function and the sum of the electron affinity and the bandgap. Assuming ideal junctions and Al work function of 4.28 eV , electron affinity of 4.48 eV and CIS bandgap of 1.1 eV, we would expect Al to make a rectifying or Schottky contact to p-type CIS, with a barrier height of around 1.3 eV [8].

Current versus voltage measurements on our Schottky barriers were used to investigate the barrier heights of our junctions. These measurements were performed with a Microtone wafer probing station and a Keithley 236 Source Measure Unit. The barrier height can be determined using

$$\phi_B = \frac{kT}{q}\ln\left(\frac{A^* T^2}{J_S}\right) \qquad (1)$$

where A^* is the effective Richardson and J_s is the saturation current density [8].

The capacitance versus voltage behavior of these Schottky barriers can be used to determine the carrier density in the semiconductor. The junction will behave like a parallel-plate capacitor with the depletion region serving as the dielectric. As a voltage is applied to the junction the depletion width will change and therefore so will the junction's capacitance. This change is proportional to the carrier density in the semiconductor. The capacitance of an ideal Schottky barrier as a function of reverse bias voltage can be expressed as

$$C = A\left[q\varepsilon_s N_d \Big/ 2\left(V_{bi} - \left(\frac{kT}{q}\right) - V\right)\right]^{\frac{1}{2}} \qquad (3)$$

where A is the junction area, ε_s is the dielectric constant, N_d is the semiconductor doping density, and V_{bi} is the built-in voltage [8]. Therefore, the slope of $1/C^2$ vs. V can be used to determine the doping density. We measure the capacitance versus voltage behavior of our junctions using a wafer probing station and a Keithley 590 CV analyser.

In addition to the Schottky barriers, we also measure the I-V charactreristics of *pn* junctions made with our films. These junctions were produced by using chemical bath deposition of n-type CdS onto our p-type CIS-based films. The deposition solution consisted of 0.2 M $Cd(CH_3OO)_2$, 0.5 M $SC(NH_2)_2$, and NH_4OH. The Cd and thiourea solutions were mixed seprately and heated to 80 °C. The ammonium hydroxide was added to the Cd solution to raise its pH to 8. The solutions were then mixed and and placed in contact with the CIS films for a duration of 5.0 minutes.

RESULTS

SEM micrographs show that the as-deposited CIS thin films on Mo were polycrystalline and dense with sub-micron sized grains (see Fig. 1). However, the uniformity of grain size and smoothness of the films was improved with the addition of the ethyl alcohol to the deposition solution (see Fig. 2). The annealed series of the films showed a coalescing of the grains (see Fig. 3). However, there was also evidence of cracking in some of the films after annealing. The CdS films grown on the CIS were extremely smooth and uniform and showed a good coverage of the CIS (see Fig. 4).

EDS analysis of our CIS films showed that the Se content varied between 50 to 53% for all the deposition potentials used. The Cu to In ratio in the as-deposited films varied almost linearly with the deposition potential (see table I). Based on these results, we would expect those films deposited a potentials less negative than -1.2 V vs. SCE would be p-type, and those deposited at more negative potentials would be n-type [5].

Figure 1. CIS as-deposited at −1.2 V vs. SCE. The scale marker is 3.0 μm.

Figure 2. CIS as-deposited at −1.2 V vs. SCE with alcohol in solution. The scale marker is 10 μm.

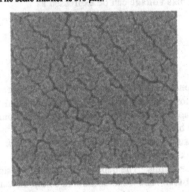

Figure 3. CIS after annealing in argon at 600 °C for 1 hour. The scale marker is 3.0 μm.

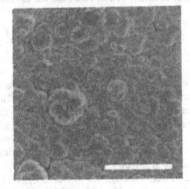

Figure 4. Chemically bath deposited CdS on CIS. The scale marker is 3.0 μm.

XRD analysis of the as-deposited films verified the existence of the main Bragg peaks for CIS in a chalcopyrite crystal structure. Unfortunately, these peaks were not intense and are quite broad. In addition, the minor peaks which conclusively verify the chalcopyrite structure (i.e., (101), (103), (105), and (213)) cannot be identified. However, after annealing these films in a flowing argon atmosphere there was a significant improvement in the crystallinity of the films (see Fig. 5). The intensity and sharpness of the main Bragg peaks increased and the minor chalcopyrite peaks became observable.

V_{dep} (V vs. SCE)	Cu/In ratio	N_a (cm^{-3})	$q\Phi_b$ (eV)
-1.0	1.30	8.6×10^{20}	0.62
-1.05	1.23	6.3×10^{20}	0.57
-1.1	1.15	1.4×10^{19}	0.53
-1.15	1.20	2.1×10^{19}	0.45
-1.2	1.01	1.1×10^{19}	0.48
-1.25	0.97	6.9×10^{18}	0.55
-1.3	0.89	8.7×10^{18}	0.53

Table I. EDS results and carrier densities and barrier heights from the electrical characterization of Al on CIS Schottky barriers.

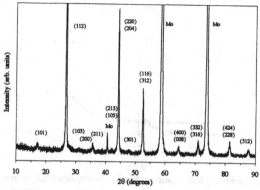

Figure 5. XRD pattern of an annealed CIS film deposited at −1.2 V vs. SCE

The I-V behavior of the Al contacts on CIS film deposited at less negative potentials than -1.2 V vs. SCE showed the rectifying behavior of a Schottky barrier. This implies that these films were p-type. The majority of the Al contacts on the films deposited at −1.25 V vs. SCE still exhibited some rectification. However, the Al contacts on the films deposited at −1.3 vs. SCE were almost entirely ohmic. The semilog plots of current density versus voltage showed the anticipated linear behavior and yielded barrier heights between 0.55 and 0.64 eV for the un-annealed films, with ideality factors much greater than one (see Table I). A value for the hole effective mass of 0.71 was used in our calculations [1]. After annealing, there was an increase on average of 0.16 eV in the barrier heights. This result is consistent with the improvement in crystallinity seen with annealing, as both effects would presumably correlate with a reduction in the number of defects in the material.

The capacitance versus voltage measurements of the Schottky barriers on the un-annealed CIS series demonstrated a linear $1/C^2$ vs. V behavior. The carrier densities based on the least-squares slope of these plots ranged from 10^{18} to 10^{20} and would indicate that this series is nearly degenerate. However, a significant decrease in the carrier was observed after annealing at 600 °C. The carrier density in the sample deposited at −1.15 V vs. SCE decreased from 2.1×10^{19} to 6.8×10^{17} cm^{-3}.

The current versus voltage measurement of the CdS on CIS pn-junctions showed good rectification (see Figure 6). Small but observable short circuit current and open circuit voltages were seen under ambient room light. The photocurrent were on the order on tens of microamps using silver paint point contacts to the CdS surface.

CONCLUSIONS

The electrodeposited $Cu_xIn_{2-x}Se_2$ thin films were shown to have stoichiometries which varied linearly with the range of deposition potentials used in this study. The film surfaces were dense and homogenous. Annealing in an Argon atmosphere was shown to increase the average grain size of the films, but some cracking did occur. X-ray diffraction studies confirmed that all of the films grown in this study had a basic chalcopyrite structure. However, there was significant improvement in crystallinity with annealing. Schottky barrier behavior was observed for Al contacts to films deposited at the voltages used in this study. The results were in agreement with these films being p-type as indicated by our EDS results.

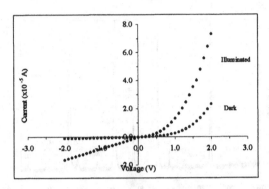

Figure 6. Chemically bath deposity CdS on electrodeposited CIS *pn*-Junction.

Analysis of capacitance versus voltage measurements on these Schottky barriers yielded carrier densities which increased with deviation from stoichiometry. A dramatic decrease in the average carrier density with respect to composition was shown with the annealing of the films.

Chemical bath deposition can be used to deposit n-type CdS onto electrodeposited p-type CIS thin films to produce a pn junction. These pn juntions exhibited good rectification and measureable photocurrents.

ACKNOWLEDGEMENTS

We wish to thank F. Montegani, the Ohio Aerospace Institute (OAI), and the NASA-ASEE program for supporting this work.

REFERENCES

1. A. Rockett and R.W. Birkmire, *J. Appl. Phys.* **70**, R81(1991).

2. J.R. Tuttle, J. Ward, A. Duda, T. Berens, M. Contreras, K. Ramanathan, A. Tennant, J. Keane, E. Cole, K. Emory, and R. Noufi, Proc. *Mat. Res. Soc. Symp.*, **426** (1996) 143.

3. R.P. Raffaelle, J.G. Mantovani, R.B. Friedfeld, S.G. Bailey, and S.M. Hubbard, *26th IEEE Photovolt. Spec. Conf* (1997) 559.

4. R.P. Raffaelle, J.G. Mantovani, S.G. Bailey, A.F. Hepp, E.M. Gordon, and R.Haraway, *Mat. Res. Soc. Symp.* **485** (1997).

5. R.P. Raffaelle, J.G. Mantovani, S.G. Bailey, A.F. Hepp, E.M. Gordon, and R.Haraway, *NASA/TM*-97-206322 (1997).

6. S.B. Zhang, S. Wei, A. Zunger, *Phys. Rev. Lett.*, **78**, 21 (1997) 4059.

7. C.X. Qui and I. Shih, *Solar Energy Mater.* **15** (1987) 219.

8. S.M. Sze, *Physics of Semiconductor Devices* (John Wiley & Sons, New York, 1981).

LARGE AREA DEPOSITION OF CADMIUM SULFIDE BY CHEMICAL BATH DEPOSITION FOR PHOTOVOLTAIC APPLICATIONS

DAVID S. BOYLE and **PAUL O'BRIEN***
Department of Chemistry, Imperial College of Science, Technology and Medicine, Exhibition Road, London, SW7 2AZ, UK.
Email addresses : d.smythboyle@ic.ac.uk and p.obrien@ic.ac.uk

ABSTRACT

There is considerable interest in the deposition of compound semiconductors by methods which involve relatively low capital expense and are technically undemanding on the experimentalist. One process to meet these criteria is Chemical Bath Deposition (CBD). Such processing methods are particularly appropriate for the production of devices for which large areas and low cost are essential, such as the BP Solar Ltd "Apollo" cells based on CdS:CdTe heterojunctions.

The electrical and optical properties of these devices have been investigated with respect to impurity profiles of 16-O, 34-S, 35-Cl and 12-C in the CdTe and CdS:CdTe interface. Device characteristics (*e.g.* V_{oc} and R_s) have been correlated with SIMS data. The distribution of the carbon contaminant appears to influence the chloride-promoted recrystallisation of CdTe. XPS analysis of CdS layers appears to indicate that the annealing process induces the formation of a chloride-rich surface layer, which may promote the n- to p- type thermal conversion of CdTe.

INTRODUCTION

Thin film n-CdS/p-CdTe solar cells such as the BP Solar "Apollo" cell [1] have been demonstrated in recent years to be relatively efficient and economically attractive photovoltaic conversion devices. It has been established that solar cell efficiencies are sensitive to the nature of the CdS:CdTe interface.[2] The optimization of each layer in a device depends on the processing of all the other adjacent layers present. The importance of CdS-CdTe interdiffusion and O and Cl doping (either deliberate or as a consequence of the materials syntheses) have yet to be established quantitatively. In the present work, we have been motivated to investigate the significance of the observation that "wet-chemical" cells (as opposed to those fabricated by "clean" techniques such as close-space-sublimation) often contain significant concentrations of

129

chemical impurities. It has been suggested that the adventitious incorporation of these impurities, *via* the CBD CdS process,[3] may be related to the relatively high cell efficiencies of these cells.[4]

Secondary Ion Mass Spectrometry (SIMS) is a useful tool for the determination of chemical impurities in semiconductor films.[5] The technique involves the detection of charged secondary ions which are produced as a result of sputtering with energetic primary ions. In dynamic SIMS the primary ion density is sufficient to erode the surface, hence depth profiling of elements is possible. A profilometer is used to determine the depth of the crater produced by ion sputtering (assuming a flat bottomed pit), hence the isotopic concentration profile through a sputtered sample may be determined.

EXPERIMENTAL

The CdS thin films (~80 nm) were grown on commercial tin-oxide glass substrates (LOF) from stirred solutions (at 343-353 K) containing cadmium salt, an amine complexant, thiourea and sodium hydroxide to obtain a solution of final pH ~12 (Table 1). After deposition the substrates were washed with de-ionised water (adherent particulate matter removed by ultrasonic agitation) and allowed to dry before annealing in air. The electrodeposition of CdTe involved the formation of CdTe layers (~1700 nm) from an aqueous solution containing Cd^{2+} and $HTeO^{2+}$ ions. A post-deposition heat treatment was employed to effect type conversation and produce an efficient p-n junction. Film thicknesses were calculated from transmission/ reflection measurements. A standard process employed by BP Solar [1] was used for the fabrication of photovoltaic modules (100x100 mm). Device performances were characterised by J-V analysis under 100 mW cm^{-2} illumination using ELH lamps behind diffusing glass. The XPS measurements were performed in the ultra-high vacuum chamber (base pressure 10^{-8} Pa) of a VG ESCALAB-Mk II (VG Scientific) using Al Kα excitation (analyser pass energy of 50 eV). The energy scale was calibrated using C_{1s} (at 284.8 eV) as a reference. SIMS analysis was performed using a CAMECA IMS 3F Ion Microscope with gas duoplasmatron (argon and oxygen) and Cs^+ (13 KeV impact energy) primary ion sources, with a Sun SPARC station data acquisition and processing system and CAMECA IMS software. Ion implanted standards of polished CdTe containing the isotopes 16-O, 34-S, 35-Cl, 12-C and 14-N were prepared at AEA Technology Ltd. The post analysis calibration of the depth scale for each SIMS analysis was achieved with a TENCOR Alfa-step stylus surface profiler.

Bath	CdCl$_2$	NaOH	ammonium hydroxide	1,2-diaminoethane	thiourea
1	2.0	14.5	250	6	2.46
2	3.0	9.2	250	8.7	2.46
3	2.0	14.5	0	6	2.46
4	2.0	5.2	0	6	7.66

Table 1: Chemical baths used to deposit CdS on TO-glass. All concentrations in mmol dm^{-3}.

RESULTS AND DISCUSSION

The XPS spectra for our CBD CdS were similar to those reported by other workers for CdS films.[6] An important observation was the confirmation of the presence of carbon, nitrogen and oxygen as impurities in the as-deposited films (Table 2). Upon annealing of films in air, the nitrogeneous contaminant was reduced to a concentration below the detection limit of the instrument; concurrently a concentration of chloride species was observed in the XPS spectrum. The presence of a chloride-rich layer on CdS has been associated with efficient type-conversion of CdTe layers upon annealing of heterojunction devices. The annealing step in air does not appear to degrade the CdS films; the absence of a photoelectron peak at 168.7 eV (S 2p for SO$_4$$^{2-}$ of cadmium sulfate [7] was an indication that formation of undesirable oxidation products was not a significant process.

The following relationship was used to relate the parameters in the SIMS studies:

$$RSF = D* \int (I/ I_{ref})$$

where RSF is the relative sensitivity factor, D is the specific ion-implantation dose in the host matrix to yield a Gaussian distribution, I is the recorded secondary ion intensity and I$_{ref}$ is the secondary ion intensity recorded for the particular isotope in the ion-implanted standard. The etch rate through the entire cell structure was relatively constant, hence the average sputter rate was calculated in terms of the crater depth and total sputter time. The ion intensity scale was converted to concentration by application of the appropriate relative sensitivity factors.[8] Initial efforts revealed that the very low negative secondary ion yield of nitrogen would not allow collection of accurate and reproducible data. Hence only the isotopic profiles for 16-O, 34-S, 35-Cl and 12-C were determined. The SIMS 14-N depth profiles in CBD CdS layers have been reported by other workers but the results were not quantitative.[4]

For all CdS/ CdTe solar cells, the general pattern of isotopic distributions from SIMS analysis were similar and the relative concentrations followed the order 16-O > 35-Cl > 12-C >

photoelectron (PE) peak	as-deposited CdS	heat-treated CdS
Cd 3d $_{5/2}$	405.1	405.1
Cd 3d $_{3/2}$	411.9	411.9
Cd (Auger)	1105.2 and 1111.2	1105.0 and 1111.5
S 2p	162.0	162.0
N 1s	395.7	
O 1s	531.9	531.8
C 1s	284.8	282.8
Na 1s		1071.5
Cl 2p $_{3/2}$		199.5

Table 2: Binding energies (eV) of PE peaks observed for CBD CdS films via Bath 1.

Cell and bath number	V_{oc} (V)	I_{sc} (mA)	R_s (Ω)	R_p (Ω)	P_{max} (mW)	Fill Factor	Efficiency (%)
1	0.776	0.917	199	6710	0.408	0.574	7.8
2	0.75	0.709	268	8380	0.292	0.55	7.2
3	0.769	0.755	320	10400	0.3	0.517	6.9
4	0.766	0.747	294	7000	0.306	0.534	7.2

Table 3 : PV cell characteristics for devices fabricated *via* four different CBD CdS window layers.

34-S through the main body of the CdTe matrix. Isotopic profiles of 12-C, 34-S and 35-Cl have similar maxima ($\sim 10^{19}$ atoms/ cm^3) located at the CdS/ CdTe interface, buried at a depth of *ca*. 1600 nm in the CdTe matrix. The cell efficiencies (6.9-7.8 %) and V_{oc} values (0.75-0.776 mV) were reasonable given the comparatively high concentrations of impurities (Table 3). High concentrations of 16-O throughout the cell structure, of the order 10^{20}-10^{21} atoms cm^{-3}, were observed. It is known that CdTe has a propensity for self-compensation, hence the effective dopant concentration may be orders of magnitude lower than the actual concentration of dopants. The interface region of CdS/ CdTe is not abrupt, as evidenced from the smooth tail in the sulfur profile extending into the CdTe substrate, a feature common to all samples.

It was instructive to compare the cell characteristics and isotopic profiles of the best and poorest performing devices in our investigations. The greatest concentration of carbonaceous material was detected for cell 3 (Figure 1), which was observed to possess a peak value of 7×10^{19} atoms cm^{-3} at a depth of 50 nm into the CdTe substrate. The PV cells associated with these films

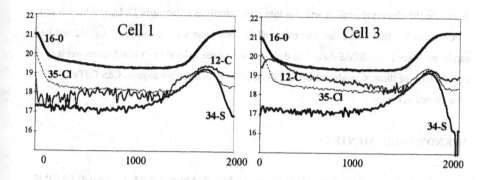

Figure 1: SIMS profiles for Cells 1 and 3. Units for y-axes are $\log_{10}$ atoms cm^{-3}, x-axes are nm.

possessed the greatest R_s and R_p values and lowest FF values (Table 3). The high R_s value would suggest that the bulk of the CdTe is poorly crystalline, resulting in a higher density of carrier recombination sites and poor cell efficiencies. For cell 1 the values for I_{sc}, V_{oc}, FF and cell efficiency were the best for all devices examined. The isotopic profiles of 12-C and 35-Cl in cells 1 and 3 indicate that the concentrations of carbon and chlorine in the CdTe bulk, beyond the interface region, are greater for cell 3 than those for cell 1. The recrystallisation of CdTe is known to be mediated by the presence of a chloride flux. We suggest that the higher concentration of carbon in cell 3 has an adverse effect upon the chloride diffusion and associated CdTe recrystallisation in the material. The relatively high V_{oc} for cell 3 appears to be inconsistent with the other cell characteristics, but the V_{oc} is more indicative of the CdS:CdTe interface region, which suggests that recrystallisation at the heterojunction is successful but incomplete in the CdTe bulk. Further studies with photoluminescence measurements on bevelled devices are planned and will provide more information.

CONCLUSIONS

In this paper we have used quantitative Secondary Ion Mass Spectrometry (SIMS) to determine the elemental profiles and absolute concentrations of isotopes 12-C, 16-O, 34-S and 35-Cl within n-CdS/p-CdTe thin film solar cell devices. Device characteristics (*e.g.* V_{oc} and Rs) can be correlated with SIMS data. The development of the chloride rich interface region is a consequence of thermal treatment of both the CdS layer and the subsequent CdS:CdTe heterojunction device.

ACKNOWLEDGEMENTS

The authors wish to thank Mr S. Hearne and Mr R. Murphy (NMRC, Ireland) for SIMS measurements, Dr Karl Senkiw (Dept. of Chem. Eng., Imperial College) for the XPS analysis, Dr. D. Johnson (BP Solar Ltd) for device fabrication and the EPSRC. This work was conducted under the Access to Large-Scale Facilities (LSF) European Community's Training and Mobility of Researchers (TMR) Programme. Paul O'Brien is the Sumitomo/ STS Professor of Materials Chemistry and the Royal Society Amersham International Research Fellow (1997-98).

REFERENCES

1.   K. Turner, J. M. Woodcock, M. E. Özsan, D. W. Cunningham, D. R. Johnson, R.J. Marshall, N.B. Mason, S. Oktik, M.H. Patterson, S.J. Ransome, S. Roberts, M. Sadeghi, J. M. Sherborne, S. Sivapathasundaram and I. A. Wells, Solar Energy Materials and Solar Cells, **35**, p.263-270 (1994).

2.   S. A. Galloway, P. R. Edwards and K. Durose, Institute of Physics Conference Series, **157**, p.579 (1997).

3.   D. S. Boyle, P. O'Brien, D. J. Otway and O. Robbe, J. Mater. Chem, submitted for publication (1998).

4.   A. Kylner, J. Lindgren and L. Stolt, J. Electrochem. Soc., **143**, p.2662 (1996).

5.   R. G. Wilson, F. A. Stevie and C. W. Magee, Secondary Ion Mass Spectrometry: A Practical Handbook for Depth Profiling and Bulk Impurity Analysis, Wiley, New York (1989).

6.   A. Kylner and M. Wirde, Jpn. J. Appl. Phys., **36**, p.2167 (1997).

7.   M. Stoev and A. Katerski, J. Mater. Chem., **6**, p.377 (1996).

8.   R. G. Wilson, Int. J. Mass. Spectrometry. Ion. Proc., **143**, p.43 (1995).

Part IV

Fundamental Studies for
Advanced Materials and Devices

MONTE CARLO PREDICTION OF ALLOY STRUCTURE

BRIAN S. GOOD*, GUILLERMO BOZZOLO** AND RONALD D. NOEBE*

*National Aeronautics and Space Administration, Lewis Research Center, 21000 Brookpark Road, Cleveland, OH 44135.

**Ohio Aerospace Institute, 22800 Cedar Point Road, Cleveland, OH 44142.

ABSTRACT

As the performance demands on metal alloys of interest to the aerospace community continue to increase, the complexity of these alloys has tended to grow as well, with corresponding increases in the cost and time needed to develop future generations of materials. While the traditional empirical approach to designing these materials has proven effective, as the complexity of the materials grows, such methods will become less efficient, limiting prospects for further improvement. The incorporation of theoretically- and computationally-based input into the design process has the potential to both improve the quality of, and reduce the time to design, new materials.

Among the most fundamental issues in new materials design, the prediction of the crystallographic structure of solids has long been a goal of researchers in the field of computer simulation. To become an effective tool for assisting in the process of alloy design, a computational scheme must make use of an energy method fast enough to allow consideration of a large number of candidate structures, and an efficient energy minimization scheme.

In this work, we discuss our application of Monte Carlo methods, using the BFS semi-empirical energy method developed in our laboratory, to the problem of prediction of the crystallographic structure of multi-component metal alloys. Using a simple Metropolis Monte Carlo method in conjunction with an accurate and transferable method for computing energetics, we have investigated the crystallographic structure of a variety of fcc- and bcc-based intermetallic compounds, and have considered the influence of alloying additions in compounds having up to five constituents. We discuss the successes and limitations of the methodology, and the prospects for obtaining accurate prediction of physical properties from our results. We also emphasize the role of this type of methodology as part of a synergistic experimental-theoretical effort for alloy design.

INTRODUCTION

Modeling at the macroscopic and mesoscopic scales has been performed for some time. It has long been realized, however, that the application of atomistic theoretical techniques to the problem of materials design holds great promise, both in its own right, and as a means of improving the quality of modeling at larger length scales. Because all physical and mechanical properties are ultimately manifestations of phenomena at the microscopic level, a detailed understanding of atomistic behavior should, in principle, provide information about macroscopic behavior as well. However, it is only recently that calculations at the atomic scale have shown the quality and detail needed to be useful in any successful materials design program.

Historically, major limitations in methodology have prevented the broad application of atomistic techniques to the design of materials. Ab initio quantum mechanical methods, while providing generally accurate information at the microscale, are too computationally intensive to be used to provide a bridge between micro- and mesoscales. The development of fast, accurate and transferable semi-empirical techniques for the computation of energies, along with new energy mini-

137

mization methods and the availability of ubiquitous high-performance computing, have allowed researchers to begin to obtain information about macroscopic behavior from microscopic calculations. Thus, the incorporation of these techniques, as exploratory and screening tools in the earliest stages of alloy design, has become practicable.

Most of the theoretical work performed in the last few years has been largely focused on the validation of the techniques used and the determination of fundamental properties for relatively simple systems. There is, as will be demonstrated in this paper, a continuing effort in developing methods to extend these calculations to time and length scales comparable to real intermetallic mechanical behavior [1]. Part of this effort is directed towards providing guidelines that can be used in the initial phases of alloy design, thus providing an efficient tool to simplify difficult and costly experiments.

The purpose of this paper is to show that, given the availability of a method capable of avoiding typical limitations of other comparable techniques [1], a great deal of information can be obtained relative to properties whose understanding is mostly based on empirical knowledge.

Within the wide range of applications of numerical simulation based on the BFS method for alloys [2], we concentrate here on the effect of alloying additions to binary ordered intermetallics. Based on the fact that mechanical and metallurgical properties of these materials can be greatly impacted by the introduction of such additions [3,4], we focus our attention on what can be learned from computational simulations at the atomic level of these systems.

In this paper we describe a scheme for the computer-aided design of metallic alloys. The scheme combines the techniques of Monte Carlo computer simulation with an accurate and transferable method for computing energetics. A Metropolis Monte Carlo algorithm is used, in conjunction with the BFS method for alloys, to create a predictive tool for materials design with a broad range of application. At present, our methodology can provide information about energetics, crystal structure, defect structures, simple physical properties, and the influence of temperature on microstructure.

MONTE CARLO METHODOLOGY

At a given temperature, the equilibrium structure of an alloy will be the one that exhibits the lowest free energy. Thus, the prediction of alloy structure involves finding the minimum-energy configuration of a quantity of atoms sufficiently large to allow characterization of the microstructure.

In the work described below, two Monte Carlo-based techniques for minimizing the energy of a computational cell are used. To compute the minimum energy structure of a cell of known composition, an "exchange Monte Carlo" techniques is used. Specifically, a computational cell of known crystal structure, lattice constant and composition is prepared. It should be pointed out that if these quantities are not known in advance, they may be obtained from static calculations using the BFS method. The cell sites are then populated by atoms whose chemical species are chosen randomly, consistent with the specified composition. The initial energy of the computational cell is computed using the BFS method, and the distribution of chemical species within the cell is adjusted until the energy is minimized. The minimization procedure, repeated until the total energy converges, is as follows:

1. A pair of atoms having different atomic species is chosen at random.
2. The energies of the atoms (and their environments) are computed and added.
3. The species of the two atoms are exchanged.
4. The energies are recomputed.
5. The exchange is accepted or rejected using the Metropolis Monte Carlo scheme [5]:
    5.a. If the exchange lowers the energy, it is always accepted.

5.b. If the exchange raises the energy by an amount ΔE the exchange is accepted with probability $exp(-\Delta E/kT)$, where k is Boltzmann's constant and T is the temperature.

The procedure is terminated when the energy of the computational cell stabilizes. It should be noted that the temperature used in the minimization is entirely arbitrary.

At high temperatures, any defects in the computational cell will be relatively mobile, so that an effective "annealing" process takes place; an atom in a metastable location will exhibit a non-negligible probability of exchange with another atom even if the cell's total energy is raised. On the other hand, the true minimum-energy state of the cell will most probably be attained at low temperature. Therefore, for some minimizations, the above procedure is performed using a sequence of decreasing temperatures. This "temperature cascade" helps to prevent defects present in the initial configuration of the cell from being "frozen-in", as can happen if the minimization is performed only at a low temperature. In addition, the cascade procedure can provide, in a crude way, insight on the relationship between heat treatment and microstructure. While such a temperature cascade does not attempt to mimic the detailed dynamics of the equilibration process, it can offer a qualitative view of the effects of rapid versus slow cooling of the system. The temperature treatment (that is, the sizes of the steps between the various temperatures in the cascade) is of critical importance in determining the final state of the system. Slow cooling results in a highly-ordered low-temperature state, while rapid cooling results in an alloy with grain structure, in which each grain has essentially the ordering of the thermodynamic ground state.

The second Monte Carlo method, "relaxation Monte Carlo", is the process by which we obtain the minimum energy geometry of a computational cell once the distribution of chemical species is determined. The process is similar to the "exchange Monte Carlo" procedure, except that the minimization occurs by imparting small random displacements to randomly-chosen atoms, rather than exchanging the chemical identities of pairs of atoms. As before, the new configuration is accepted or rejected according to the Metropolis criterion [5].

Ideally, the state of the computational cell should be minimized by a combined exchange-relaxation Monte Carlo scheme, where the energies used in the Metropolis criterion are fully relaxed exchange energies. Even with a method as efficient as BFS, however, such a scheme is computationally impractical for cells as large as the ones used here. This exclusion may be justified to a certain extent by considering the results of a recent BFS study of segregation in metallic alloys [6]. In this work, heats of segregation were computed with and without atomistic relaxation. In all but one case, the inclusion of relaxation did not alter the identity of the segregating species. Furthermore, the magnitudes of the relaxed and unrelaxed heats of segregation were not very different in most cases.

THE BFS METHOD

Since its inception a few years ago, the BFS method [2] has been applied to a variety of problems, including bulk properties of solid solutions of fcc and bcc binary alloys [2], the temperature dependence of segregation profiles in binary alloys [6,7], surface alloys [8], and the design of 4- and 5-element alloys [9,10], providing a foundation for the work presented in this paper. In what follows, we provide a brief description of the operational equations of BFS used in the energy calculations for the Monte Carlo method. The reader is encouraged to seek further details in previous papers where a detailed presentation of the foundation of the method, its basis in perturbation theory and a discussion of the approximations made are clearly shown [2,9].

The BFS method provides a simple algorithm for the calculation of the energy of formation of an arbitrary alloy (the difference between the energy of the alloy and that of its pure elements). In BFS, the energy of formation is written as the superposition of individual contributions of all the atoms in the alloy

$$\Delta H = \sum_i \varepsilon_i \tag{1}$$

For each atom, we partition the energy into two parts: a strain energy and a chemical energy contribution. The first takes into account the atomic positions of the neighboring atoms of atom i, regardless of their chemical identity. For its calculation, we use the actual geometrical distribution of the atoms surrounding atom i, computed as if all of its neighbors were of the same species as atom i. In this sense, the BFS strain energy differs from the commonly defined strain energy in that the actual chemical environment is replaced by that of a monoatomic crystal. Its calculation is then straightforward, even amenable to first-principles techniques. In our work, we use Equivalent Crystal Theory (ECT) [11] for its computation, due to its proven ability to provide accurate and computationally economical answers to most general situations. In all cases considered in this work, a rigorous application of ECT is reduced to that of its two leading terms, which describe average density contributions and bond-compression anisotropies. We neglect the three- and four-body terms dealing with the bond angle and face-diagonal anisotropies. The chemical environment of atom i is considered in the computation of the BFS chemical energy contribution, where the surrounding atoms are forced to occupy equilibrium lattice sites corresponding to the reference atom i. Building on the concepts of ECT, a straightforward approach for the calculation of the chemical energy is defined, properly parameterizing the interaction between dissimilar atoms.

Thus defined, the BFS strain and chemical energy contributions take into account different effects (i.e., geometry and composition), computing them as isolated effects. A coupling function g restores the relationship between the two terms. It is defined in such away as to properly consider the asymptotic behavior where chemical effects are negligible for large separations between dissimilar atoms. Summarizing, the contribution to the energy of formation of atom i is then

$$\varepsilon_i = \varepsilon^S + g\varepsilon^C \tag{2}$$

The strain energy contribution is obtained by solving the ECT perturbation equation

$$NR_1^p e^{-\alpha R_1} + MR_2^p e^{-\left(\alpha + \frac{1}{\lambda}\right)R_2} = \sum_j r_j^p e^{-(\alpha + S(r))r_j} \tag{3}$$

where N and M are the number of nearest- and next-nearest neighbors respectively, and where p, l, α and λ are ECT parameters that describe element i (see Ref. 11 for definitions and details), r denotes the distance between the reference atom and its neighbors, $S(r)$ describes a screening function [11] and the sum runs over nearest and next nearest neighbors. This equation determines the lattice parameter of a perfect, equivalent crystal where the reference atom i has the same energy as it has in the geometrical environment of the alloy. R_1 and R_2 denote the nearest- and next-nearest neighbor distances in this equivalent crystal of lattice parameter a^S.

Once the lattice parameter of the (strain) equivalent crystal, a^S, is determined, the BFS strain energy contribution is computed using the universal binding energy relation of Rose et al. [12], which contains all the relevant information concerning a single-component system:

$$\varepsilon^S = E_C^i (1 - (1 + a^{S*}) e^{-a^{S*}}) \tag{4}$$

where E_C is the cohesive energy of atom i and where the scaled lattice parameter a^{S*} is given by

$$a^{S*} = q\frac{(a^S - a_e)}{l_i} \tag{5}$$

where q is the ratio between the equilibrium Wigner-Seitz radius and the equilibrium lattice parameter a_e.

The BFS chemical energy is obtained by a similar procedure. As opposed to the strain energy term, the surrounding atoms retain their chemical identity, but are forced to be in equilibrium lattice sites of an equilibrium (otherwise monoatomic) crystal i. The BFS equation for the chemical energy is given by

$$NR_1^{p_i}e^{-\alpha_i R_1} + MR_2^{p_i}e^{-\left(\alpha_i + \frac{1}{\lambda_i}\right)} = \sum_k \left(N_{ik}r_1^{p_i}e^{-\alpha_{ik}r_k} + M_{ik}r_2^{p_i}e^{-\left(\alpha_{ik} + \frac{1}{\lambda_i}\right)r_2} \right) \tag{6}$$

where N_{ik} and M_{ik} are the number of nearest- and next-nearest neighbors of species k of atom i.
The chemical environment surrounding atom i is reflected in the parameters α_{ik}, given by

$$\alpha_{ik} = \alpha_i + \Delta_{ki} \tag{7}$$

where the BFS parameters Δ (a perturbation on the single-element ECT parameter α_i) describe the changes of the wave function in the overlap region between atom i and k. Once Eq. (6) is solved for the equivalent chemical lattice parameter a^C, the BFS chemical energy is given by

$$\varepsilon^C = \gamma_i E_C^i (1 - (1 + a^{C*})e^{-a^{C*}}) \tag{8}$$

where $\gamma_i = 1$ if $a^{C*} > 0$ and $\gamma_i = -1$ if $a^{C*} < 0$. The scaled chemical lattice parameter is given by

$$a^{C*} = q\frac{(a^C - a_e)}{l_i} \tag{9}$$

Finally, the BFS chemical and strain contributions are linked by a coupling function g which describes the influence of the geometrical distribution of the surrounding atoms in the relevance of the chemical effects, given by

$$g_i = e^{-a^{S*}} \tag{10}$$

where the scaled lattice parameter a^{S*} is defined in Eq. (5).
In this work we used the BFS interaction parameters Δ determined following the procedure outlined in Ref. 13. These parameters are obtained from first-principles, all-electron, density functional calculations of the elemental constituents and ordered binary compounds of these elements. The particular implementation used is the Linear-Muffin-Tin Orbitals (LMTO) method [13] in the Atomic Sphere Approximation. As mentioned above, in order to provide parameters to the BFS method, we need to calculate the equilibrium properties of the elemental solid for the same symmetry of the compound to be studied, since BFS is referenced to the ground state properties of the system in that symmetry. Once these parameters are computed, they remain the same for any other calculation involving any of these elements in an alloy of the same crystal structure, requiring no further adjustment or replacement.

STRUCTURE OF TERNARY ALLOYS

In what follows, we discuss Monte Carlo BFS results for NiAl-based alloys of increasing complexity. Each step in our study is taken from conclusions drawn from previous steps, illustrating the utility of this computationally efficient method as our exploratory tool.

1. NiAl: As a test of the validity of the BFS method in dealing with B2 ordered intermetallics and the accuracy of the parameterization used, we first discuss the results of the BFS calculations for the defect structure of NiAl across its B2 phase field.

In a B2 ordered structure, Ni and Al atoms occupy sites in interpenetrating cubic sublattices. By analyzing a large number of possible atomic distributions, one can extract information on the lowest energy states indicating the existence of two separate regimes: pure substitutional Ni-rich

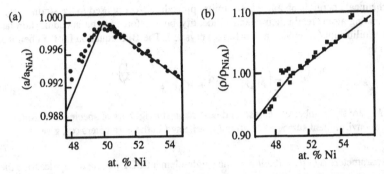

Fig. 1: (a) Lattice parameter and (b) density of binary alloys as a function of Ni concentration, normalized to their stoichiometric values. The solid symbols denote experimental results from different investigators summarized in Ref. 3. The lines denote the BFS predicted properties, corresponding to the structure with the lowest energy state for each composition.

alloys with no structural vacancies (at T=0), and a defect structure consisting of mostly Ni vacancies on the Al-rich side of stoichiometry. Fig. 1.a and 1.b compare these BFS predictions with available experimental data [3,14]. To demonstrate the effectiveness of the Monte Carlo methodology, we have performed a BFS Monte Carlo minimization of a 1024 atom computational cell of stoichiometric NiAl, using a temperature cascade of ten steps, the last of which was T = 10 K. At each temperature, 500 iterations of 1024 exchanges were performed. The initial random-alloy state of the cell is shown in Fig 2.a, while the final, ordered state, is shown in Fig. 2.b. The cells have been expanded in the z-direction as an aid to visualization.

Further insight into the minimization procedure may be gained by considering the behavior of the BFS energy of the cell as a function of the number of iterations, as shown in Fig. 3.a. It can be seen that there is a precipitous drop in energy during the earliest stages of the minimization, and that, after this drop, the remainder of the Monte Carlo procedure may be considered essentially a refinement process, with a slight downward trend in the energy. In spite of the rapid drop, it is evident from Fig. 2.b that no defects have been frozen into the structure. A careful examination of the energy in Fig 3.a reveals, however, that the

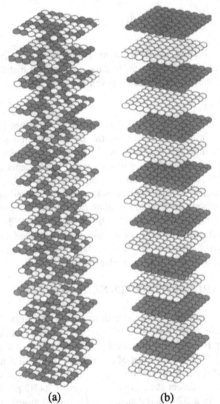

Fig. 2: Initial (high temperature) and final states (after stabilization) of a stoichiometric NiAl alloy.

142

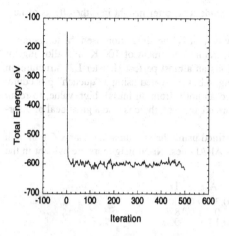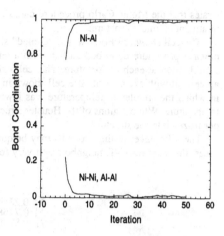

Fig. 3: (a) Energy of the computational cell as a function of minimization history and (b) bond coordination matrix elements as functions of minimization history.

attainment of a perfect B2 structure (Fig. 2.b) is in a sense fortuitous. The perfect B2 state was attained several times prior to the termination of the Monte Carlo procedure, with fluctuations having energies slightly above the B2 energy evident between these events. The fact that the cell was in the perfect B2 state at the termination of the minimization is somewhat coincidental. Still, the cell configurations during the iterations immediately preceeding the final state exhibit nearly perfect B2 ordering, with a small number of antisite defects.

We additionally characterize the state of the computational cell by the bond coordination matrix C_{ij}, defined as the probability that a nearest neighbor of an atom of chemical species i will be of species j. The C_{ij} are obtained by performing an average over one or more states of the computational cell as the minimization proceeds. Because of the symmetry of the B2 structure, the resulting matrix C_{ij} is symmetric, and may be characterized by only two numbers, the Ni-Ni and Al-Al element, and the Ni-Al element, as shown in Fig. 3.b. The Ni-Ni/Al-Al element goes rapidly to zero, while the Ni-Al element tends to unity at the same time, as expected for the B2 structure.

Alloying additions can have a dramatic effect on material properties such as the yield strength, hardness, diffusivities, creep strength, and environmental resistance. Of fundamental interest in considering a candidate alloying addition is the site occupation, defect concentration, formation of additional phases that could result, and the ensuing changes in the properties of the base material. While this issue has been thoroughly investigated for simple alloys using analytical BFS calculations [14], we now concentrate on the results of numerical simulations. The goal of this analysis is to determine, by simple numerical simulations, basic features related to the microstructure and properties of high order alloys in order to guide the experimental work on these materials.

2. NiAl+Ti: We first consider $Ni_{50} (Al,Ti)_{50}$ alloys, known to form Heusler precipitates beyond 5 at.% Ti [14]. Previous work based on a comparison of different ordering schemes of these alloys has demonstrated that the BFS method correctly predicts these features [14]. Results of two exchange Monte Carlo simulations for Ni_2AlTi are shown in Fig. 4, along with the corresponding coordination matrices. A comparison of the degree of disorder in the two cells demon-

strates that the Monte Carlo procedure can, in a crude way, offer insight into the effects of heat treatment on microstructure.

The cell shown in Fig. 4.a was "cooled" slowly, that is, the simulation used the temperature cascade procedure described earlier, with a temperature decrement of 100 K, and allowing for equilibration at each temperature. This cell exhibits an almost perfect Heusler $L2_1$ structure, in agreement with experiment. The cell shown in Fig. 4.b was created using a "quench" procedure, in which the simulation temperature was lowered abruptly from an initial, high value, to room temperature. While regions of the Heusler structure are evident, there is also a great deal of disorder frozen into the structure.

The differences in the two cells may be quantified using the coordination matrix C_{ij}. In both cases, there are no Ti-Ti neighbors and very few Al-Al ones. Ni-Ni neighbors are evident in the

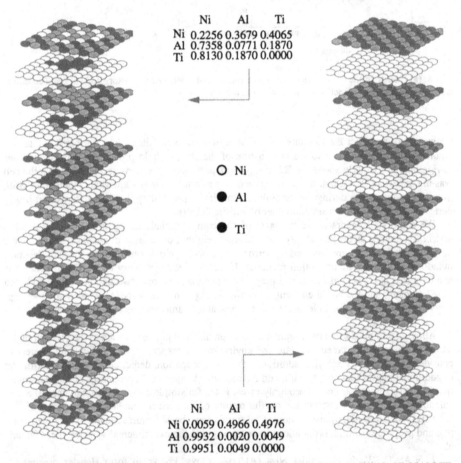

	Ni	Al	Ti
Ni	0.2256	0.3679	0.4065
Al	0.7358	0.0771	0.1870
Ti	0.8130	0.1870	0.0000

O Ni

● Al

● Ti

	Ni	Al	Ti
Ni	0.0059	0.4966	0.4976
Al	0.9932	0.0020	0.0049
Ti	0.9951	0.0049	0.0000

Fig.4: Final structures of a Monte Carlo/Metropolis/BFS simulation on a 1024 atom cell of a Ni-25Al-25Ti alloy. Both (initially random) final states are obtained by lowering the temperature in (a) rapid cooling process, and (b) by slowly lowering the temperature in 100 K intervals and allowing the cell to equilibrate. The inset includes the coordination matrix for the final state of the simulation (see text).

quenched cell, but are essentially gone in the other (cascade) cell, where all Al and Ti nearest neighbors are Ni.

3. NiAl+Cr: Based on previous work using the BFS method in determining the solubility limit of Cr in NiAl (~2 at. % Cr) [10] and having determined an appropriate temperature treatment for the computational cell, we now consider cascade processes on two Ni-Al-Cr alloys. Fig. 5 depicts computational cells for two NiAlCr alloys, (5.a) Ni-25Al-25Cr, and (5.b) Ni-33Al-34Cr. Two cells are shown for each alloy. In each case, the cell on the left represents the state of the alloy after equilibration at the next-to-last temperature in the cascade, while the cell on the right is the final equilibrated state. The Ni-25Al-25Cr cell (Fig. 5.a), exhibits three distinct phases: an α-Cr phase, a Ni-rich NiAl phase, and a Ni interphase "coating" around the Cr precipitate. The higher temperature states show several antistructure atoms and Al atoms in solid solution in the Cr pre-

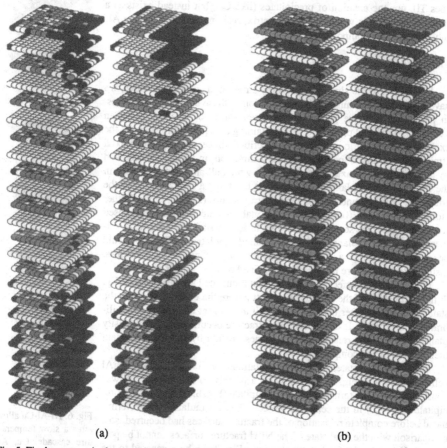

(a)   (b)

Fig. 5: Final two steps in the temperature cascade of a (a) Ni-25Al-25Cr and a (b) Ni-33Al-34Cr alloy. The first column indicates the atomic distribution once the energy of formation stabilizes (i.e., little difference with its final, T= 0 value). The second column indicates the final (ideal) state of the simulation. The third and fourth columns show similar results for an equiatomic Ni-Al-Cr alloy. Cr, Al and Ni atoms are represented by black, grey and white spheres, respectively.

145

cipitate, as well as some Cr atoms embedded in the NiAl matrix. This is, perhaps, a better representation of the microstructure of this alloy, within the framework of the Monte Carlo algorithm used, to reach the ground state of the system. While the actual, ideal thermodynamic ground state is difficult to obtain experimentally, the higher temperature state provides information both on the ground state as well as a more realistic representation of the actual alloy. The Ni-33Al-34Cr alloy, on the other hand, exhibits a simpler, two-phase structure consisting of an approximately stoichiometric NiAl matrix with α-Cr precipitate. Again, in the higher temperature state, there is a noticeable amount of Al atoms residing in the Cr precipitate.

4. NiAl + Cu: Fig. 6 shows a computational cell for Ni-40Al-10Cu. This alloy exhibits neither the formation of an ordered ternary phase (as does Ti), nor the creation of precipitates (like Cr), but instead exists as a solid solution of Cu within the NiAl matrix, residing primarily on the Al-sublattice for this composition.

FRACTURE

In addition to the simulation of alloy structure as described above, we have done some preliminary characterization of these computed structures. In particular, we have performed an investigation of fracture, using an extremely simplified procedure. After attaining an equilibrium state via the exchange Monte Carlo procedure as described above, we apply a uniaxial strain to the computational cell (in the z-direction, in this case), followed by a Monte Carlo relaxation procedure in which all atoms except those in the top and bottom planes are allowed to relax. At a critical value of the strain, the cell is found to fracture catastrophically (Fig. 7). Upon further atomistic relaxation (but without additional uniaxial strain), the exposed fracture surfaces are seen to contain a high-density of faceted segments primarily along {110} planes. Clearly, a detailed simulation of fracture should be performed using a dynamical simulation technique. Nevertheless, insight into the fracture process can be gained by this simple procedure.

Five states of a perfect B2 NiAl computational cell are displayed in Fig. 7: the initial state of the cell, the state just before the onset of fracture, the state immediately after onset, a state well after onset and the final equilibrated relaxed state. It can be seen that fracture occurs along high-density planes, and that the relaxed surface planes created by the fracture retain their correct mixed composition.

Initial and final states of similar simulations for Ti and Cr alloyed NiAl cells are shown in Fig. 8. In each case, a perfect computational cell was constructed with the ternary Heusler precipitate (Fig. 8.a) and the α-Cr precipitate (Fig. 8.b) in the center of the cell. These simulations were terminated before complete relaxation of the fracture surfaces had occurred, so a comparison with the final state of the NiAl fracture surfaces cannot be performed. However, the final states shown in Fig. 8 can be compared to the next-to-last NiAl fracture cell in Fig. 7. In the case of embedded Heusler precipitates (Fig. 8.a), the central region maintains its integrity during the fracture process, with two fracture surfaces evident along the precipitate-matrix interface. A somewhat similar behavior is seen for the Cr precipitate

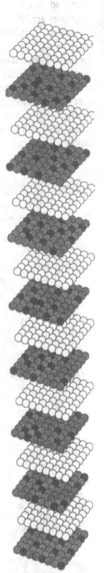

Fig. 6: Ni-Al-Cu alloy after a slow temperature cascade. Cu, Al and Ni atoms are represented with black, grey and white spheres, respectively.

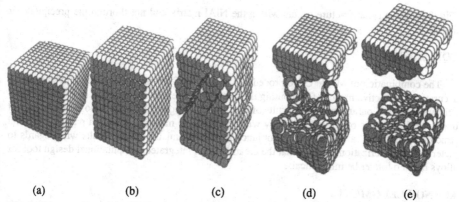

(a)   (b)   (c)   (d)   (e)

Fig. 7: Fracture of a perfect NiAl cell. States of the computational cell at (a) the beginning of the simulation, (b) after uniaxial strain, just before the onset of fracture, (c) just after the onset of fracture, (d) during relaxation of the newly formed fracture surfaces, and (e) after fracture relaxation is essentially complete. Ni and Al atoms are denoted with white and grey spheres, respectively.

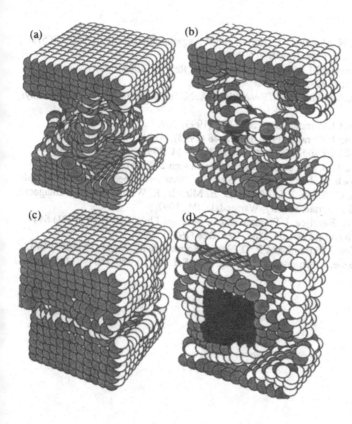

Fig. 8: (a) Intermediate state during the fracture process of a NiAl alloy with an embedded Heusler, Ni$_2$AlTi, precipitate. Al, Ni and Ti atoms are denoted with white, grey and black spheres, respectively. (b) A cut of the computational cell showing the internal structure of the cell shown in (a), (c), (d) same for NiAl+Cr.

(Fig. 8.b), except that fracture occurs within the NiAl matrix and not thorugh the precipitate or interface.

CONCLUSIONS

The combination of Monte Carlo procedures and the BFS energy method has proven to be an accurate and effective means of computing the microstructure of NiAl-based alloys. The transferability of the BFS parameters (and results of other work) suggest that there is no reason to expect that the effectiveness of the methodology will be limited to materials having a particular crystal structure. This broad applicability, in conjunction with promising initial results with regards to materials characterization, suggest that the creation of an integrated computational design tool for alloys may no longer be impractical.

ACKNOWLEDGMENTS

We would like to thank N. Bozzolo for fruitful discussions, C. Amador for providing us with the pure element parameters needed for these calculations and F. Honecy for assisting with the numerical simulations. This work was funded by the NASA HITEMP and PPM program offices.

REFERENCES

1. C. T. Liu, J. Stringer, J. N. Mundy, L. L. Horton and P. Angelini, Intermetallics 5 (1997) 579.
2. G. Bozzolo and J. Ferrante, Phys. Rev. B50 (1994), 5971.
3. R. D. Noebe, R. R. Bowman, M. V. Nathal, Internat. Mat. Rev. 38 (1993) 193.
4. D. Miracle, Acta Met. 41 (1993) 649.
5. N. Metropolis, A. W. Rosenbluth, M. N. Rosenbluth, A. N. Teller and E. Teller, J. Chem. Phys. 21 (1953) 1087.
6. G. Bozzolo, B. Good and J. Ferrante, Surf. Sci. 289 (1993) 169.
7. B. Good, G. Bozzolo and J. Ferrante, Phys. Rev. B 48 (1993),18284
8. G. Bozzolo, R. Ibanez-Meier and J. Ferrante, Phys. Rev. B 51 (1995), 7207.
9. G. Bozzolo and J. Ferrante, J. Computer-Aided Mater. Design 2 (1995) 113.
10. G. Bozzolo, R. D. Noebe, J. Ferrante and A. Garg, in ``Structural Intermetallics 1997," eds. M. V. Nathal, D. Darolia, C. T. Liu, P. L. Martin, D. B. Miracle, R. Wagner and M. Yamaguchi, The Minerals, Metals and Materials Society, Warrendale, PA, 1997.
11. J. R. Smith, T. Perry, A. Banerjea, J. Ferrante and G. Bozzolo, Phys. Rev. B 44 (1991) 6444.
12. J. H. Rose, J. R. Smith and J. Ferrante, Phys. Rev. B28 (1983), 1835.
13. G. Bozzolo, C. Amador, J. Ferrante and R. D. Noebe, Scripta Metall. Mater. 33 (1995) 1907.
14. G. Bozzolo, R. D. Noebe, J. Ferrante and A. Garg, Mat. Sci. Eng. A 239-240 (1997) 769.

"GENETICALLY ENGINEERED" NANOSTRUCTURE DEVICES

GERHARD KLIMECK, CARLOS H. SALAZAR-LAZARO, ADRIAN STOICA, AND
THOMAS CWIK
Jet Propulsion Laboratory, California Institute of Technology, Pasadena, CA 91109

ABSTRACT

Material variations on an atomic scale enable the quantum mechanical functionality of devices
such as resonant tunneling diodes (RTDs), quantum well infrared photodetectors (QWIPs),
quantum well lasers, and heterostructure field effect transistors (HFETs). The design and
optimization of such heterostructure devices requires a detailed understanding of quantum
mechanical electron transport. The Nanoelectronic Modeling Tool (NEMO) is a general-purpose
quantum device design and analysis tool that addresses this problem. NEMO was combined with
a parallelized genetic algorithm package (PGAPACK) to optimize structural and material
parameters. The electron transport simulations presented here are based on a full band
simulation, including effects of non-parabolic bands in the longitudinal and transverse directions
relative to the electron transport and Hartree charge self-consistency. The first result of the
genetic algorithm driven quantum transport calculation with convergence of a random structure
population to experimental data is presented.

INTRODUCTION

The NASA/JPL goal to reduce payload in future space missions while increasing mission
capability demands miniaturization of measurement, analytical and communication systems.
Currently, typical system requirements include the detection of particular spectral lines,
associated data processing, and communication of the acquired data to other subsystems. While
silicon device technology dominates the commercial microprocessor and memory market,
semiconductor heterostructure devices maintain their niche for light detection, light emission,
and high-speed data transmission. The production of these heterostructure devices is enabled by
the advancement of material growth techniques, which opened
up a vast design space. The full experimental exploration of
this design space is unfeasible and a reliable design tool is
needed.

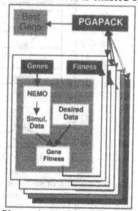

Military applications have similar system requirements to
those listed above. Such requirements prompted a device
modeling project at the Central Research Laboratory of Texas
Instruments (which transferred to Raytheon Systems in 1997).
NEMO was developed as a general-purpose quantum
mechanics-based 1-D device design and analysis tool from
1993-97. The tool is available to US researchers by request on
the NEMO web site [1]. NEMO is based on the non-
equilibrium Green function approach, which allows a
fundamentally sound inclusion of the required physics:
bandstructure, scattering, and charge self-consistency. The
theoretical approach is documented in references [2, 3] while
some of the major simulation results are documented in
references [4-6]. This paper highlights the recent work on
genetic algorithm based device parameter optimization.

Figure 1: Architecture of a
genetic algorithm-based NEMO
simulation.

QUANTUM DEVICE PARAMETER OPTIMIZATION
USING GENETIC ALGORITHMS

Heterostructure device designs involve the choice of material compositions, layer thicknesses, and doping profiles. Material parameters such as band offsets, effective masses, dielectric constants etc. influence the device simulation results in addition to the structural design parameters. The full exploration of the design space using purely experimental techniques is unfeasible due to time and financial constraints. For example, it takes a well-equipped research laboratory approximately five working days [7] for the growth, processing and testing of a particular resonant tunneling diode design. NEMO can provide quantitative [4-6] current voltage characteristics (I-V's) within minutes to hours [8] of CPU time for a single set of device and material parameters. With this quantitative simulation capability the design parameter space can be explored expediently once an automated system for the design parameter variation is implemented. This paper presents the combination of NEMO with a parallelized genetic algorithm package (PGAPACK) [9] as indicated in Figure 1. The architecture lends itself to the optimization of any parameters that enter a NEMO simulation. To evaluate how good a particular parameter set is, a fitness function must be developed as discussed in the next section.

SIMULATION TARGET AND FITNESS FUNCTION

In this work the RTD is used as a vehicle to study the effects of structural and doping variations on the electron transport. I-V's of two devices that are part of a well-behaved test matrix of experimental data published in reference [5] are used as a design target. The raw I-V data (see the example in Figure 2) contains the contact series resistance and oscillations in the negative differential resistance (NDR). The oscillation in the NDR is attributed to external circuit effects [10] and cannot be simulated within NEMO. The step-like feature in the NDR is cut out of the raw data to generate a "clean" set of experimental data. The contact series resistance can be estimated from the peak current of a series of nominally identical devices [5] with different cross

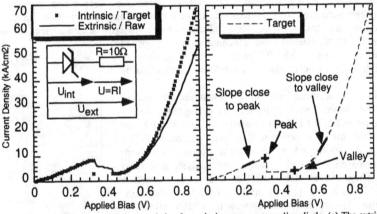

Figure 2: Generation of the target I-V characteristic of a typical resonance tunneling diode. (a) The extrinsically measured I-V (solid line) includes a series resistance and oscillations in the negative differential conductance region (0.32V-0.43V). The series resistance can be estimated from a series of devices with different cross sections. The intrinsic I-V is the target for the optimization (crosses). (b) Features that enter into the evaluation of the fitness of simulated data. Of particular interest are the peak and valley voltage and current and the slopes close to the peak and the valley.

sections. The voltage drop over the contact resistance can be subtracted out of the extrinsic voltage scale to yield the intrinsic voltage scale (see inset of Figure 2a)

The fitness of the simulated data is measured against such target I-V. There are four particular features that are explicitly evaluated for each simulated I-V: peak and valley current and voltage, and the slope close to the peak and the valley (see Figure 2b). Differences between the target and the simulation in these four features and the absolute and relative error for all simulated data points enter into the fitness function with a weighted average. The target fitness evaluated against itself results in a value of 1. Disagreements between simulation and target result in fitness values between 0 and 1.

TRANSPORT MODEL

The electron transport simulations are based on a single band model, which incorporates [3] effects of non-parabolic bands in the longitudinal and transverse directions relative to electron transport. The model parameters are derived from a tight binding sp3s* multiband model. This single band model captures the relevant transport physics such as complex band wrapping in the barriers and the non-parabolicity of the conduction band. The computation of the non-parabolic single band model executes about 60 times faster than the computation of the full band sp3s* model (for structures considered here). This dramatic increase in speed allows inclusion of Hartree charge self-consistency with non-parabolicity in the transverse direction. The double integral in total energy, E, and transverse momentum, k, to obtain the electron density at each site i (Eq. (1)) is carried out explicitly [2] in the inner loop of the charge self-consistency. The current density, I, is evaluated self-consistently with the electron density, n, in the double momentum and energy integration.

$$n_i \propto \int kdk \int K_i(k,E)dE \qquad (1)$$

$$I \propto \int kdk \int T(k,E)(f_L(E) - f_R(E))dE \qquad (2)$$

where T is the transmission coefficient, $f_{L/R}$ are the fermi functions in the left and the right reservoirs, and K is the electron density kernel [2].

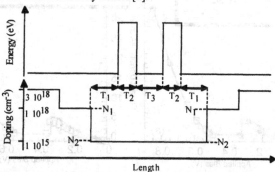

Figure 3 Conduction band edge and doping profile of a typical resonant tunneling diode. The central device region is typically undoped. The low doped spacer thickness, the barrier thicknesses and the well thickness are labeled T_1, T_2, and T_3, respectively. The low spacer doping and the central device doping are labeled N_1 and N_2, respectively. These five parameters are varied.

SET-UP OF NUMERICAL EXPERIMENT

In the numerical experiment described in Figure 3, five parameters (2 doping concentrations, N_1, N_2, and 3 thicknesses, T_1, T_2, T_3) are varied within the genetic algorithm in order to achieve the best fit to an experimental I-V curve. The simulation is started from a random population of 200 parameter sets. The doping population is logarithmically distributed around the nominal values by factors of 10 ($N_1\_[1\times10^{17}, 1\times10^{19}]$, $N_2\_[1\times10^{14}, 1\times10^{16}]$). The layer thickness population is uniformly distributed around the nominal value by 10 monolayers ($T_1\_[1,17]$ for device 1, $T_1\_[10,30]$ for device 2, $T_2, T_3\_[6,26]$). In each generation 63 of the worst genes [11] are dropped out of the population and new genes are generated [9] from the rest by mutation and crossover. Mutation allows the parameters to leave the original parameter range.

SIMULATION RESULTS

Two I-V's from slightly different structures serve as a target of the genetic algorithm optimization. Both structures were specified to the grower to have 16 monolayers (ml) of barriers (T_2) and well (T_3), no intentional doping in the central device ($N_2 = 1\times10^{15}$ cm^{-3}), $N_1 = 1\times10^{18}$ cm^{-3} doping in the low doping spacers, and 3×10^{18} cm^{-3} in the high doping contacts (see Figure 3). The nominally only difference in the two devices is in the no-doping spacer length T_1 of 7 vs. 20 ml. The simulation is started from the random populations as described in the previous section. The genetic algorithm converges for both I-V's to the nominal structure values, well within the experimental uncertainty as shown in Figure 4. Again it is found that the well widths must be increased in the simulation by a few monolayers versus the nominal values to achieve the best agreement with experimental data [5]. Different relative weights will result in different "optimal" structures as shown in Figure 4b. The convergence was achieved by evaluation of about 1000 parameter sets within a discretized parameter space of about 2×10^6 parameter sets [12].

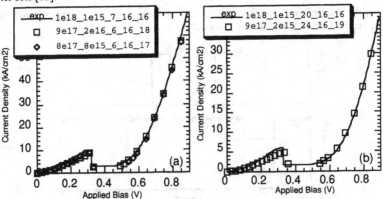

Figure 4: Current voltage characteristics of two different InGaAs/InAlAs resonant tunneling diodes. The nominal structures have barrier (T_2) and well (T_3) thicknesses of 16 monolayers (ml), and doping a doping profile of 10^{18} cm^{-3} (N_1) and 10^{15} cm^{-3} (N_2). The devices (a) and (b) differ nominally in their no-doping spacer thicknesses (T_1) of 7 and 20 ml, respectively. The solid lines show experimental data published in reference [5], where the noise in the valley current region was eliminated. The curves are labeled by the 5 parameters N1_N2_T1_T2_T3.

FUTURE WORK

This work is the first step to integrate NEMO within a high performance parallel computational environment. A desired curve can now be entered as the target of the simulation and the genetic algorithm is expected to obtain the optimal parameter set. Future work will utilize this method to analyze the vast material and structure parameter space. It is planned to evaluate other optimization techniques such as simulated annealing and directive approaches as well. These optimization techniques will be made available within a graphical user interface which enables the selection of parameters to be varied, the setting of parameter ranges and the setting of optimization parameters, such as population sizes, and mutation and crossover rules.

SUMMARY

We present the first NEMO simulations driven by a genetic algorithm to optimize structure design parameters such as layer thicknesses and doping profiles. The convergence of the initially random population of devices to experimental specified device parameters is demonstrated for two different devices. The agreement between experimental device specification and simulation parameters is within the experimental error margin. The distinct current-voltage features such as the shape of the peak and the turn-on in the valley current are very well reproduced with the transport model. These simulations are performed for the first time in Hartree charge self-consistency within the framework of a double integration over energy and momentum instead of the typical Tsu-Esaki single energy integration. The transport simulation are performed within a novel non-parabolic single band model which is derived from a more complete sp3s* tight binding model. This single band model captures the relevant transport physics such as complex band wrapping in the barriers and the non-parabolicity of the conduction band in the longitudinal and transverse transport direction.

ACKNOWLEDGEMENTS

The research described in this paper was carried out by the Jet Propulsion Laboratory, California Institute of Technology, under a contract with the National Aeronautics and Space Administration.

BIBLIOGRAPHY

1. NEMO, Nanoelectronic Modeling, in http://www.raytheon.com/rtis/nemo/.
2. R. Lake et al., J. Appl. Phys., **81**(12), 7845 (1997).
3. R. Lake et al., phys. stat. sol. (b), **204**, 354 (1997).
4. G. Klimeck et al., Appl. Phys. Lett., **67**(17), 2539 (1995).
5. G. Klimeck et al., IEEE DRC, 1997: p. 92.
6. R. C. Bowen et al., J. Appl. Phys., **81**, 3207 (1997).
7. A. C. Seabaugh, Texas Instruments, private communication, 1997.
8. The actual CPU time needed for a single I-V simulation depends strongly on the choice of material systems, bandstructure models, temperature scattering models, and bias points. The individual I-V characteristics presented here take about 30 minutes to compute on a single 200MHz R10000 CPU of an SGI Origin.
9. D. Levine, http://www-unix.mcs.anl.gov/~levine/PGAPACK/index.html, Parallel Genetic Algorithm Library.
10. J. N. Schulman, Second Workshop on Characterization, Future Opportunities and Applications of 6.1Å III-V Semiconductors, Aug. 24-26, 1998, Naval Research Laboratory,

Washington, DC, http://estd-www.nrl.navy.mil/code6870/code6870.html. H. C. Liu, J. Appl. Phys. **64**, 4792 (1988). . H. C. Liu, J. Appl. Phys. **53**, 485 (1988).

11. PGAPACK is implemented with MPI where N-1 of N processors are slaves to one master processor. The master takes care of the collection of data from the slaves. In a cluster of 64 CPU's we therefore renew only 63 genes in every generation.

12. The mutation operations may drive a particular parameter outside its original range. Therefore the full parameter space may not be limited.

SILICON QUANTUM DOTS IN SILICA

D. O. Henderson[a], M. H. Wu[a], R. Mu[a], and A. Ueda[a]
C. W. White[b] and A. Meldrum[b]
[a]Chemical Physics Laboratory, Physics Department
Fisk University, Nashville, TN
[b]Oak Ridge National Laboratory, Oak Ridge, TN

ABSTRACT

Silicon ions were implanted into fused silica substrates at doses of 1×10^{21}, 2×10^{21}, 5×10^{21}, and 1×10^{22} ions/cm^3. The implanted substrates were annealed at 1100°C for one hour in a reducing atmosphere (95% Ar+5% H$_2$). Optical absorption spectra recorded after the annealing treatment showed absorption onsets at 3.86, 3.73, 2.86 and 2.52 eV for substrates implanted with 1×10^{21}, 2×10^{21}, 5×10^{21}, and 1×10^{22} ions/cm^3, respectively. Static photoluminescence (PL) measurements indicated red emission between 1.72 and 1.61 nm with a slightly increasing red shift with ion dose. Time resolved PL at room temperature revealed slow (~50 μs) and fast (~20 μs) lifetimes which increased with decreasing temperature. TEM studies showed that the particle size increased with increasing ion dose. Typical particle sizes ranged between 2 and 5 nm indicating quantum confinement of the exciton, which can account for the blue shift in the absorption edge with decreasing ion dose. However, the maxima in the PL spectra for all ion doses are relatively independent of the ion dose and are strongly shifted from the absorption spectra. This suggests that radiative recombination occurs from a common luminescent center, possibly a surface or interfacial state in the SiO$_x$ layer surrounding the nanocrystal.

INTRODUCTION

Nanocrystals (NCs) are rapidly emerging as a class of materials that possess potential for a wide range of applications and devices including nonlinear optics, ultra-fast-all-optical switches, ultra-high-density memory storage devices, nanostructured lasers, chemical catalysts, and photovoltaics. From a fundamental perspective, NCs have also attracted the attention of the scientific community because they offer the challenge to study how material properties evolve as a function of their size and degree of dimensional confinement.

Semiconductor NCs, in particular have recently been under intense study because they offer the opportunity to investigate quantum confinement effects. Quantum confinement effects are activated when the NC size is comparable to the exciton radius. Quantum size effects can be classified into three regimes based on the NC size: 1) weak confinement when the mean particle radius a is ~3 times the exciton Bohr radius a_{exc}, 2) intermediate confinement when a is approximately equal to a_{exc}, and 3) strong confinement when a is a few times smaller than a_{exc}. In the effective mass approximation, the Hamiltonian for localized electrons and holes is expressed as

$$H = -\frac{\hbar^2}{2 m_e}\nabla_e^2 - \frac{\hbar^2}{2 m_h}\nabla_h^2 + V(r_e) + V(r_h) - \frac{e^2}{\varepsilon |r_e - r_h|} \tag{1}$$

where the subscripts e and h refer to the electron and hole coordinates, respectively, $V(r_i)$ is the potential experienced by the localized electron or hole, m_i is the electron or hole effective mass, ε is

155

the low frequency dielectric constant and the last term describes the electron-hole Coulombic interaction. A splitting of the conduction and valence bands occurs when the confinement energy term in eqn. 1 dominates the overall interaction. The energy of the lowest optical transition for an infinite potential well is given by[1]

$$\Delta E = E - E_g = \hbar^2 \frac{\pi^2}{2\mu a^2} - 1.8 \frac{e^2}{\epsilon a} \tag{2}$$

where μ is the reduced mass of the electron-hole pair, E_g is the bulk crystal gap, a is the radius of the NC and the second term is the correction for Coulombic interaction. Eqn. 2 clearly shows that as the NC size decreases, there in an increase in the bandgap energy.

PL spectra of silicon nanoparticles prepared by various techniques have been investigated extensively.[2-10] Silicon NCs prepared by ion implantation into SiO_2 layers at a dose of 5×10^{16} ions/cm^2 and annealed at 900°C for times ranging from 0 to 3 hours[10] produced photoluminescence from the blue to the red (1.6-2.9 eV) depending on the annealing time. As compared to various theoretical models, the PL observed in the ion implanted samples remains ~1eV lower for the smallest crystallites. This discrepancy was accounted for by suggesting the existence of surface tail states. Ludwig fabricated NC silicon using a spark processing technique. PL was observed in the blue, green and red, which depended on the atmosphere under which the NCs were produced.[10] Blue PL with a tail extending into the red was observed for processing in flowing dry air, while green PL was observed for NCs formed in stagnant air. It was concluded that the source of the PL originated from Si NCs and not from defects in the surrounding SiO_2 matrix. Vepřek and Wirschem prepared NC Si/amorphous-SiO_2 thin films by plasma chemical vapor deposition.[10] They observed PL in the red and did not observe any particle size dependence on the emission wavelength although calculated values indicated a bandgap of ~4 eV for 1.2 nm size particles. The discrepancy was accounted for by the model of Koch et al.[10] which states that photogenerated carriers are trapped in surface states or at the Si/SiO$_2$ interface where they reside for several microseconds before undergoing radiative recombination. However, it appears that the nature of the luminescent center responsible for the large red shift has yet to be identified. Theoretical calculations have provided some insight and suggest that the trap may be silicon dimer bonds that are passivated by hydrogen or silicon oxide.[11]

Low temperature and time resolved PL of Si NCs have also been studied. In general, the PL quantum yield increases with decreasing temperature, with the exception of the anomaly observed for porous silicon. The low temperature PL studies presented by Ludwig[10] show PL in the blue and green. The blue PL quantum yield increases as the temperature decrease as does the green PL, but the green PL also shows a red shift as the temperature decreases. Ludwig concluded that because the temperature dependence of the PL does not follow known relationships for bandgap or excitonic transitions, localized states (such as amorphous Si:H where radiative transitions occur between band tail states) should play a similar role in spark processed NC Si. Based on the assumption that such states are present in the insulating matrix of spark processed NC Si, it was suggested that these states would prevent photoexcited electrons from diffusing away, which in turn would prevent nonradiative decay paths and would account for the high quantum efficiency.

The time resolved PL on spark processed Si investigated by Ludwig[10] show PL in the blue, green with a tail extending into the red. The lifetimes were typically a few nanoseconds for the blue and green PL and the red tail was analyzed as having a lifetime of 10-30 ns.

We have extended the investigations on the static and low temperature time resolved PL for samples prepared by Si ion implantation into silica. The samples were implanted at various energies and concentrations to approximate a linear concentration vs. depth profile in an attempt to reduce the size dispersion of the Si nanocrystals formed after annealing in a reducing atmosphere.

EXPERIMENT

The samples used in the experiments reported here were produced by implantation of Si⁺ ions into Corning 7940 fused silica. Each sample was implanted with ions at several different energies[12] to obtain flat depth profiles. Four samples, with Si concentrations of 1 x 10²¹/cm³, 2 x 10²¹/cm³, 5 x 10²¹/cm³ and 1 x 10²²/cm³ were studied. All samples were annealed at 1100°C for one hour in an atmosphere of 95% Ar + 5% H₂. As reported by White and co-workers, thermal annealing results in formation of nanocrystals with average sizes that increase with the Si concentration. TEM measurements reported to give average nanocrystal sizes ranging from 2 - 5 nm.[12]

The excitation beam for the PL measurements was provided by the frequency tripled output of a 30 ps pulse width Nd:YAG laser, which was used directly to excite the samples or to pump an optical parametric generator/amplifier system. Excitation energy densities in all experiments were kept below 10 μJ/cm² so that each nanocrystal absorbed an average of less than one photon of each laser pulse. PL spectra were measured by a photomultiplier tube (PMT) after spectral selection in a computer controlled ¼ m monochromator. Signals from the PMT were digitized by a boxcar averager and averaged by a personal computer. The maximum gate width of the boxcar (16 μs) was used to obtain PL spectra. Time resolved spectra were

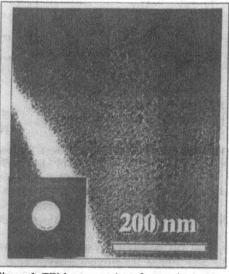

Figure 1. TEM cross-section of a sample implanted with 2x10²² ions/cm³ and annealed at 1100°C for i h in a 95% Ar + 5% H₂ atmosphere. Inset shows the electron diffraction pattern.

collected by recording the signal from the PMT with a 500 MHz bandwidth digital oscilloscope. Samples were mounted in a closed cycle helium cryostat to perform temperature dependent measurements.

RESULTS

TEM cross-sections of the samples implanted with 2x10²² and 1x10²² ions/cm² indicate the size of the NCs are typically in the range of 2-5 nm, while the sample containing 2x10²² ions/cm² are larger with particle sizes up to 8 nm in diameter (fig. 1). The electron diffraction studies confirm the particles nanocryrstalline silicon.

The optical transmission spectra are shown for the implanted samples that were annealed at 1100°C in an atmosphere of 96% Ar + 4% H₂ in figure 2. It is apparent that there is a consistent increase in the absorption onset as

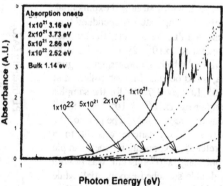

Absorption onsets
1x10²¹ 3.16 eV
2x10²¹ 3.73 eV
5x10²¹ 2.86 eV
1x10²² 2.52 eV
Bulk 1.14 ev

Figure 2. Optical absorption spectra of Si implanted in SiO₂.

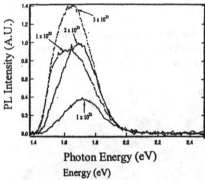

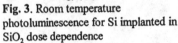

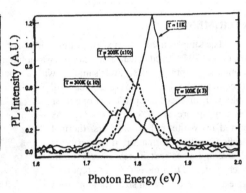

Fig. 3. Room temperature photoluminescence for Si implanted in SiO₂ dose dependence

Fig 4. PL for SiO₂ implanted with 1×10^{21} Si ion/cm³ temperature dependence

the ion dose decreases (NC size decreases). The band edges are at 2.52, 2.86, 3.73 and 3.92 eV for the ion doses of 1×10^{22}, 5×10^{21}, 2×10^{21} and 1×10^{21} ions/cm², respectively. The region between 700-1200 nm has a transmittance between ~90 and 95% which results from front-surface and back surface-reflection losses from the silica substrate.

Room temperature PL for the same set of samples are illustrated in figure 3. The maxima in the PL are observed at 1.72, 1.69, 1.64 and 1.62 eV for the ion doses of 1×10^{22}, 5×10^{21}, 2×10^{21} and 1×10^{21} ions/cm³. The intensity is also observed to scale with ion dose and the lineshapes are approximately the same with no fine structure observed for all doses. As compared to the transmission spectra which show a strong dose dependent blue shift of 1.4 eV (highest dose to lowest dose), the PL show a shift of 0.1 eV. The low temperature PL shown in figure 4 for a sample containing 2×10^{21} ions/cm² reveals that the intensity of the PL increases by nearly a factor of 20 for the measurement made at 11 K as compared to the room temperature measurement.

The dose dependence on lifetime at room temperature is shown figure 5. The PL was collected at 1.72 eV. The curves were fit with a stretched exponential and the lifetimes were found to decrease with increasing NC size (increasing ion dose) by nearly a factor of 3. However, the lowest dose could not be fit with a stretched exponential. For the temperature dependent time resolved PL, we consider the samples containing 1×10^{22}, 2×10^{21} and 1×10^{21} ions/cm³ which are presented in figure 6a, 6b ad 6c. The PL was collected at 1.65, 1.68 and 1.72 eV for the samples containing 1×10^{22}, 2×10^{21} and 1×10^{21} ions/cm², respectively. The plot for the lowest dose sample was fit to a stretched exponential; the sample containing 2×10^{21} ions/cm² was fit with a double stretched exponential, while the highest dose sample showed a less reliable fit. The lifetimes at all

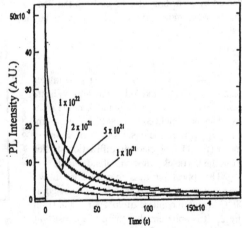

Fig.5 PL lifetime dependence for Si implanted in SiO₂ and annealed at 1100°C in 95% Ar + 5% H₂ for 1 h.

temperatures are shorter for the highest dose sample as compared to samples with lower doses and differ by a factor of ~20 at 10 K and by a factor of ~3 at 250 K. It is clear from the plot for the lowest dose sample, the lifetime shows a strong dependence on temperature in the 11-100 K range where the lifetime decreases by a factor of ~8, whereas for the highest dose the lifetime decreases by a factor of ~1.3.

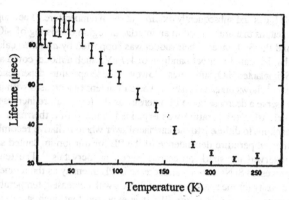

DISCUSSION

Particle sizes determined by TEM and the size dependent blueshift observed in the onset of the optical absorption spectra are in qualitative agreement with the predictions of quantum confinement theory. Also, the relative blueshifts appear to be in reasonable quantitative agreement with the calculations for bandgap shifts proposed by various authors, in particular with those of Delley and Stiegmeyer.[14] However, the contributions from molecular chain like structures absorbing in the blue cannot be excluded.

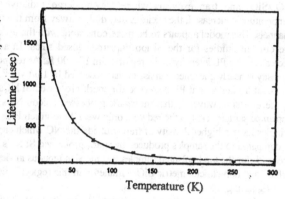

The room temperature PL ranges between 1.72 and 1.62 eV for all ion doses and does not appear to reflect quantum confinement effects when compared to the corresponding absorption spectra which show onsets ranging from 3.92 to 2.52 eV. Clearly, for all implanted substrates there is no band edge luminescence. The insensitivity of the PL peak position with respect to the absorption onset appears to

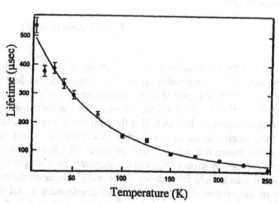

indicate that the PL is originates **Figure 6.** Temperature dependence of PL lifetimes: a) 1×10^{22}, from a common luminescent center. b)2×10^{21}, c) 1×10^{21} ions/cm^2

A similar observation (i.e. the PL was independent of particle size) was made by Kanemitsu et al.[2] who reported broad PL at 1.65 eV for particles with average diameters of 7, 10 and 13 nm that were prepared by laser breakdown of

SiH$_4$ gas and subsequently oxidized at room temperature. They suggested that the PL originates from excitons that are trapped in an interfacial region consisting of SiO$_x$ between the silicon quantum dot and the SiO$_2$ layer. Their model was supported by *ab initio* calculations for a thin film of siloxane that indicated a direct bandgap of 1.7 eV. Such a model could account for the PL observed for the Si implanted SiO$_2$ substrates. However, the temperature dependence of the PL reported by Kanemitsu et al.[2] shows unusual behavior where PL intensity increases with increasing temperature, whereas we observe a decrease in the PL intensity as the temperature increases. The increasing PL intensity with increasing temperature was explained in terms of a thermally activated process which causes the excitons to diffuse into the interfacial layer where radiative recombination occurs. On the other hand, the temperature dependence of the PL for our ion implanted samples appears to follow the trend observed for spark processed silicon nanocrystals.[15] The temperature dependent PL for spark processed Si NCs shows an increase in PL intensity as the temperature decreases. The increased PL intensity for the spark processed Si NCs with decreasing temperature suggests that strongly localized states play a role in the PL. It is expected that such states may exist in the insulating layer of SiO$_x$/SiO$_2$ and trap excitons where they undergo radiative recombination. However, as the temperature is increased, the excitons may diffuse away from the trap and relax through nonradiative channels. This model appears to be more consistent with the temperature dependent PL reported in the current studies for the Si ion implanted fused silica. In addition, we observe a very abrupt decrease in the PL intensity in the region from 10-100 K. At temperatures greater than 100 K the PL intensity is nearly quenched, similar to that observed by Ludwig.[15] Despite the similar behavior in the temperature dependent PL observed for spark processed Si NCs and Si implanted SiO$_2$, they differ in the emission wavelengths: the spark processed silicon NCs emit in the blue-green and the Si implanted samples emit in the red with only weak blue-green PL. This suggests that the Si implanted samples have a higher density of traps outside the NC which capture the carriers leading to red PL as compared to the samples produced by spark processed Si NCs. Furthermore, analysis of the blue-green PL for the ion implanted samples is somewhat tenuous in view of the fact that ion beam induced defects and non-stoichiometric SiO$_x$ luminesce in this region which makes it difficult to separate the PL originating from NCs and defects.

The room temperature time resolved PL spectra were fit with a stretched exponential function expressed as[13]

$$I_{PL}(t) = I_0 \exp[(-t/\tau)^\beta] \tag{3}$$

where I_0 is the initial value of the PL intensity, β is the stretching parameter, and τ is the relaxation time constant. The stretched exponential is often used to describe the dynamics of inhomogeneous systems which is expected to be appropriate for the ion implanted samples where there is size dispersion that results from the nature of the nucleation and growth rates of the NCs during the annealing treatment. It is clear that the lifetimes of the NCs decrease as the NC size increases. This is most likely related to the number of traps in the interfacial region between the NC and the SiO$_x$/SiO^2 layers. It is expected that the larger NCs have more trapping sites in the interfacial region than the relatively smaller NCs and consequently have a higher probability of trapping a carrier pair, which in turn leads to faster decay. The size dependence on the relaxation times is also consistent with the model proposed by Qin et al.[9] They developed a model which shows that the relaxation rate decreases as the particle size decreases. The mechanism for relaxation occurs by tunneling from the interior of the NC into a surrounding SiO$_x$ layer whereupon the carriers undergo radiative recombination at a trap. This process provides a relaxation channel that, in part, overcomes the problem of the phonon bottleneck that is expected to occur for relaxation within the NC itself. How well this model applies to the results presently currently is difficult to assess quantitatively, but the general prediction that the relaxation rates increase with increasing particle size is consistent with our results. Another factor that may explain the faster relaxation rate of the large NCs as compared to the relatively smaller ones is that the energy level spacing for the larger NCs is more compact than for the small nanocrystals. Thus, for a fixed excitation energy, there are more states available for the excited carriers to tunnel into the SiO_x matrix as compared to the small nanocrystals. Also, the amplitude of the wavefunction for the larger NCs becomes more localized near the edges of the confining barrier which enhances the probability of tunneling into the SiO_x layer. On the other hand, the smaller NCs excited with the same energy promote carriers into states where the wavefunction amplitude is less strongly localized at the edges of the potential well. Consequently, it is expected that there would be a decrease in the probability of tunneling into the surrounding matrix. This would account for the faster relaxation times in the large NCs as compared to the small ones. Thus, contributions from the increase of traps and the localization of the wavefunction at edges of the potential barrier for the larger NCs as compared to the smaller ones can account for the faster decay times in the large nanocrystals.

The temperature dependent decay curves are also consistent with faster relaxation in the large NCs compared to the relatively smaller ones. However, the sample containing the highest concentration of Si exhibits an unusual decay curve that shows a relatively temperature independent lifetime in the 10-50 K temperature range. Apparently, in this temperature range, processes which could reduce the PL lifetime are quenched, but it is not clear as to what mechanism is responsible for this behavior. The samples containing the highest concentration of Si also show a relatively weak temperature dependence on the relaxation time as compared to the lower concentrations. This can also be accounted for by considering the amplitude of the wavefunction for carriers that are excited in the larger NCs as compared to small NCs as well as the number of trapping centers outside the NC. As stated previously, given the same pump energy for the large and small NCs, carriers in the large NCs will occupy a higher quantum level than the small NCs. The amplitude of the wavefunction for the large NCs will be more localized near the edge of the potential barrier which would increase the tunneling frequency into the surrounding SiO_x layer as compared to the small NCs. Also, because the large NCs have larger number of trapping centers there is greater efficiency for being trapped after tunneling occurs. Consequently, the carriers in large NCs have less residence time within the NC itself whereas the carriers in the smaller nanocrystals spend a longer time in the interior of the NC before tunneling and being trapped at a luminescent center outside the NC. Based on this assumption, the carriers in the smaller NCs would be more likely scattered by phonons as compared to the large NCs. Since scattering of carriers by phonons is a temperature dependent process that occurs within the NC, we suggest that the small NCs that have longer lifetimes within the NC would show a stronger temperature dependence of their lifetimes due scattering by phonons as compared to the large NCs.

CONCLUSIONS

Silicon NCs were formed by ion implantation into SiO_2 substrates followed by thermal annealing at 1100°C in a reducing atmosphere. TEM confirmed the presence of Si NCs. The optical absorption spectra show an increase in the absorption onset as the ion does decreases which is attributed to quantum confinement effects. The room temperature PL show luminescence at ~1.7 eV and is nearly independent of the Si NC size. We suggest that the apparent absence quantum size effects in the PL spectra indicates that the carriers are trapped at a common luminescent center that may reside at the Si NC/SiO_x interface. The decrease in the radiative lifetimes for larger NCs originates from the higher concentration traps that reside outside the NC compared to the small ones. It is postulated that the strong temperature dependence on the lifetimes of the carriers for the small NCs arises from the longer residence times inside the nanocrystal as compared to the large NCs (due to the higher number of traps at the Si NC/SiO$_x$ interface), which in turn leads to scattering by phonons. The scattering by phonons then reduces the lifetimes as the temperature increases.

ACKNOWLEDGMENTS

The authors at Fisk would like to acknowledge the financial support by NASA grant no. NCC3-575 (Consortium for the Advancement of Renewable Technology, CARET), DOE (grant no. DE-F605-94ER45521 and the research at ORNL is sponsored by the Division of Materials, U.S. Department of Energy under contract DE-AC05-84OR21400 with Lockheed Martin Energy System, Inc.

REFERENCES

1. L. E. Brus, J. Chem. Phys. **80**, 4403 (1984).
2. Y. Kanemitsu, T. Ogawa, K. Shiraishi, and K. Takeda, Phys. Rev. B **48**, pp. 4883-4886 (1993).
3. J. Diener, M. Ben Chorin, D. I. Kovalev, S. D. Ganichev, and F. Koch, Phys. Rev. B **52**, pp. R8617-R8620 (1995).
4. R. P. Chin, Y. R. Shen, and V. Pretova-Koch,Science **270**, pp. 776-778 (1995).
5. L. Rebohle, J. von Borany, R. A. Yankov, W. Skorupa, I. E. Tyschenko, H. Fröb, and K. Leo, Appl. Phys. Lett. **71**, pp. 2809-2811 (1997).
6. K. Kim, Phys. Rev. B **57**, pp. 13072-13076 (1998).
7. D.J. Lockwood, Z. H. Lu, and J. M. Baribeau, Phys. Rev. Lett. **76**, pp. 539-541 (1996).
8. L. Tsybeskov, MRS Bulletin **23**, pp.33-38 (1998).
9. G. Qin and G.G. Qin, J. Appl. Phys. **82**, pp. 2572-2579 (1997).
10. R. E. Hummel and P. Wißmann, *Handbook of Optical Properties Volume II, Optics of Small Particles, Interfaces, and Surfaces*, pp. 83-143, CRC, New York (1997) and references therein.
11. G. Allan, C. Delerue and M. Lannoo, Phys. Rev. Lett. **76**, pp.2961-2964 (1996).
12. C. W. White, J. D. Budai, S. P. Withrow, J. G. Zhu, S. J. Pennycock, R. A. Zuhr, D. M. Hembree, D. O. Henderson, R. H. Magruder, M. J. Yacaman, G. Mondragon, and S. Prawer, Nuclear Instruments and Methods **B127/128**, 545 (1997).
13.Yu-Shen Yuang, Yang-Fang Chen, Yang-Yao Lee, and Li-Chi Lui, J. Appl. Phys. **75**, pp.3041-3044 (1994).
14. See pp. 84 in reference 10.
15. See pp. 115 in reference 10.

HIGHLY REFLECTIVE AND SURFACE CONDUCTIVE SILVER-POLYIMIDE FILMS FABRICATED VIA REDUCTION OF SILVER(I) IN A THERMALLY CURING POLY(AMIC ACID) MATRIX

R. E. SOUTHWARD,[1] D S. THOMPSON,[2] D. W. THOMPSON,[2] J. L. SCOTT,[2] and S. T. BROADWATER[2]
[1]National Research Council Postdoctoral Associate,
Materials Division, NASA-Langley Research Center, Hampton, VA 23681;
[2]Department of Chemistry, College of William and Mary, Williamsburg, VA 23187.

ABSTRACT

Highly reflective and surface conductive silvered polyimide films have been prepared by the incorporation of the (1,1,1-trifluoro-2,4-pentanedionato)silver(I) complex into a dimethyl-acetamide solution of the poly(amic acid) formed from 3,3´,4,4´-benzophenone tetracarboxylic acid dianhydride (BTDA) and 4,4´-oxydianiline (ODA). Thermal curing of the silver(I)-containing poly(amic acid) films leads to cycloimidization with concomitant silver(I) reduction yielding a reflective and conductive silvered film surface.

INTRODUCTION

The synthesis of metal clusters frequently involves colloid-stabilizing polymers such as polyvinyl alcohols, poly(amide imide)s, and other macromolecules (1). We invoked this rubric to synthesize silver-metallized polymeric films with high reflectivity and conductivity. Our initial approach (2) involved thermally (150-225°C) induced reduction of positive valent silver (Ag(I)) complexes dissolved in polymeric films resulting in matrix-stabilized metal clusters. Upon heating to 300-340 °C colloid stabilization is sacrificed, and metal clusters undergo phase separation yielding composite materials which are either reflective or surface conductive, but not both. Our most reflective films (R = 80%) are not conductive, and our conductive films (0.5 ohm/sq for a 70 nm Ag layer) have low reflectivity.

We now report a Ag(I)-polymer system that yields films with both high reflectivity and conductivity. There are several applications for such metallized films including lightweight optical mirrors and sunshields for the NASA Next Generation Space Telescope project (3), large scale inflatable radiofrequency antennas for the management of electromagnetic signals in space (4), and thin film reflectors and concentrators for solar thermal propulsion and dynamic power generation (5). Reflective surfaces on polymeric supports offer substantial advantages in weight, flexibility, elasticity, and fragility.

Specifically, the research herein focuses on the deposition of metallic Ag on polymeric films by a single stage, internal, self-metallization process. Efforts center on Ag because of its unrivaled reflectivity and conductivity. We chose aromatic poly(amic acid)-polyimide polymers because of their excellent chemical and thermal stability. Our synthetic protocol is illustrated in Scheme 1. Silver(I) acetate (AgOAc) and 1,1,1-trifluoro-2,4-pentanedione (TFAH) are allowed to react in dimethylacetamide (DMAc) to give in solution the (1,1,1-trifluoro-2,4-pentane-dionato)silver(I) complex, AgTFA. A DMAc solution of the poly(amic acid) form of BTDA-/ODA is added to the DMAc solution of AgTFA. A film is then cast. Thermal curing of the Ag(I)-poly(amic acid) film effects reduction of Ag(I) coupled with imidization of the amic acid. As the poly(amic acid) imidizes silver(0) atoms/clusters migrate in part to the surface to give a conductive and reflective film. Herein details of the synthetic chemistry and physical characterization for these films are reported.

The usual techniques for the production of metallized films include physical vapor deposition via thermal evaporation or sputtering, chemical vapor deposition (CVD), electrodeposition, and electroless chemical reduction from solution in which the substrate is immersed. A major problem with these external deposition procedures is poor adhesion of metal to polymer (6). Also, CVD rarely works well with polymeric films since they can seldom be heated without degradation to the temperatures needed to effect reduction of a positive valent metal from a volatile molecular precursor. The internal metallization protocol of this paper yields metal atoms and near-atomic metal clusters which diffuse and aggregate to give reflective and conductive metallic surfaces with uncompromised adhesion at the metal-polymer interface.

Mat. Res. Soc. Symp. Proc. Vol. 551 © 1999 Materials Research Society

EXPERIMENTAL

Materials. All chemicals were obtained from commercial sources. BTDA/ODA poly(amic acid) solution was prepared with a 1% offset of dianhydride at 15% solids (w/w) in DMAc. The resin was stir red for 5 h. The inherent viscosity was 1.7 dL/g at 35° C.

Preparation of BTDA/4,4′-ODA metallized films. Ag-containing solutions were prepared by first dissolving silver(I) acetate in DMAc containing TFAH. The 15% poly(amic acid) solution was then added to give the desired Ag to polymer ratio. Doped poly(amic acid) solutions were cast as films onto soda lime glass plates using a doctor blade set at 500-650 μm to obtain cured films 20-25 μm thick. After remaining in an atmosphere of slowly flowing dry air for 18 h, the films were cured in a forced air oven. The cure cycle involved heating over 20 min to 135 °C and holding for 1 h, heating to 300 °C over 4 h, and holding at 300 °C for 7 h.

RESULTS and DISCUSSION

Reflectivity, conductivity, and surface properties. Table 1 displays characterization data for three Ag films cured at 300 °C for 7 h. The first entry is an undoped BTDA/ODA control. The second entry is a silvered film of area (12 x 20 cm) cast from the same Ag-doped resin that was used to prepare small glass slides (27 x 46 mm) of metallized polymer for the reflectivity versus cure plot of Figure 1. The last two entries are for films prepared from a single resin solution. The films differ only in that one film was cast directly onto a glass plate whereas the other was cast more thinly onto an undoped BTDA/ODA polyimide film. Casting the Ag-doped resin onto a parent base conserves silver and ensures that

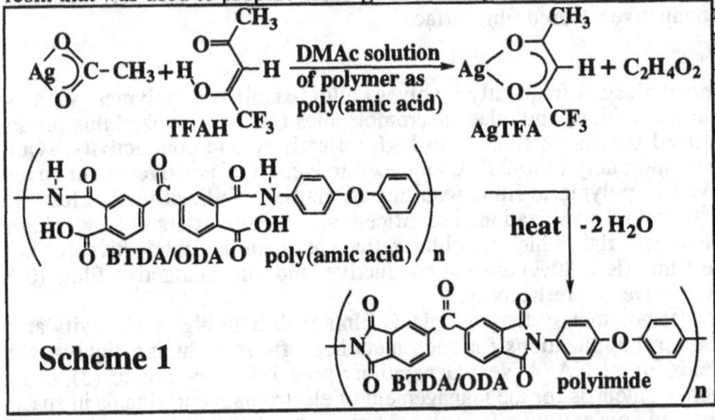

Scheme 1

Table I. Selected characterization data for AgTFA-BTDA/ODA metallized films.

Percent Silver (Calc)[a]	Percent Reflectivity of Silvered Films[b] at 531 nm as a function of angle					Tg, °C (DSC)	CTE (ppm/ K)	Surface Resistivity[c] (polished film) (ohm/sq)
	20°	30°	45°	55°	70°			
0 Control	Not applicable					275	42.8	7 x 10^{15} air side 2 x 10^{17} glass
13.0	98	97	97	95	91	276	34.3	<0.1
12.8	98	98	98	99	92	270	33.0	<0.1
12.8	95	95	94	92	91	273	32.8	<0.1

mechanical and thermal properties of these "film-on-film" composites closely resemble those of the parent polyimide. The TEM of Figure 2 shows the relative thickness of the metallized and parent layers (1:7) and also shows that there is no diffusion of Ag into the polyimide base. The reflectivities and conductivities of all films are impressive.

We followed the evolution of the silvered surface by casting a AgTFA-BTDA/ODA solution onto a series of glass slides and measuring reflectivity of individual films as they were withdrawn at selected temperatures over the cure cycle. Figure 1 shows a plot of reflectivity

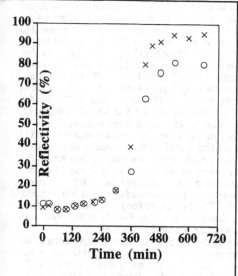

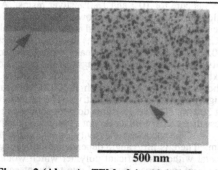

Figure 2 (Above). TEM of the 12.8 % film of Table 1 cast on a parent BTDA/ODA base showing the polymer-polymer interfacial region.

Figure 1 (Left). Reflectivity versus thermal cure for AgTFA-BTDA/ODA films with 10.7 (O) and 13.0 % (X) silver. Time zero for the abscissa is at 135 °C after 1 h; the temperature rises from 135 to 300 °C over 240 min and then remains constant at 300 °C.

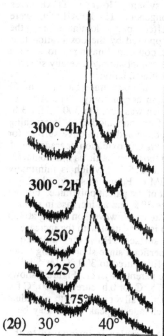

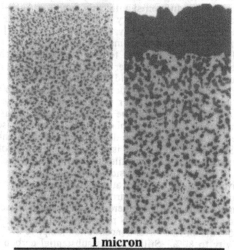

Figure 4 (Above). TEMs of two 10.7% Ag films of Figure 1: (Left) film heated to 275° for 0 h; (Right) film heated to 300° for 5 h. Left is not conductive; Right has ρ = 5 ohm/sq.

Figure 3 (Left). XRD patterns in the Ag 111 and 200 region for selected 10.7 % Ag films of Figure 1. Temperature at which films were removed from oven is on each curve.

versus cure for 10.7 and 13.0% silver resins. After heating at 135 °C for 1 h the silver(I)-doped films have the identical yellow appearance of undoped BTDA/ODA films, and XRD shows no peaks typical of face centered cubic silver metal. Surface luster becomes apparent only near 275 °C; however, XRD reflections for silver are seen as early as 175 °C as shown in Figure 3 for the 10.7% series. The X-ray peaks are broadened between 175 and 300 °C which suggests that silver formed in the early stages of the cure is dominated by nanometer-sized crystallites which exhibit Scherrer broadening. TEMs for films heated to 275° for 0 h and 300 °C for 7 h are shown in Figure 4. After 3 h at 300 °C the films show a Ag layer which is 150-200 nm and is highly reflective and conductive. The reflectivity for the 10.7% films levels near 80% after 4 h at 300 °C. The 10.7% films do not become conductive until cured for 5 h at 300 °C. Films with less than ca. 10% silver do not become conductive under any of the conditions examined.

The specular reflectivity values for the three films of Table 1 are greater than 90% at 531 nm. There is little angle dependence indicating a reflective surface that is predominately silver metal without significant polymer which would increasingly absorb radiation at 531 nm as the angle of incidence increased due to the effective increase in the absorption pathway. Figure 5 shows the angle dependence of reflectivity as a function of cure time. For the 13.0% film cured to 300 °C for 2 h, XPS data show that there is residual polymer at the surface. This film exhibits the greatest loss in reflectivity with increasing angle. The film cured for 7 h, which has a nearly pure silvered surface, exhibits the lowest angle dependence. This loss of reflectivity with increasing angle may be due to highly absorbing polyimide remaining near the surface.

In our previous work with BTDA/ODA-AgOAc-1,1,1,5,5,5-hexafluoro-2,4-pentandione (HFAH) none of the metallized surfaces was conducting even though excellent reflectivity was obtained. Even when metallized films were cured to 340 °C, conductivity was not induced (2). The TFA ligand of this paper differs only slightly from the HFA ligand by replacement of fluorines with hydrogens in one methyl group. We did not expect the Ag(I)-TFAH system to give films which differed significantly from the Ag(I)-HFAH system. However, TFAH gives metallized films which are very different from their HFAH congeners. The HFAH films were ca. 80% reflective and were never conductive. Clearly, ligand effects play a dominant role in the metallization of polyimide films. This conclusion is also supported by the observation that silver(I) nitrate and silver(I) tetrafluoroborate, which have non-coordinating anions, give degraded and brittle films under similar conditions. Thus, one cannot simply take any silver(I) compound with a poly(amic acid) and obtain metallized films of high quality.

The STM for the conductive 12.8% film of Table 1 cast on a polyimide base (Figure 6) shows an continuous irregular topography which would sustain conductivity. The AFM of Figure 6 for the non-conductive 10.7% sample of Figure 1, which was cured to 300 °C for 3 h, shows a regular array of globular particles which are isolated from one another by intervening polymer. We suggest that the development of the highly reflective and conductive surface for the 12.8% AgOAc-TFAH film is due to a combination of sintering and oxidation degradation of near-surface polyimide to volatile products, both of which bring the silver particles into contact.

XPS data for the air-side surface of the film cured to 275 °C, which is minimally reflective and has little silver at the surface as seen in the TEM of Figure 4, show only 2.8% silver and a carbon, oxygen, nitrogen ratio which is close to that of the parent polyimide. For the film cured to 300 °C for 0 h there is an increase in reflectivity to 28% and an increase in surface silver to 4.0%; the C, O, and N ratio remains near that of the parent. However, on curing at 300 °C for 5 h the air side silver concentration jumps to 27% which is consistent with the increase in reflectivity to 80%. Still, there is substantial carbon at the surface. However, ion milling experiments show that the surface carbon is predominately superficial since short milling times reveal a silver layer that is >98% pure. The glass side of the film cured to 300 °C for 5 h has much less silver and an organic composition closer to that of the parent polymer. Figure 5 shows Ag 3d XPS spectra for three 10.7% films of Figure 1 cured to 275 °C for 0 h, cured to 300 °C for 0 h, and cured to 300 °C for 5 h. All samples show symmetrical silver peaks with a gradual narrowing with increasing cure temperature and time. The change in peak shape may be a function of changes in metal particle sizes with cure time.

Synthetic aspects of film preparation. The AgTFA complex is introduced to the poly(amic acid) resin via an in situ synthesis. Even though AgTFA is not isolated, we can be

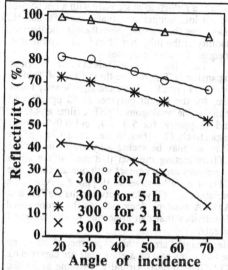

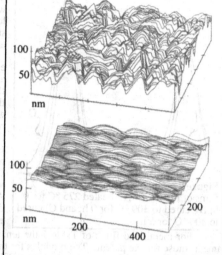

Figure 5. Reflectivity versus angle for 13.0 % AgTFA-BTDA/ODA films cured at 300 °C for varying times. Sheet resistance values are: 7 h cure - 0.1 ohm/sq, 5 h cure - 18ohm/sq, 3 h cure - 225 ohm/sq, and 2h cure - not conductive.

Figure 6. STM of the 12.8 % conductive film (air-side) of Table 1 (R=95% and ρ = <0.05 ohm/sq) cast on a BTDA/ODA base. Bottom: AFM of a 10.7 % non-conductive film (air-side) cured at 300 °C for 3h (R = 61%).

assured that this complex is formed. First, when the poly(amic acid) is added to AgOAc alone, immediate polymer gelation occurs. The pathway for gelation is deprotonation of the aromatic carboxyl groups of the poly(amic acid) via transfer to the more basic acetate ions of AgOAc. Carboxylate groups of the polymer then coordinate to Ag(I) ions to form the extended gel network. It is well known that silver carboxylate complexes are dimeric involving the coordination of two carboxylate groups (8). If TFAH is added to AgOAc before the addition of the poly(amic acid), no gelation occurs, and the insoluble AgOAc is rendered soluble. Second, support for AgTFA formation comes from the AgOAc-HFAH system. Using HFAH rather than TFAH with AgOAc in BTDA/ODA gives a similarly clear homogeneously doped resin solution. Again, we assume that the AgHFA complex is formed in situ. When films of the in situ AgHFA-BTDA/ODA system are thermally cured, silvered surfaces with 80% reflectivity are produced. Would one obtain the same metallized BTDA/ODA film if one began with a pure solid AgHFA complex? This question can be answered because the AgHFA complex can be isolated as the (η^4-1,5-cyclooctadiene) adduct, [(COD)(HFA)Ag], whose crystal structure is known. When this solid complex is dissolved in DMAc with BTDA/ODA and a film is thermally cured, metallized films are produced with identical properties to those prepared via the in situ AgOAc-HFA route (7). This is strong evidence that AgHFA occurs in both systems. We found that TFAH/Ag(I) ratios of 1.35:1, rather than 1:1, gave doped solutions which were less viscous and easier to manipulate.

Thermal and mechanical properties of the metallized films. The glass transition temperatures of the metallized films do not vary from that of undoped BTDA/ODA by more than a few degrees. This suggests that the bulk polymer structure is not compromised by the reduction of silver(I) and the formation of silver(0). However, the formation of metal clusters in the bulk of the polymer as well as on the surface diminishes the high temperature thermal-oxidative stability of the hybrid film. While in a nitrogen atmosphere the temperature at which there is ten percent weight loss is not vastly different from that of the undoped polymer, in air

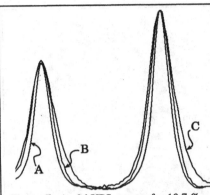

Figure 7. Ag 3d XPS spectra for 10.7 % films of Figure 1: A) heated 275 °C for 0 h; B) cured to 300 °C for 0 h; and C) cured to 300 °C for 5 h.

there is a reduction in stability with a ten percent weight loss temperature that is *ca.* 150-200 °C lower than the control. Nonetheless, the thermal stability of the mirrored films in air is more than adequate for most purposes.

The linear coefficients of thermal expansion (CTE) for the three films of Table 1 are 33-34 ppm/K. These are significantly lower than for the parent polymer at 43 ppm/K and than for the analogous AgHFA films at 43, 44, and 43 ppm/K for 5.0, 7.4, and 9.9 % silver, respectively (2). Thus, in the conductive AgTFA films we may be seeing a hybrid value for the CTE reflecting the fact that the surface silver aggregates are in contact with one another. The CTE of metallic silver is 19 ppm/K. Again, the variation in CTE values between the AgTFA and AgHFA systems demonstrates the pronounced differences that can arise due to subtle ligand effects.

For metallized films of Table 1 the tensile strengths are, within experimental error, the same as those for the parent. The modulus for the two films cast on a glass plate are elevated ca. 5 % while the value for the film cast on a BTDA/ODA base is virtually the same as for the undoped polyimide. All of the films are flexible and can be tightly creased without rupture.

CONCLUSIONS.

We have synthesized silvered BTDA/ODA polyimide films from single phase AgOAc-1,1,1-trifluoro-2,4-pentanedione-BTDA/ODA solutions cast and cured on glass plates. Depending on concentration and thermal conditions metallized films can be fabricated with excellent specular reflectivity, surface conductivity, outstanding metal-polymer adhesion, and appropriate mechanicals. While the formation of metallic silver lowers thermal stability in air, the metallized polymers still have a wide thermal use range. We have also shown that BTDA/ODA films cast on a parent BTDA/ODA polyimide base yield highly reflective and conductive composite films. This "film-on-film" approach minimizes silver required for the formation of a reflective surface and assures composite properties which are essentially those of the parent polyimide (9).

REFERENCES

1. "Clusters and Colloids: Theory to Applications," G. Schmid, Ed., NCH, Weinheim, 1994.
2. a) R. E. Southward, D. W. Thompson, and A. K. St. Clair, *Chem. Mater.*, **9**, 501 (1997); b) R. E. Southward, D. S. Thompson, D.W. Thompson, and A. K. St. Clair, *ibid.*, **9**, 1691 (1997).
3. a) G. Horner, D. Stoakley, and A. K. St. Clair, *Bull. Amer. Astron. Soc.*, **189**, 60.06(1996); b) R. Angel, B. Martin, S. Miller, D. Sandler, D. Burns, and D. Tenerelli, *Bull. Amer. Astron. Soc.*, **189**, 60.07 (1996); c) J. J. Triolo, J. B. Heaney, and G. Hass, *SPIE*, **21**, 46 (1977).
4. R. E. Freeland and G. Bilyou, *43rd Congress of the International Astronautical Federation, IAF-92-0301*, Washington, D.C., 1992.
5. a) P. A. Gierow, *Proceedings of the ASME-JSME-JSES Solar Energy Conference*, Reno, NV, pp. 1-7 (1991); b) K. Ehricke, ARS paper 310-56, Mtg. of the Am. Rocket Soc., Cleveland, Ohio, June 18-20 (1956); c) D. A. Gulino, R. A. Egger, and W. F. Bauholzer, *NASA Technical Memorandum 88865* (1986)..
6. G. Rozovskis, J. Vinkevicius, and J. Jaciauskiene, *J. Adhes. Sci. Technol.*, **10**, 399 (1996).
7. A. F. Rubira, J. D. Rancourt, M. L. Caplan, A. K. St. Clair, and L. T. Taylor, *Chem. Mater.*, **6**, 2351 (1994).
8. a) Coggin, P.; McPhail, A. T. *J. Chem. Soc., Chem. Commun.*, **1972**, 91; b) Blakeslee, A. E.; Hoard, J. L. *J. Am. Chem. Soc.*, **1965**, 78, 3029.
9. DWT acknowledges and thanks the Petroleum Research Fund for partial support of this work.

Part V

Microgravity Materials Processing: Fundamental Studies

THERMOPHYSICAL PROPERTIES OF FLUIDS MEASURED UNDER MICROGRAVITY CONDITIONS

H.-J. Fecht
University of Ulm, Faculty of Engineering,
Albert-Einstein-Allee 47, D-89081 Ulm, Germany
and
R.K. Wunderlich
Technical University Berlin, Institute of Metals Research
Hardenbergstr. 36, PN 2-3, D-10623 Berlin, Germany

ABSTRACT

The analysis of nucleation and growth processes relies mostly on circular arguments since basic thermophysical properties necessary, such as the Gibbs free energy (enthalpy of crystallization, specific heat), the density, emissivity, thermal conductivity (diffusivity), diffusion coefficients, surface tension, viscosity, interfacial crystal / liquid tension, etc. are generally unknown with sufficient precision and therefore often deduced from insufficient linear interpolations from the elements. The paucity of thermophysical property data for commercial materials as well as research materials is mostly a result of the experimental difficulties arising from the unwanted convection and reactions of melts with containers at high temperatures. An overview will be given on the results of thermophysical property measurements during several different space flights using containerless processing methods. Furthermore, a perspective on a future measurement program of thermophysical properties supported by the European Space Agency is described. In this regard, the International Space Station is considered as the ideal laboratory for high precision measurements of thermophysical properties of fluids which help to improve manufacturing processes for a number of key industries.

INTRODUCTION

Materials properties, such as mechanical strength, creep resistance, ductility, wear resistance as well as magnetic and electronic characteristics, are determined by the atomic structure and composition and the number and kind of defects produced during the synthesis of materials. Therefore, the structural control on an atomic, nanoscopic and microscopic level, in particular during the liquid-to-solid phase transformation, is very important for quality control and the design of advanced materials for specific technological applications. In order to apply computational design tools for describing (i) the heat flow balance and fluid dynamics and (ii) the resulting microstructure of a casting reliable thermophysical and related properties are required as input parameters.

Numerical simulation programs of industrial casting and solidification processes have been established as a valuable tool in improving process reliability, quality control and high geometric shape accuracy. However, for further quality optimization the availability and accuracy of the thermophysical property data used as input parameters are generally not sufficient. Such data are not existent for most commercial alloys or there are only values available showing an intolerable error width.

THERMOPHYSICAL PROPERTIES

The paucity of thermophysical property data for commercial materials as well as materials of fundamental interest is a result of the experimental difficulties arising from the unwanted reactions of melts with containers at high temperatures. For instance, viscosity measurements reported for pure aluminium and iron vary by ± 100 % and ± 50 %, respectively, around the mean. On the other hand, the knowledge of these properties is essential for the understanding and subsequent modeling of metallurgical processes, thermodynamic phase equilibria and phase diagram evaluation. Thus, accurate input data are needed in the stable liquid and at different levels of liquid undercooling. This will expand our knowledge of the universality of the thermophysical properties in the metastable state and can be applied to metallic liquids of broad scientific and technological interest.

Some of these data can be obtained more or less accurately by conventional methods. High precision measurements however on chemically highly reactive materials and fluids in the undercooled regime at the temperatures of interest require the use of containerless processing and non-contact diagnostic tools [1]. By eliminating the contact between the melt and a crucible accurate surface nucleation control and the synthesis of materials free of surface contamination become possible. Alternative methods of active nucleation control are based on falling droplet and encased fluid samples [2].

INDUSTRIAL INTEREST

The need for thermophysical property data to improve process control and product quality has been recognized by European industry in recent years. This includes primary metals producers, secondary refiners as well as end users. The thermophysical data are needed to (i) gain a better understanding of the (solidification) process, (ii) to solve problems encountered with the process in order to reduce the number of defects in the materials and thus improve product quality and (iii) to minimize waste and energy costs. Examples for relevant industrial processes are:
- Casting processes (Fe-, Ni-, Ti-, Al-, Mg- alloys, refractories, MMC's)
- Crystal growth of poly- and single-crystalline materials (turbine blades and disks,semiconductors)
- Glass production (metallic and non-metallic)
- Rapid prototyping
- Spray forming and powder production
- Surface modification by spraying techniques
- Welding (conventional, laser, electron).

Furthermore, a questionnaire distributed within European industry indicates the urgent need of high quality data of themophysical properties. In general, the companies are dissatisfied with the amount of data available for commercial materials. On the other hand, it is reckognized that such data are needed for:
- Computer modeling of solidification processes
- Gaining a better understanding of the processes
- Better process control and
- Improved product quality.

As such, the following properties which are generally unknown for commercial materials are considered as important in the corresponding hierarchy:
1. Melting range
2. Fraction solid / fraction liquid
3. Heat capacity, enthalpy, total hemispherical emissivity
4. Density
5. Surface (interfacial) tension
6. Thermal conductivity and diffusivity

7. Viscosity
8. Spectral emissivity, optical properties
9. Diffusion coefficients
10. Electrical conductivity.

SPACE RELEVANCE

The precise measurements of thermophysical data, especially in the liquid state, are considered as a prime goal in order to improve the modeling of industrially relevant solidification processes. Although the standard earth-based techniques appear to work for several materials they can not be confidently applied to reactive melts at high temperatures due to contamination and exothermic reactions from the crucibles required to hold liquid samples against gravity. New scientific methods as well as apparative developments are now available which allow to eliminate containers. Further increase of accuracy will be possible by carrying out these measurements in the microgravity environment of space under ultra high vacuum conditions.

Scientific experiments have been carried out in the containerless processing facility TEMPUS during the Spacelab missions IML-2 and MSL-1. It has been demonstrated that the reduction of positioning forces in microgravity leads either to a significant improvement of the accuracy or makes the measurement possible at all. During 200 hours of measurement time using non-contact diagnostic tools thermophysical data have been obtained in a temperature range and with a precision so far unattainable. It can be easily foreseen that these new experiment techniques can be extended to measure thermophysical properties of liquid materials which are of commercial interest and open a new research field regarding high precision thermophysical property measurements and their application for high precision numerical modeling of industrial solidification processes.

SPACE EXPERIMENTS DURING IML-2 AND MSL-1

The transient thermal response of a specimen to a variable heat input is determined by thermophysical properties such as the specific heat, thermal conductivity, thermal conductances to a heat bath and measuring devices and by the enthalpy and kinetics of phase transformations. Modulation calorimetry applied in a suitable range of modulation frequencies allows to separate the influence of the different thermal relaxation times on the thermal response of a specimen exposed to modulated heating power input and, as such, a direct determination of the specific heat from measured temperature variations. This preempts that the amplitude of power modulation in the temperature range of interest can be accurately measured and that the relevant thermal relaxation times differ sufficiently from the modulation time scale. Typically the latter condition is met in levitated metallic specimens because of the large difference in the rates of radiative heat loss and thermal transport within the sample [3].

In the present investigation the specific heat in the stable and undercooled melt of eutectic Zr-36at%Ni, Zr-24at%Fe and more complex Zr-Al-Ni-Cu-(Ti,Nb) alloys with melting points typically between 1100 and 1300 K have been investigated in microgravity experiments. These alloys form metallic glasses by rapid quenching methods and constitute the basis of the more complex 'bulk' glass formers. However, no accurate experimental values of the specific heat in the stable and undercooled melt are available near the eutectic temperature.

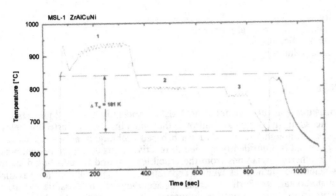

Fig. 1:   Typical experiment run during MSL-1 with processing in the stable and undercooled melt and recalescence.

The experiments were performed in the containerless processing device TEMPUS specifically designed for microgravity thermophysical experiments. Microgravity conditions are necessary for this experiment in order to avoid excessive heating effects associated with leviatation under 1g conditions thus allowing processing under ultrahigh vacuum conditions and a high degree of liquid undercooling. This experimental procedure further ensures stable quiescent sample conditions by the absence of the large stirring forces resulting from levitation under under 1-g conditions.

RESULTS

Thermophysical properties of liquid Zr-based alloys with near-eutectic compositions have been measured under containerless sample processing, ultra-high-vacuum and microgravity conditions applying a new A.C. modulation technique. As an example, some of the most interesting experimental results obtained during the MSL-1 mission are presented here. Primary ingots for processing in MSL-1 were prepared from low oxygen metals in an arc melter under a purified argon atmosphere. The final MSL-1 batch specimen had a typical oxygen contend of 400 ppm increasing to 600 ppm after processing in TEMPUS. The specimen was melted by a fast heating pulse and heated to equilibrium temperature in the stable melt while performing modulation calorimetry (1). Subsequently the sample was cooled by a fast cooling pulse to a temperature in the undercooled melt performing again modulation calorimetry at equilibrium temperature (2). This step was repeated (3) at another temperature plateau in the undercooled melt with subsequent undercooling and recalescence. With regard to nucleation the specimen exhibited excellent stability in the undercooled melt allowing for the first time extended specific heat and thermal expansion measurements in this regime. However due to increasing sample deformation as function of processing time in the liquid the cp measurement (3) had to be interrupted. The deformation is caused by an increasing rotational speed of the specimen and was quantitatively analyzed with the high resolution camera.

The raw temperature signal (2) with modulation frequency $\omega = 0.08$ Hz is shown in Fig. 2a. The signal is further processed by application of a drift correction to compensate for a change in the bias temperature before reaching temperature equilibrium at $t \approx 500$ sec. ΔT_{mod} is then obtained from subsequent Fourier analysis as shown in Fig 2b. In addition, ΔT_{mod} is calculated from a running average over modulation cycles after application of a 0.80 Hz low pass FFT filter to remove the high frequency modulation components caused by harmonic oscillations in the potential of the positioning field. ΔT_{mod} values obtained by both methods agree within better 0.1 K.

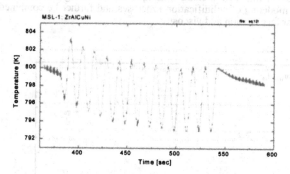

Fig. 2a: Raw temperature signal from cp measurement from Fig. 1

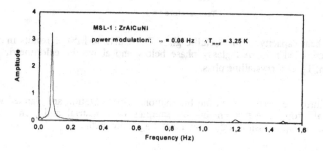

Fig. 2b: Fourier transform of temperature signal shown in Fig. 2a after performing a drift correction

As a particular result specific heat values of the $Zr_{65}Al_{7.5}Cu_{17.5}Ni_{10}$ specimen obtained in the space experiment for the stable and undercooled liquid are shown in Fig. 3. In addition the corresponding values in the crystalline phase and those obtained by heating the glassy phase above the calorimetric glass transition in a laboratory DSC using the step heating method are shown. The temperature range of the of the latter measurement is limited by the onset of crystallization. These results seem to indicate a nonmonotonous behavior of the liquid specific heat exhibiting a maximum in the undercooled regime.

Furthermore, the enthalpy of fusion ΔH_f can be evaluated from measurements of the duration of the recalescence or solidification plateau at small undercooling (< 10K). With the knowledge of the specific heat the total hemispherical emissivity in the solid and liquid state can be determined with high accuracy. Measurements of the change in rf-heater frequency and current during the phase transition together with a further analysis based on the time dependent

175

change of the coupling coefficients between sample and positioning coil (heater coil off) allows to obtain the fraction solid versus fraction liquid. Further analysis of the phase lag between modulation heating power and corresponding temperature signal gives the apparent thermal conductivity. As such, this new method is unique in measuring the thermophyical properties relevant for the modeling of solidification processes and further be combined with measurements of density, surface tension and viscosity.

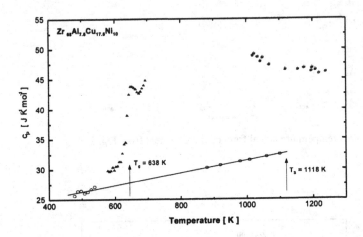

Fig. 3: Specific heat capacity of the metallic glass ZrAlCuNi. ●: MSL-1 results in stable and undercooled melt. Δ : glassy phase below and above the calorimetric glass transition, T_g. O : crystalline phase

In the future, further experiments on the International Space Station are planned which allow thermophysical property measurements on samples of industrial and technological interest using containerless processing and non-contact diagnostic methods.

ACKNOWLEDGEMENTS

The financial support from ESA (Topical Team Thermophysical Properties), DLR (DARA grant 50 WM 94-31-4) and DFG (G.F. Leibniz program) are gratefully acknowledged.

REFERENCES

[1] H.-J. Fecht, W.L. Johnson, *Rev. Sci. Instr.* **62** (1991) 1299.

[2] D.M. Herlach, B. Cochrane, I. Egry, H.-J. Fecht und L.A. Greer: „Containerless Processing in the Study of Metallic Melts and their Solidification", *Int. Mat. Rev.* **38** (1993) 273.

[3] R.K. Wunderlich, D.S. Lee, W.L. Johnson, H.-J. Fecht, *Phys. Rev. B* **55** (1997) 26.

MICROGRAVITY PROCESSING OF BIOPOLYMER/METAL COMPOSITES FOR NLO APPLICATIONS

C.M. CUTTLE*, M.V. CATTANEO**, J.D. GRESSER**, D.L. WISE**, D.O. FRAZIER***,
F. ARANDA****, and D.J. TRANTOLO**
*Northeastern University, Chemical Engineering, Boston, MA 02115
**Cambridge Scientific, Inc., 195 Common Street, Belmont, MA 02478-2909
***National Aeronautics and Space Administration - George C. Marshall Space Flight Center, AL
****Hanscom AFB, Bedford, MA

ABSTRACT

The overall objective of this project is the development of NLO-active materials with optical clarity and mechanical strength. These materials are intended for laser eye protection. By combining χ^2 and χ^3 optical properties, the intensity of incident laser radiation may be efficiently reduced. Using an in-plane poling technique, aligned films of liquid crystal poly(benzyl-L-glutamate), PBLG, were made which showed higher second harmonic generation (SHG) values compared to quartz. Silver sols in the 10-90 nm diameter size range were complexed with tricyanovinyl aniline, TCVA, resulting in composite PBLG/Ag Sol films with higher than at least an order of magnitude of χ^3 values materials such as polydiacetylenes and nitroanilines. These polymeric NLO materials offer definite advantages in terms of easy processability into films for the manufacture of the optical elements necessary for laser eye protection.

INTRODUCTION

The overall objective of this project is the development of NLO-active materials with optical clarity and mechanical strength. These materials are intended for laser eye protection. By combining χ^2 and χ^3 the intensity of incident laser radiation may be efficiently reduced. The target materials must not only possess reasonable NLO (i.e., χ^2 and χ^3) susceptibilities, but also be optically clear, thermally stable, and have high impact strength. To develop this material, a χ^3-active TCVA (tricyanovinyl aniline)/silver sol was processed with a χ^2-active polymeric host to develop a NLO-active polymer/metal ("polymet") composite.

Biopolymers may be aligned in an electric field which, when applied parallel to the surface of the polymer solution, results in an aligned film of reasonable area. The relationship between alignment and nonlinear optical (NLO) activity indicates the minimum degree of alignment required for χ^2 activity as measured by second harmonic generation (SHG). Small increases in alignment give rise to large increases in NLO activity. NLO activity, though, asymptotically approaches a limiting value within a given biopolymer system [13].

The dodecyl group of N-methyl-N-dodecyl-4-tricyanovinyl-aniline, TCVA, provides a surfactant-like property, which helps stabilize the complex between the dye and the silver sol (Figure 1). Adsorption of TCVA onto a stabilized silver sol was achieved by mixing a TCVA solution in acetone $(5.5 \times 10^{-3} M)$ with the aqueous silver sol $(2 \times 10^{-4} M)$ in the presence of polyvinyl alcohol, PVA. When the TCVA is combined with the silver sol and a PVA solution, the hydrophobic dodecyl group of the TCVA molecule attaches itself to the silver sol molecule as is shown in Figure 2.

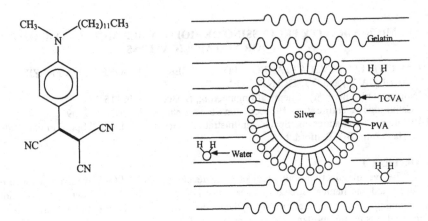

Figure 1. Molecular structure of the TCVA, tricyanovinyl-aniline, TCVA.

Figure 2. TCVA-PVA complex

EXPERIMENT

Materials

- Silver nitrate, ACS: Bradford Scientific, Inc., Epping, NJ, Cat. AG-530, Lot 317-1
- Poly-Vinyl-Alcohol (PVA), PolySciences, Cat. 15132, Lot 85610, MW 78,000, 88% mole hydrolyzed
- N-methyl-N-dodecyl-4-tricyanovinyl aniline (TCVA)
- Polycarbonate (Lexan™), Cat. 9034, 1/32" thick, Commercial Plastics, Inc.(Somerville, MA.)
- Gelatin (type A) from porcine skin, 300 Bloom strength (Sigma Chemical Co.)

Preparation of Polymet Composites

Colloidal dispersions of silver metals have previously been prepared by reduction of silver nitrate in ethanol 5×10^{-3}M [12]. An average silver particle size of about 25 nm and a distribution ranging from 5 nm to about 150 nm was obtained when the silver sol was prepared by this method. However, the mean particle size of silver in the PBLG film was greater than 100 nm as shown by Transmission Electron Micrography (TEM) (Figure 3). The large size particles are believed to be a result of agglomeration of smaller particles during the film formation process. The particle size of the silver needs to be less than 25 nm in order to avoid the refractive index (RI) mismatch between the host matrix and the secondary phase, which produces a loss of second order susceptibility, χ^2 [1][6].

To reduce the agglomeration, a different procedure for forming a stable silver sol was carried out using polyvinyl alcohol (PVA) as a stabilizing agent. Briefly, 100 milliliters of aqueous $AgNO_3$ (5×10^{-3}M) was added stepwise to 300 milliliters of vigorously stirred ice-cold aqueous $NaBH_4$ (2×10^{-3}M). Simultaneously, a 50 milliliter solution of 1% PVA was added stepwise during this reduction. The mixture was then boiled for ½ hour to decompose any excess $NaBH_4$.

Hydrogels were then formed by suspending the TCVA/PVA/Ag Sol in a 10% aqueous gelatin solution with an aligned PBLG film [11]. This composite was placed between two Polycarbonate sheets (Lexan 9034) and sealed using an inner silicon seal as shown in Figure 4.

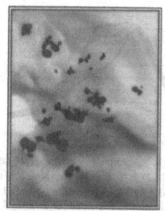

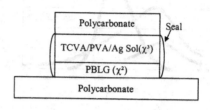

Figure 3. Transmission Electron Micrograph (Edge View) of Ag Sol particles prepared in Ethanol and suspended in PBLG/TCVA/Ag Film.

Figure 4. Layered Film Assembly of Composite Films

RESULTS

TEM of the sol with and without TCVA showed particles ranging in size from 10 nm to 90 nm (Figures 5 and 6). This Ag Sol absorbed light at $\lambda_{max} = 400\,nm$, indicating a particle radius of 10 nm (Figure 7A) [5].

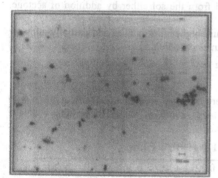

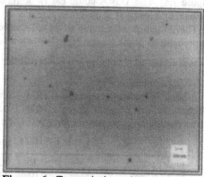

Figure 5. Transmission electron micrograph of PVA-stabilized Ag Sol particles prepared in water.

Figure 6. Transmission electron micrograph of PVA-stabilized Ag Sol particles prepared in water and coated with TCVA.

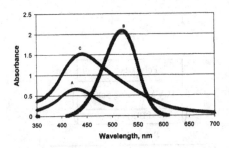

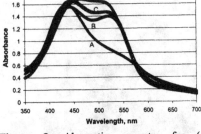

Figure 7. Absorption spectra for (A) 1×10^{-3}M PVA-stabilized silver sol only, (B) 5.5×10^{-5}M TCVA in acetone, and (C) 1.1×10^{-5}M TCVA and in the presence of 2×10^{-4}M Ag Sol after dilution of 1:5 dilution of A and B in water.

Figure 8. Absorption spectra for (A) 1×10^{-3}M TCVA and in the presence of 2×10^{-4}M Ag Sol after dilution of 1:5 dilution in water, (B) same as A after 1:5 dilution in Ethanol, (C) same as A after 1:5 dilution in acetone, and (D) same as A after 1:5 dilution in Dioxane.

Adsorption of TCVA onto the PVA-stabilized silver sol was achieved by simply mixing a 5.5×10^{-5}M solution of TCVA in acetone (Figure 7B) with a 2×10^{-4}M aqueous PVA-stabilized silver sol solution (Figure 7A). The complex formation between the TCVA and the silver sol was achieved instantaneously. Addition of the TCVA to the sol resulted only in some broadening of the spectrum of the sol (Figure 7C) with hardly any adsorption due to free dye in the solution (Figure 7B). The TCVA could easily be recovered from the sol either by addition of acetone, dioxane, or ethanol as shown in Figure 8.

The formation of a stable silver sol should ensure higher film quality and partially alleviate the problem of phase separation in the presence of an electric field.

Measurement of NLO Optical Susceptibilities: χ^2

- χ^2 *Determination*
  Second harmonic generation (SHG) values were obtained from the analysis of films with different TCVA concentrations. Table 1 summarizes some of the SHG results with the corresponding order parameter for a PBLG-only film.

Table I. SHG Signal of Pure PBLG Films

Order param.	0.35	0.37	0.32	0.49	0.36	0.53	quartz
SHG signal @ $\lambda = 1.5\mu m$	0.01	0.04	0.01	0.06	0.25	0.08	4.5
SHG signal @ $\lambda = 1\mu m$	0.18	.85	0.80	0.80	1.45	0.28	0.45

At 1 μm incident radiation the SGH signal (which is proportional to the χ^2 value) for PBLG was 1.45, which is higher than the quartz.

- χ^3 *Determination*
  χ^3 values for the films were determined by degenerate four wave mixing (DFWM). Figure 9 shows the schematic for the DFWM arrangement.

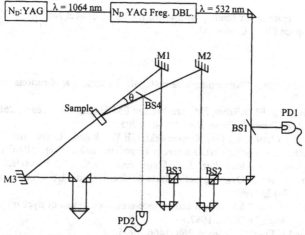

Figure 9. Schematic of Degenerate Four Wave Mixing (DFWM).
[M = Mirror, BS = Beam Splitter, PD = Photodetector]

The intensity of the conjugate beam is calculated using the following relation

$$I = \left(\frac{\omega}{2\varepsilon_o cn^2}\right)|\chi^3|^2 d^2 I_1(\omega)I_2(\omega)I_3(\omega)... \tag{1}$$

When the sample is compared to a reference sample of CS_2 under identical conditions, the χ^3 value of the sample can be calculated as follows.

$$\frac{\chi_{sample}^3}{\chi_{CS_2}^3} = \left(\frac{n_{sample}}{n_{CS_2}}\right)^2 \frac{d\alpha}{1-e^{-\alpha}} e^{\alpha d/2} x \left(\frac{I_{sample}}{I_{CS_2}}\right)^{1/2} \frac{d_{CS_2}}{d_{sample}}... \tag{2}$$

Composition PBLG	Wt Ratio TCVA/Ag	Average Thickness	Field kV/cm	%T @532 nm	χ^3 x 10^{-10} esu
100	1.5	30.5 μm	1.0	19.0	11.0

Literature values of χ^3 x 10^{-10} esu: Benzene = 0.048; Nitrobenzene = 0.0167;
Aniline = 0.003; o, m, p-Nitroanilines = 0.033= –0.197, Polydiacetylene = 0.37 – 1.6.

Figure 10. NLO characterization — χ^3 Values of TCVA/PBLG Polymets

CONCLUSIONS

Silver sols were produced which could form stable complexes with χ^3-active molecules such as N-methyl-N-dodecyl-4-tricyanovinyl aniline (TCVA). These χ^3 materials were then incorporated into a gelatin matrix and sealed in the presence of χ^2-active poly-L-benzyl glutamate (PBLG) films produced by an in-plane poling technique. The composite materials, made by adding χ^2 and χ^3 components, showed higher NLO properties than quartz and polydiacetylene.

ACKNOWLEDGMENTS

The authors are grateful for funding of this work under Contract No. NAS8-97006 from the NASA Marshall Space Flight Center, AL.

REFERENCES

1. L Beecroft and C Ober. "Nanocomposite materials for optical applications", Advance ACS Abstract (1997)
2. JH Golden, H Deng, FJ DiSalvo, JMJ Frechet and PM Thompson. *Science*, **268**, 1463, 1995.
3. A Hagfelt and M Ghratzel. *Chem. Rev.*, **95**, 49, 1995.
4. P Kitipichai, R LaPeruta, Jr., GM Koronowski, GE Wnek and I Gorodisher. "Synthesis and optical characterization of new NLO-active polyurethane and a silver colloidal suspension in a select polyurethane", *Mat. Res. Soc. Symp. Proc.,* Vol. 247, **Electrical Optical and Magnetic Properties of Organic Solid State Molecules**, eds. LY Chining, AF Garito, DJ Sandman pp. 117-123, 1992.
5. PC Lee, and D Meisel. "Adsorption and surface-enhanced raman of dyes on silver and gold sols", *J. Phys. Chem.*, **86**, 3391, 1982.
6. H Matsuda, and K Fukuda. *Science*, **268**, 1466, 1995.
7. MG Mogul, JD Gresser, DL Wise, GE Wnek and DJ Trantolo. "Second harmonic generation and laminar structures in poly (γ-benzyl-L-glutamate) films aligned in electric fields" in **Photonic Polymers: Fundamentals, Methods and Applications**, DL Wise, GE Wnek, DJ Trantolo, JD Gresser, TM Cooper, eds., Marcel Dekker, Inc., NY, 1998.
8. AMT Nguyen and AF Diaz. *Adv. Mater.*, **11**, 858, 1994.
9. Reetz, *Science*, **267**, 367, 1995.
10. S Schneider, P Halbig, H Grau, U Nickel, "Reproducible preparation of silver sols with uniform particle size for application in surface-enhanced raman spectroscopy"
11. DJ Trantolo, JD Gresser, DL Wise, MG Mogul, TM Cooper, and GE Wnek. "Electric field alignment of biopolymers for nonlinear applications," in **Optical and Photonic Applications of Electroactive Polymers**, SPIE, 2528, 219, 1995.
12. DJ Trantolo, MG Mogul, DL Wise, DO Frazier, JD Gresser. *Space processing of materials*, SPIE, **2809**, 106, 1996.
13. Wiese, H. and Horn, D., "Fiber optic quasielastic light scattering in concentrated dispersions: the on-line process control of carotenoid micronization", *Ber. Bunsenges. Phys. Chem.,* **97**, 1589-1597, 1993.

THE EFFECTS OF GROUND AND SPACE PROCESSING ON THE PROPERTIES OF ORGANIC, POLYMERIC, AND COLLOIDAL MATERIALS

DONALD O. FRAZIER*, MARK S. PALEY*, BENJAMIN G. PENN*, HOSSIN A. ABDELDAYEM*, DAVID D. SMITH*, WILLIAM K. WITHEROW*, AND WILLIAM E. CARSWELL*, MARIA I. ZUGRAV**
*NASA Marshall Space Flight Center, Space Sciences Laboratory, Huntsville, AL 35812
**Consortium for Materials Development in Space, University of Alabama in Huntsville, Huntsville, Alabama 35899

ABSTRACT

In recent years, a great deal of interest has been directed toward the use of organic materials in the development of high-efficiency optoelectronic and photonic devices. There is a myriad of possibilities among organic materials, which allows flexibility in the design of unique structures with a variety of functional groups. The use of nonlinear optical (NLO) organic materials as thin-film waveguides allows full exploitation of their desirable qualities by permitting long interaction lengths and large susceptibilities allowing modest power input.[1] There are several methods in use to prepare thin films such as Langmuir-Blodgett (LB) and self-assembly techniques,[2-4] vapor deposition,[5-7] growth from sheared solution or melt,[8,9] and melt growth between glass plates.[10] Organic-based materials have many features that make them desirable for use in optical devices, such as high second- and third-order nonlinearity, flexibility of molecular design, and damage resistance to optical radiation. However, processing difficulties for crystals and thin films has hindered their use in devices.

We discuss the potential role of microgravity processing of a few organic and polymeric materials. It is of interest to note how materials with second- and third-order NLO behavior may be improved in a diffusion-limited environment and ways in which convection may be detrimental to these materials. We focus our discussion on third-order materials for all-optical switching, and second-order materials for frequency conversion and electro-optics. The goal of minimizing optical loss obviously depends on processing methods. For solution-based processes, such as solution crystal growth and solution photopolymerization, it is well known that thermal- and solutal-density gradients can initiate buoyancy-driven convection. Resultant fluid flows can affect transport of material to and from growth interfaces and become manifest in the morphology and homogeneity of the growing film or crystal. Likewise, buoyancy-driven convection can hinder production of defect-free, high-quality crystals or films during crystal and film growth by vapor deposition.

INTRODUCTION

Third-Order Materials for Optical Switching

Optical-fiber communication systems have undergone stunning growth over the past decade. The technologies that have arisen in support of these systems have been incredibly fortuitous. The operating wavelength of erbium-doped amplifiers, for instance, serendipitously coincides with the minimum loss wavelength of fused silica fibers. But, even as fiber optic networks have been implemented on a universal scale, electronic switching is still the main routing method. Fibers have dramatically increased node-to-node network speeds (electronic

183

switching will limit network speeds to about 50 Gb/sec). However, it is apparent that terabit-rate speeds will soon be needed to accommodate the 10–15 percent/month growth rate of the Internet and the increasing demand for bandwidth-intensive data such as digital video.[11]

All-optical switching using NLO materials can relieve the escalating problem of bandwidth limitations imposed by electronics. Several important limitations need to be overcome, such as the need for high $\chi^{(3)}$ materials with fast response and minimum absorption (both linear and nonlinear), development of compact laser sources, and reduction of the switching energy. The goal of minimizing optical loss obviously depends on processing methods. For solution-based processes, such as solution crystal growth, electrodeposition, and solution photopolymerization, it is well known that thermal- and solutal-density gradients can initiate buoyancy-driven convection. Resultant fluid flows can affect transport of material to and from growth interfaces and become manifest in the morphology and homogeneity of the growing film or crystal. Likewise, buoyancy-driven convection can hinder production of defect-free, high-quality crystals or films during crystal and film growth by vapor deposition.

Second-Order Materials for Electro-Optic Applications

Applications of materials with second-order nonlinearity include frequency conversion, high-density data storage, and electro-optic modulators and switches. The first demonstration of second harmonic generation (SHG) was in quartz,[12] and it has traditionally been observed in inorganic crystals. A decade later it was demonstrated that the second-order nonlinearity may be several orders of magnitude larger in organic crystals possessing delocalized π-electron systems in which intramolecular charge transfer occurs between electron donor and acceptor substituents.[13] While organic materials may offer larger nonlinearities than inorganic crystals, the utilization of organic crystals is limited by the small number of molecules with large hyperpolarizabilities that have a noncentrosymmetric crystalline state (11 of the 32 crystal classes possess inversion symmetry and cannot be used as $\chi^{(2)}$ materials). Moreover, the maturity of inorganic crystal growth is relatively advanced, while that of organic crystal growth has not had the necessary time for comparable development.

Whereas the second-order nonlinearity in inorganic systems is a bulk effect ascribable to crystalline structure, the primary contribution to bulk nonlinearity for organic systems is due to the ensemble of nonlinearly responding molecules. Van der Waals forces between mers (molecular units) are small and the induced dipoles result most directly from the external field. This provides an added degree of flexibility for organic materials since the required asymmetry does not require crystalline structure but instead may be achieved in amorphous geometry. For example, instead of relying on the art of crystal growth, electric field poling of polymers containing the nonlinear chromophore may be used to induce macroscopic asymmetry. Alternatively, self-assembly or liquid-crystal ordering can achieve the required asymmetry, possibly encouraged in the environmental quiescence offered by reduced convection during microgravity processing.

Three promising classes of organic compounds for optical thin films and waveguides are polydiacetylenes (PDA's), which are conjugated zig-zag polymers, phthalocyanines (Pc's), which are large ring-structured porphyrins, and dicyanovinyl anilines (DCVA), which are donor/acceptor aromatic compounds. Epitaxial growth on ordered organic and inorganic substrates under various processing conditions have been useful for preparing highly oriented PDA and Pc films.[14–16] The degree of significance relating processing conditions to uniformity in thickness, degree of orientation, and optical properties for a specific processing technique is the general focus of work in this area.

A study on the effect of processing conditions relevant to thin-film deposition by various techniques is particularly difficult because of the possibility that convection may play a major role. It is a goal of some researchers to produce good quality anisotropic films; therefore, an important, yet understudied, requirement should be to assess the role of gravity during processing. This may be particularly true for the vapor deposition of diacetylenes where subsequent polymerization in the crystal is topochemical and occurs readily only when neighboring monomer molecules are sufficiently close and suitably oriented.[17] Likewise, this requirement is equally viable for the vapor deposition of Pc's in view of the results of microgravity experiments by 3M Corporation involving the preparation of thin films of copper Pc (CuPc).[15][16][18–26] Indeed, a variety of microstructural forms was obtained in thin films of CuPc, dependent on processing methods and conditions. Small changes in processing parameters caused large changes in molecular orientation within the film. Microgravity-grown CuPc had several desirable features which indicate that the growth of organic films in low gravity (low-g) may result in better quality films for optical and electrical applications.[25][26] It is also interesting that thin-ordered films of DCVA have resulted only from physical vapor transport processing in microgravity.

ORGANIC-BASED THIN FILMS FOR NONLINEAR OPTICAL APPLICATIONS

Gravitationally Sensitive Processing Techniques for Polydiacetylene Films

PDA's (fig. 1) are highly conjugated organic polymers that are of considerable interest because of their unique chemical, optical, and electronic properties.[27–30] This class of polymers has received extensive attention as organic conductors and semiconductors, as well as NLO materials. The high mobility of the π-electrons in the polymer backbone allows them to have large optical/electrical susceptibilities with fast response times. Single crystals of poly (2,4-hexadiyne-1,6-ditosylate), also known as PTS, possess one of the largest (possibly the largest) nonresonant third-order optical nonlinearities ever measured, on the order of 10^{-9} esu.[31] Because the origin of the nonlinearity is electronic, PDA's can have very fast response times, on the order of femtoseconds. They can be highly ordered, even crystalline, which is important for optimizing their electronic and optical properties, and they can readily be formed into thin films, which is the preferred form for many applications. Changing the functionality of the side groups, thereby making it possible to tailor their properties to meet specific needs, can vary the physical, chemical, and mechanical properties of PDA's. Thus, there is a great deal of interest in the use of these polymers for technological applications.

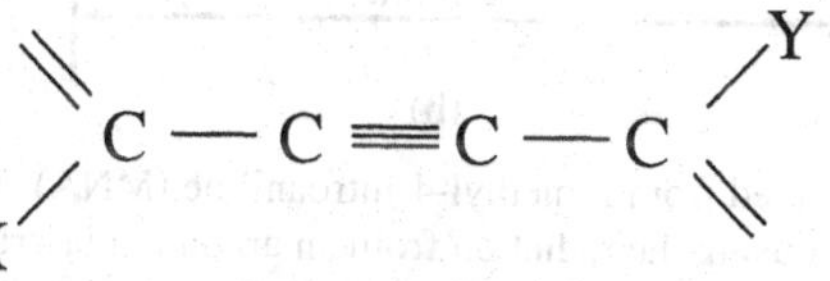

Figure 1. Structure of polydiacetylene repeat unit.

A novel technique, recently discovered, for growing PDA thin films involves exposing a transparent substrate in contact with diacetylene monomer solution to ultraviolet (UV) light.[32] A polymer film deposits on the side of the substrate in contact with monomer in solution, and there are distinct gravitational effects, which influence film quality. Good quality thin films elude growth from solutions absent of uniform flow fields and homogeneous temperature distributions near the substrate surfaces. Perhaps of immediate relevance, knowledge gained from low-g experiments may enable the optimization of growth conditions on Earth. From a device perspective, the UV technique makes construction of extremely complex waveguides possible. Utilizing a computer-controlled x-y translation stage programmed to trace out a desired pattern, researchers demonstrated that UV radiation (364 nm) from an argon-ion laser could trace out a test pattern[33] (fig. 2a). It is possible to construct a Mach-Zehnder interferometer with an optimized curvature using this technique.[34] After mounting a test cell containing diacetylene monomer solution on a translation table, a focused UV laser beam passing through the UV transparent surface of the test cell traced the desired paths to form the polymer-based optimized Mach-Zehnder waveguide (fig. 2b). Based on optical microscopy, and on refractive index measurements using waveguide mode analysis, these thin films of PDAMNA, a derivative of PDA containing 2-methyl-4-nitroaniline (MNA) moieties, have good optical quality, superior to that of films grown using conventional crystal-growth techniques. Considering the simplicity of photodeposition, this technique could make the production of PDA thin films for applications such as NLO devices technologically feasible. Hence the NLO properties of the PDAMNA films were investigated, specifically, their third-order NLO susceptibilities. Typically, $\chi^{(3)}$ values for PDA's can vary by several orders of magnitude, depending on the degree of resonance enhancement and other factors.[29 35] Preliminary measurements with the PDAMNA films using a Ti-Sapphire laser at 810 nm give $\chi^{(3)}$ values on the order of 10^{-11} esu, with response times on the order of femtoseconds.[36] There are no indications of either one- or two-photon absorption at this wavelength. To ascertain the true potential for device applications (the figures of merit), thorough measurements of light scattering, linear absorption, two- and three-photon absorption, damage thresholds, and so forth must be carried out.[37] However, it is quite promising that this UV technique makes construction of extremely complex waveguides possible.

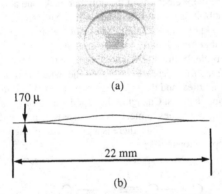

Figure 2. A PDA film derived from 2-methyl-4-nitroaniline (MNA) circuit photodeposited onto UV transparent substrates using the radiation from an argon ion laser: (a) Demonstration pattern on a quartz disk; (b) enlarged image of an actual Mach-Zehnder interferometer on a glass microscope slide.

It is well known that gravitational effects, such as buoyancy-driven convection, can affect heat and mass transport processes in solution.[38] Photodeposition of PDA films from solution is no exception. We shall first discuss how buoyancy-driven convection can arise during photodeposition of PDAMNA films from solution, and then describe how this convection can affect the morphology, microstructure, and properties of the films obtained. Both the monomer solution and the film generate heat due to absorption of UV radiation. The radiative heating, along with the thermal boundary conditions of the walls of the thin-film growth chamber, will give rise to a complex temperature pattern in the solution. Due to the lack of thermodynamic equilibrium, the solution will possess temperature and concentration gradients, and, therefore, density gradients. These gradients, under the influence of gravity, can induce convective-fluid flows in the solution (buoyancy-driven convection).

The onset of thermal convection is determined by a stability parameter known as the Rayleigh number, Ra, defined as[38]

$$Ra = \frac{\alpha g d^3 \Delta T}{\nu \kappa} \qquad (1)$$

where α is the coefficient of thermal expansion of the solution; g is the acceleration due to gravity; ΔT is the temperature difference across distance, d, in the solution; ν is the kinematic viscosity; and κ is the thermal diffusivity. For photodeposition of PDAMNA films, the value of ΔT (over a distance of <1 mm) can vary from a only a few tenths of a degree to several degrees, depending on the intensity of the UV radiation. In order to grow thicker films (>1 μ), higher intensity radiation is necessary, making large temperature gradients unavoidable. The intensity and flow pattern of convection can be predicted when the Rayleigh number is known. For instance, for an infinite fluid layer in the horizontal direction with a temperature gradient in the vertical direction (colinear with gravity), convective motion will occur in the form of rolls with axes aligned horizontal when $Ra > 1,708$ (the critical Rayleigh number), while no convection will occur if $Ra < 1,708$.[39] The exact critical value can only be determined by numerical solution of the fluid flow in the chamber. In the case of horizontal temperature gradients (orthogonal to gravity), all values of the Rayleigh number lead to convection, and the magnitude of the velocity of the fluid flow is proportional to the square root of the Rayleigh number.

Density gradients can also arise in the solution due to variations in the concentrations of the chemical species present in the solution. Variations in the monomer concentration are caused by depletion of monomer from the solution at the surface of the growing film and in the bulk. Also generation of dimers, trimers, and other soluble byproducts in the bulk solution may result in additional concentration density gradients. Such solutal gradients, along with the temperature gradients, can give rise to double-diffusive convection. This complicated convective motion is usually analyzed with the aid of the solutal Rayleigh number, in addition to the thermal Rayleigh number.[38] The solutal Rayleigh number, Ra_S, is defined as

$$Ra_s = \frac{\beta g d^3 \Delta C}{\nu D} \qquad (2)$$

Where β is the coefficient of concentration expansion, ΔC is the concentration difference across distance, d, in the solution, and D is the diffusion coefficient. Double-diffusive convection flows can be far more complex than simple thermal-convection flows. Hence, we see that convection can arise by several means during PDA film photodeposition from solution. The extent of convection, and its intensity and structure, can only be understood through accurate numerical modeling of the fluid motion and thermodynamic state of the system.

One significant effect of convection can be seen when PDAMNA films grown in 1-g are viewed under an optical microscope: they exhibit small particles of solid polymer embedded throughout. These form when polymer chains in the bulk solution collide due to convection and coalesce into small solid particles on the order of a few hundredths of a micron in size. Because these particles are so small, almost colloidal in nature, they do not sediment out readily and thus remain suspended in the bulk solution. Convection then transports these particles to the surface of the growing film where they become embedded. These particles are defects that can scatter light and thus lower the optical quality of the films.

To study the effects of convection on the occurrence of these particles in the films, the growth chamber was placed in different orientations with respect to gravity in order to vary the fluid flow pattern.[40] PDAMNA films were grown with the chamber vertical (irradiating from the top) and with the chamber horizontal (irradiating from the side). In the case when the chamber is vertical and the solution is irradiated from the top, the axial temperature gradient is vertical with respect to gravity, and the bulk solution is stably stratified because warmer, less dense solution is above cooler, more dense solution. Thus in this orientation, convection should be minimized. In the case when the chamber is horizontal and the solution is irradiated from the side, the axial temperature gradient is horizontal with respect to gravity, which makes the density gradients less stable. Hence convection should be much more pronounced in this orientation. Numerical simulations of the fluid flow are consistent with these expectations.[40] This result is reflected in the distribution of solid particles observed in the PDAMNA films grown in the two different orientations. Films grown with the chamber horizontal clearly contain a greater concentration of particles than films grown with the chamber vertical (fig. 3). This is consistent with expectations based on the relative amounts of convection in the two orientations; films grown under increased convection contain more particles than those grown under less convection. Also, wave-guiding experiments with these films demonstrate that the films containing more particles exhibit greater light scattering than those containing fewer particles do.

Note that even the film grown in the vertical orientation, where convection is minimized, still contains particles. Thus while convection is lessened, in this case, it is not eliminated. There are two reasons for this. First, there are still radial thermal-density gradients in the horizontal direction even when the chamber is vertical because the solution near the side walls is cooler than that near the center; these can give rise to convection. Also, because the substrate is transparent to UV light, it is not heated directly by the radiation—the only means by which it receives heat is via conduction. Thus, initially, there will be some heat flow from the warm solution to the cooler substrate, producing a very shallow, unstable thermal-density gradient in the immediate vicinity of the substrate/solution interface, which sits above the stably stratified bulk solution. Any convection initiated in this unstable layer may penetrate deeper into the stable layer below, giving rise to the phenomenon of penetrative convection.[41] The bottom line is that even under optimum conditions in 1-g, convection is still present during PDA thin-film photodeposition from solution, causing particles in the films.

Not only can convection affect the transport of particles which are polymerized from the bulk solution, it can also transport colloidal particles which are purposely introduced into the system to alter the optical properties. The study of the effect of composite geometry on the NLO properties of materials is an active area of research. One of the most commonly studied systems involves small metal inclusion particles surrounded by a continuous host medium. The linear optical properties of these composites are described by the theory of Maxwell Garnett.[42] In recent years the model has been extended to include nonlinear materials.[43 44] Unfortunately, assimilation of metal particles into highly nonlinear solid-state materials is difficult using traditional chemical techniques because most highly nonlinear polymeric host candidates require organic solvents which are compatible with the colloid. Ion implantation is a realistic alternative but is not readily available in most laboratories. Moreover, the distribution of metal in the axial direction follows a

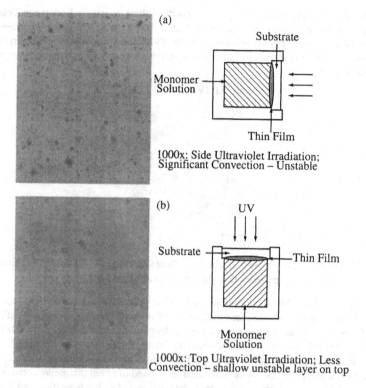

(a)

Substrate

Monomer Solution

Thin Film

1000x: Side Ultraviolet Irradiation; Significant Convection – Unstable

(b)

UV

Substrate

Thin Film

Monomer Solution

1000x: Top Ultraviolet Irradiation; Less Convection – shallow unstable layer on top

Figure 3. PDAMNA films grown in two different orientations: (a) Films grown with the chamber horizontal; (b) films grown with the chamber vertical.

Gaussian profile. Recently, however, Brust et al.[45] functionalized gold particles with thiol groups which serve to protect the particles from solvent degradation and attack. Interestingly, the colloidal metal can actually be dried and stored without agglomeration. The thiol groups in this case have little effect on the optical properties of the colloid. These thiol-capped metal particles can then be resuspended in many organic solvents and incorporated into the PDAMNA films through photopolymerization. Hence, this recipe offers a simple way to introduce small metal particles into highly nonlinear polymers. The resulting films, however, suffer from gradients in the metal concentration, which are clearly visible under reflected room light. Moreover, the films have a much higher concentration of metal than would be expected from a diffusional process. Hence, these gradients most likely arise from the process of convection. Convection often tends to destroy the homogeneity of particle-doped systems, whether it is colloidal metal films or doped porous glass. In this case, the metal particle dopants serve not only to modify the NLO properties of the system, but also to elucidate the role of convection in the formation of polymeric thin films. Additionally, convection can also affect film deposition at the molecular level; specifically, the molecular orientation of the PDA chains.[46] Preliminary studies conducted using atomic force microscopy and x-ray photoelectron spectroscopy indicate that films photodeposited onto quartz (in 1-g) for very short duration of time (a few minutes) exhibit good polymer chain alignment in the direction normal to the substrate. Films grown for longer duration show significantly poorer chain alignment. Thus, in the early stages of deposition, there appears to be some tendency for orientation. During an experiment recently conducted aboard the Space Shuttle *Endeavour* (CONCAP IV-3) in which photodeposition of PDAMNA films from solution was carried out in microgravity,[40] the best space-grown film clearly exhibits fewer particles than the best ground-based films (fig. 4).

CONCAP IV-3 PTFG Cell 9

100 μm

Figure 4. PDAMNA films grown in space. The best space-grown films clearly exhibit fewer particles than the best ground-based films.

More recently, PDA's have been investigated as potential second-order NLO materials. Theoretical calculations have indicated that certain PDA's could possess extremely high second-order NLO susceptibilities; e.g., molecular hyperpolarizabilities on the order of $1,000 \times 10^{-30}$ esu.[47] In order to make use of this second-order nonlinearity, it is necessary to orient the polymers into acentric structures, either crystals or thin films. This is not trivial; many compounds, which have desirable properties at the molecular level, tend to orient themselves centro-symmetrically in the bulk to minimize electrostatic interactions. SHG has been observed from certain asymmetrical liquid-crystalline diacetylene monomers (although, interestingly, not from the corresponding polymers)[48] from both LB and self-assembled PDA monolayer and multilayers[49,50] and even from a spin-coated PDA film.[51] Lastly, powder SHG efficiencies comparable to that of 2-methyl-4-nitroaniline have been obtained from vapor-deposited polycrystalline films of a PDA possessing MNA as a side group.[47] If the crystallites are partially aligned by growing the films quasi-epitaxially onto prealigned poly (tetrafluoroethylene) substrates, the SHG efficiency increases by almost one order of magnitude.

Photodeposition from solution can potentially provide a convenient means of obtaining PDA films for a variety of applications. However, once formed, PDAMNA films (and for that matter, most PDA's) are quite intractable materials; they are rigid and brittle, cannot be melted without decomposing, and are usually insoluble. In order to develop materials that can be more readily processed, we have begun investigating blends (mixtures) and copolymers of PDA's with other materials. One class of polymers that offers excellent optical transparency and is easily processible is polymethacrylates, such as poly(methy methacrylate); PMMA, i.e., Plexiglas™. Currently we are synthesizing, characterizing, and preparing films of both blends and copolymers of PDAMNA with PMMA. In this copolymerization study, we address problems of malleability, alignment, and stability. Previous work on other methacrylate/NLO dye systems has shown that homopolymers are often difficult to work with, but copolymers are not.[52]

Another promising application of photodeposition of PDAMNA films is holographic information storage. Holograms can be made readily from PDAMNA using a standard side-band holographic setup, as has been demonstrated in our laboratory. The laser source was an argon ion laser with a wavelength of 365 nm. A beam splitter split the output beam into a reference beam and an object beam. The object was a transparent plastic with the letters NASA printed on it. The two beams were incident on a test cell containing a quartz window and the DAMNA monomer solution. Interference took place at the quartz window/solution interface to create the complex diffraction grating that makes up a hologram, and PDAMNA was deposited on the quartz window to record the hologram. A helium-neon laser with a wavelength of 632.8 nm was then used to reconstruct the PDAMNA holograms. The reconstructed hologram reproduced the object wave front, recreating the NASA lettering. Because photodeposition is not reversible, the holograms produced are permanent.

We want to briefly mention one other promising use for PDAMNA blends and copolymers besides thin films—namely, NLO fibers. Because these DAMNA-polymethacrylate-based materials are soluble in organic solvents, and in some cases can be melted without decomposition, they are readily amenable to being drawn into fibers by either melt or solution extrusion techniques. Fiber optics are playing an increasingly important role in the telecommunications industry, and the ability to readily produce a high quality optical fiber that has good NLO properties would be of value.

Finally, one of the most promising methods potentially benefiting from microgravity processing is physical vapor transport. The standard techniques for obtaining PDA thin films involve the growth of crystalline diacetylene monomer films or the deposition of LB films, followed by topochemical polymerization of these films in the solid state to yield ordered PDA films. This ability to undergo solid-state polymerization is a very intriguing property of diacetylenes; in principle, one can start with a single crystal monomer and obtain a single crystal polymer.[7,53] However, the process is not trivial. The formation of high-quality, crystalline diacetylene monomer films or LB films can be very tedious and difficult, and, furthermore, not all monomers polymerize readily in the solid state.[8] A more commonly employed technique for obtaining diacetylene monomer films is vapor deposition. One of the chief limitations to vapor deposition of high quality monomer films (e.g., single crystalline films with good molecular orientation and few defects) has been a lack of understanding of how the processing conditions affect monomer film growth. It is certainly well known that parameters such as temperature, pressure, concentration, and so forth can affect vapor transport processes. Until recently, the influence of gravity was often tacitly ignored.

The effects of gravity, such as buoyancy-driven convection, can greatly influence heat and mass transport during growth by physical vapor transport, and thereby influence all of the aforementioned growth parameters. Thus discerning the effects of gravity (or the lack thereof) could play a critical role in optimizing the growth of high-quality PDA films by vapor deposition. A method that we have attempted for assessing convection in a gas phase is to perform the computation at low pressure. This relieves the need for specific materials constants.[54,55] It is important to note that buoyancy effects are possible only if the molecular mean free path is short enough relative to cell dimensions. In this case, molecular flows are not in the free molecular flow regime. A mathematical model has been developed to determine buoyancy-driven heat transfer in an ideal gas under a variety of orientations relative to gravitational accelerations.[54] The model demonstrates that convection can occur at total pressures as low as 10^{-2} mm Hg in cells having relatively high length-to-width ratios. A preliminary experimental test of the model involved deposition of the diacetylene monomer, 6-(2-methyl-4-nitroanilino)-2, 4 hexadiyn-1-ol (DAMNA) (fig. 5), at an evacuation pressure of 10^{-2} mm Hg for 30 minutes. Cell construction was as depicted in figure 6 with source temperature of 120 °C and sink temperature of 30 °C.

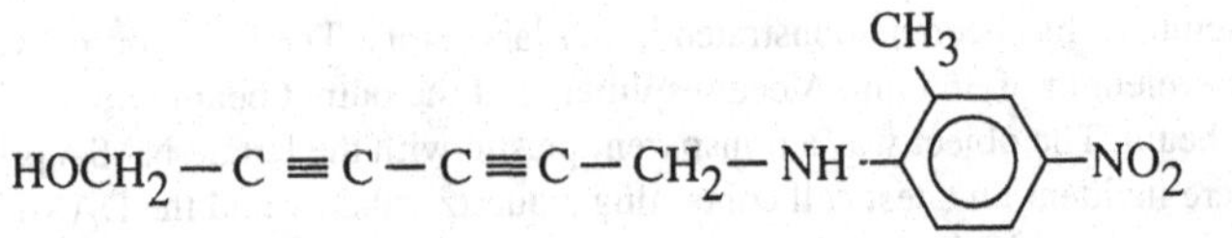

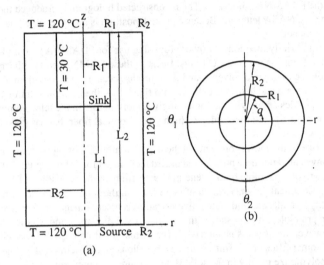

Figure 5. Diacetylene monomer (DAMNA).

(a)

(b)

Figure 6. Vapor-deposition cell for physical vapor transport of DAMNA.

Comparison of figures 7 and 8 and figures 7 and 9 shows the differences in induced flow fields in the $\theta_1 = 0°$–$180°$ and $\theta_2 = 90°$–$270°$ planes, respectively, driven by vertical and tilted cavities. There are two recirculation flows induced by vertical and tilted orientations. For the vertical orientation, the flow depicted in figure 7 is axisymmetric, hence also representative of flows in the θ_2 plane. These flows are Benard-type flows driven by counter-rotating cells (clockwise on the left and counter-clockwise on the right) due to an incipient instability in the narrow cylindrical cell cavity. For the tilted cell, a resultant asymmetric flow in the θ_1 plane approaches an antisymmetric flow profile resulting from differential heating between vertical and opposite walls of the cavity. The two recirculation flows for this orientation appear in the θ_2 plane (fig. 9). This asymmetric three-dimensional flow profile is quite complex and represents a greater degree of convection in the cell cavity for the tilted orientation than for the vertical orientation.

Experimentally, it is helpful to utilize Beer-Lambert's relationship for transmission of radiation through a medium to test gravitationally sensitive flow pattern predictions.[54] We would expect that the flow pattern in figure 9 (cell tilted at 45° relative to gravity vector) would affect film quality over most of the deposition surface differently than the flow pattern depicted in figure 7 (cell oriented vertically). A peliminary experiment suggests this to be the case with respect to film thickness. Table I contains representative absorbance intensities from scanning an approximately 1-cm-diameter spot in similar vicinities of films formed during vapor deposition

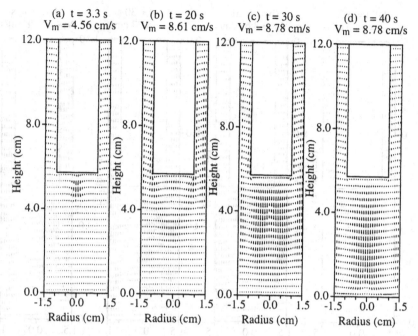

Figure 7. Axisymmetric flow in the $\theta_l = 0°-180°$ plane in a cell in which deposition occurs with the cell oriented vertically.

in vertically and obliquely oriented cells onto 15-mm diameter disks. Visually, the films from the obliquely oriented cells are a deeper yellow color than those from vertically oriented cells. We qualitatively assume generally thicker films from the relative appearances. The surfaces of these monomer films are translucent and microcrystalline. At the present time, it is also preferable for us to discuss relative film thickness from beam attenuation in qualitative terms (although more quantitative than visual observation), while observing that the spectroscopic irradiation spot sizes are large enough to average over about 45 percent of the film surfaces. The same relative result occurred in that beam attenuation was greater at each maximum absorbance wavelength repeatedly. More convection in cells having the tilted orientation is apparently responsible for correspondingly greater film thickness.

Gravitationally Sensitive Processing Techniques for Phthalocyanines

The phthalocyanines are excellent candidates for use in developing NLO devices because of their two-dimensional planar π-conjugation, better chemical and thermal stability than most other organic materials, and ease of derivatizing through peripheral and axial positions (fig. 10). Large[56-65] and ultra-fast[66-68] third-order nonlinearities have been demonstrated for Pc's. Thin films of Pc for fabrication of waveguides can be obtained by physical vapor transport (PVT) because of their exceptional thermal stability and ease in sublimation. Matsuda et al.[59] observed that Pc's with axial ligands, for example, vanadyl phthalocyanine (VOPc), have higher $\chi^{(3)}$ values than

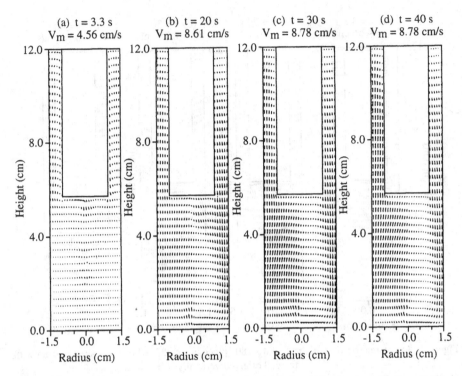

Figure 8. Flow in the $\theta_l = 0°–180°$ plane in a cell in which deposition occurs with the cell tilted 45° to the vertical axis.

most usual unsubstituted Pc's. The $\chi^{(3)}$ values of thin films of Pc vacuum deposited at 10^{-4} Pascal onto fused quartz are shown in table II. The maximum values for the unsubstituted Pc's at 1.9 μm were $1.5×10^{-12}$ esu for CuPc and $0.8×10^{-12}$ esu for nickel phthalocyanine (NiPc). In comparison, chloro-indium Pc and vanadyl Pc had $\chi^{(3)}$ values of $1.3×10^{-10}$ esu and $3×10^{-11}$ esu, respectively. Ho et al.[56] grew films of chloro-gallium (GaPc-Cl) and fluoro-aluminum (AlPc-F) Pc's onto fused silica flats at 150 °C and 10^{-6} torr. The $\chi^{(3)}$ values for GaPc-Cl and AlPc-F were $5×10^{-11}$ and $2.5×10^{-11}$ esu at 1,064 nm for thickness of 1.2 μm and 0.8 μm, respectively.

Wada et al.[69] measured a $\chi^{(3)}$ of $1.85×10^{-10}$ esu at 1.907 μm for a 51.4-nm-thick film of VOPc vacuum deposited onto quartz. The $\chi^{(3)}$ values for two different phases in vanadyl and titanyl phthalocyanines (TiOPc's) were measured by optical third-harmonic generation at wavelengths of 1,543 nm and 1,907 nm by Hosoda et al.[70] The transformation of as-prepared Pc films from phase I to phase II was performed by thermal annealing and was accompanied by a red shift in absorption spectra and an increase in $\chi^{(3)}$ values of 2–3 times. $\chi^{(3)}$ values for as-prepared films of VOPc and TiOPc were $3.8×10^{-11}$ esu and 10^{-11} esu, while the annealed films had values of $8.1×10^{-11}$ esu and $4.6×10^{-11}$ esu, respectively. Recently, SHG was reported for vacuum-deposited CuPc films which possess inversion symmetry. Chollet et al.[71] measured a $d_{eff} ≈ 2×10^{-19}$ esu at 1.064 μm fundamental wavelength for films with thickness ranging from 50 to 500 nm. These films were prepared at a pressure of 10^{-6} torr and source temperature of 120 °C. The films were homogeneous and partly oriented with a relatively large distribution of molecular axes, oriented

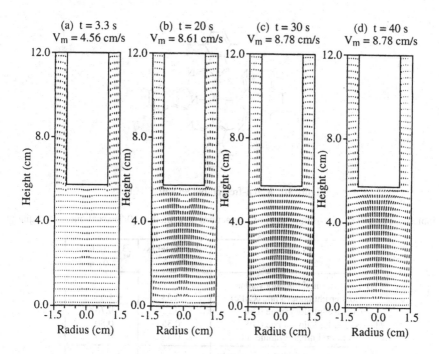

Figure 9. Flow in the $\theta_2 = 90°-270°$ plane in a cell in which deposition occurs with the cell tilted 45° to the vertical axis.

Table I. Representative absorbance intensities from scanning an approximately 1-cm-diameter spot in similar vicinities of DAMNA films formed during vapor deposition.

λ (nm)	A_{max} (cell vertical film)	A_{max} (cell tilted film)
422	0.200	0.385
214	0.395	0.546

almost perpendicular to the substrate. This order was confirmed by SHG measurements, which was attributed to quadrupolar or dipolar origins. Other researchers have observed relatively strong SHG during low-pressure deposition of CuPc by PVT.[72 73]

The research of Debe et al.[74-76] indicates that better quality organic thin films for use in NLO devices might be obtained by closed-cell PVT in microgravity. The advantage of thin-film growth in microgravity is that it provides the opportunity to eliminate buoyancy-driven convection during higher pressure processing. Recent reports[74-76] of the Space Shuttle mission STS–51 of August/September 1985 include results of experiments in which PVT epitaxially deposited CuPc onto highly oriented seed films of metal-free Pc (H_2Pc). The substrate was a 1.4-cm-diameter solid-copper disc. The substrate seed film was prepared on Earth by vacuum sublimation of H_2Pc onto a copper disc at a temperature range known to produce highly oriented films.[76] The high-pressure microgravity-grown CuPc films had several desirable features, which indicate

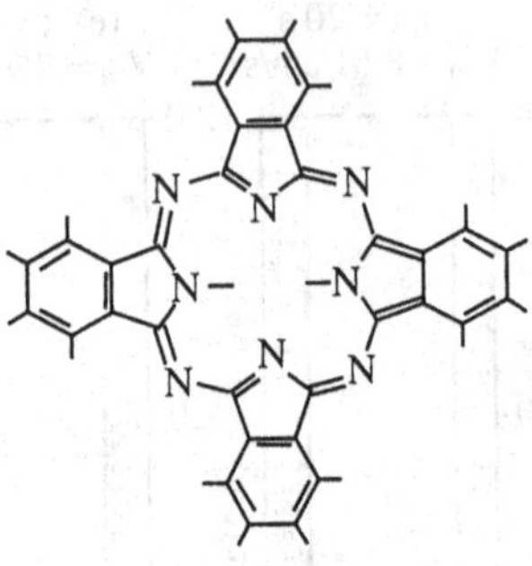

Figure 10. Metal-free phthalocyanine (H₂Pc).

Table II. The $\chi^{(3)}$ values of thin films of Pc vacuum deposited at 10^{-4} Pascal onto fused quartz.

Compound	Film Thickness (μm)	$\chi^3 \times 10^{-12}$ esu (1.9μm)
a) Unsubstituted Phthalocyanine		
Copper Phthalocyanine (CuPc)	0.53	1.5
Cobalt Phthalocyanine (CoPc)	0.22	0.76
Nickel Phthalocyanine (NiPc)	0.35	0.80
Platinum Phthalocyanine (PtPc)	0.41	0.60
b) Unsubstituted Phthalocyanine with Axial Ligands		
Vanadyl Phthalocyanine (VOPc)	0.28	30
Titanyl Phthalocyanine (TiOPc)	0.26	27
Chloro-aluminum Phthalocyanine (ClAlPc)	0.26	15
Chloro-indium Phthalocyanine (ClInPc)	0.14	130

that the growth of organic films in low-g may result in better quality films for NLO applications. 3M researchers report these films to be randomly oriented when epitaxially deposited onto H₂Pc films when processing occurs in 1-g (fig. 11b), but highly oriented and densely packed following microgravity processing by the same method (fig. 11a).[25 26 74-76]

As stated earlier, electric field poling of polymers containing the nonlinear chromophore may be used to induce macroscopic asymmetry. However, in the case of the ring-structured macromolecule, phthalocyanine, poling is not generally an option to maximize ordering in deposited films. It is, therefore, beneficial to achieve, if possible, self-assembly to induce required asymmetry by other means. Whenever self-assembly might occur in molecules not prone to poling, exploitation of conditions favorable toward asymmetry could prove beneficial. In the case of Pc, for example, $\chi^{(3)}$ enhancement and possible $\chi^{(2)}$ inducement might result from self-assembly during microgravity processing. The benefit may lie in either transfer of knowledge to improve ground-based processing or development of benchmark devices for setting realistic technology goals.

Abdeldayem et al.[77], recently observed intrinsic optical bistability in vapor-deposited thin films of metal-free Pc, ranging in thickness from 40 to 800 nm, using continuous wave (CW) and

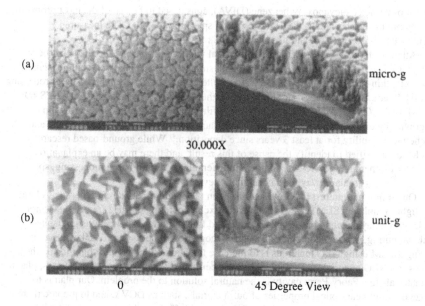

(a)

micro-g

30,000X

(b)

unit-g

0   45 Degree View

Figure 11. Copper phthalocyanine films epitaxially vapor deposited onto copper substrate: (a) μ-g deposition of CuPc epilayer; (b) 1-g deposition of CuPc epilayer. (Courtesy of 3M Corporation.)

chopped He-Ne lasers at 633 nm. Source and substrate temperatures were maintained at 300 °C and 5 °C, respectively, while vapor-vacuum deposition occurred at 10^{-6} torr onto quartz disks. Bistability in the film was attributed to changes in the level of absorption and refractive index caused by thermal excitation. Technology potential was demonstrated for a relatively low speed AND logic gate with millisecond response time in a 230-nm-thick metal-free phthalocyanine film. The estimated third-order nonlinear susceptibility measurements by four-wave mixing using pulsed Nd:YAG laser at 532 nm was on the order of 10^{-8} esu. This relatively large value is attributed to both resonant and thermal mechanisms that might be present in the system at this wavelength.

Gravitationally Sensitive Processing of N,N-dimethyl-p-(2,2-dicyanovinyl) aniline (DCVA)

DCVA is a donor/acceptor-substituted aromatic compound that, because of its donor/acceptor structure, has a highly polarizable dissymmnetric π-electron cloud. Although such compounds may exhibit high values of molecular optical nonlinearities, β, only those compounds that form acentric crystals (space group $P2_1$),[78] as is the case for DCVA, are expected to have high χ^2 values required for consideration in electro-optic devices. Orientation of the molecular dipole moment in DCVA crystals is close to optimal for manifestation of a large nonlinear response in the solid state. Measurement of the vector part of χ^2 along the dipole moment direction, using nonconventional electric field-induced second harmonic generation (EFISH) in 1,4-dioxane, gives a value of 145.5 ±9.1×10^{-51} cm^3/V^2. Static second-order polarizability, β, of DCVA is calculated to be 88.57×10^{-51} cm^3/V^2. These values are several times larger than the corresponding values for o-methoxy-dicyanovinylbenzene (DIVA), considered to be one of the most promising of the organic materials and has the acentric crystal space group, $P2_1$.[79] The measured χ^2 for DIVA in $CHCl_3$ is 18.71×10^{-51}, and its calculated β is 12.35×10^{-51} cm^3/V^2.[80]

Attempts to grow films of DCVA by physical vapor transport on Earth have not been successful to date. Efforts to do so yield small bulk single crystals and polycrystalline islands. Surprisingly, thin-ordered films of DCVA have resulted from physical vapor transport processing aboard the Space Shuttle Endeavour during two flights (STS-59; CONCAP IV-2 and STS-69; CONCAP IV-3). Oriented and ordered DCVA thin films that were reproducibly grown in microgravity by effusive ampoule physical-vapor transport have shown thermodynamic and photochemical stability for at least 3 years since deposition.[81] While ground-based research efforts have intensified to identify the cause of this result, and there may be an explanation consistent with the microgravity environment transferable to ground-based processing, space-flight preparations are underway for a benchmark device.

One important interest in NLO materials is in their light modulation properties. A large spatial light modulator (SLM) array (1000×1000 pixels) operating at MHz frame rates is not possible with today's technology. Such arrays would find important applications in displays, optical computing, optical interconnects, pattern recognition, robotics, etc. The key issue in designing a SLM using NLO materials is to increase the interaction length of light inside the materials. As NLO materials can only be spun at a certain thickness on typical substrates, placing NLO materials in a Fabry-Perot cavity is the natural solution to the problem. Our plan is to optimize the nonlinear optical properties of our materials, such as DCVA, and to prepare them in a form suitable for building spatial-light modulators. Our efforts will include microgravity processing by physical-vapor transport on appropriate substrates to exploit the aforementioned anomaly with respect to DCVA, and to employ the photodeposition process on modified DAMNA onto similarly prepared substrates. The space environment may minimize scattering centers in photodeposited films while possibly inducing orientation of the polymer chains.

CONCLUSIONS

Organic materials such as conjugated polymers, although less developed than semiconductor systems, are highly promising because they have generally larger nonlinearities, are more facile for molecular tailoring, and are more malleable than their inorganic counterparts. It is reasonable to consider that microgravity can play an important role in the formation of organic and semiorganic crystals for second- and third-order applications. In the diffusion-limited regime of space, larger and more defect-free crystals may be grown, hence improving optical transparency and conversion efficiency. By different processing techniques, improved optical quality and molecular alignment have been observed in PDA and Pc thin films processed in microgravity or reduced convection environments. Indeed, one material to date, DCVA, with a potentially very high electro-optic coefficient, promises to reveal through space-based processing methods for improving on processing techniques. Perhaps of more immediate relevance, knowledge gained from low-g experiments may enable the optimization of growth conditions on Earth.

REFERENCES

1. B.K. Nayar and C.S. Winter, Optical and Quantum Electronics, **22**, 297 (1990).
2. G.M. Carter, Y.J. Chen, and S.K. Tripathy, Appl. Phys. Lett., **43**, 891 (1988).
3. F. Kajzar, J. Meissier, J. Zyss, and I. Ledoux, Opt. Commun., **45**, 133 (1983).
4. F. Kajzar, and J. Messler, Thin Solid Films, **11**, 132 (1988).
5. M.K. Debe and K.K. Kam, Thin Solid Films, **186**, 289 (1990).
6. D.O. Frazier, B.G. Penn, W.K. Witherow, and M.S. Paley, SPIE Crystal Growth in Space and Related Diagnostics, **1557**, 86 (1991).
7. G.Z. Wegner, Naturforsch, **246**, 824 (1969).
8. M. Thakur, and S. Meyler, Macromolecules, **18**, 2341 (1985).
9. M. Thakur, G.M. Carter, S. Meyler, and H. Hryniewicz, Polymer Preprints, **27(1)**, 49 (1986).
10. I. Ledoux, D. Josse, P. Vidakovic, and J. Zyss, Optical Engineering, **27(1)**, 49 (1986).
11. M.N. Islam, Phys. Today, 34 (May 1994).
12. P.A. Franken, A.E. Hill, C.W. Peters, and G. Weinreich, Phys. Rev. Lett., **7**, 118 (1961).
13. B.L. Davydov, L.D. Derkacheva, V.V. Dunina, M.E. Zhabotinskii, V.F. Zolin, L.G. Koreneva, and M.A. Samokhina, Opt. Spectrosc., **30**, 274 (1971).
14. M.S. Paley, D.O. Frazier, S.P. McManus, S.E. Zutaut, and M. Sangahadasa, Chem. Mater., **5**, 1641 (1993).
15. M.K. Debe, Prog. Surf. Sci., **24(1-4)**, 1 (1987).
16. C.J. Liu, M.K. Debe, P.C. Leung, and C.V. Francis, Appl. Phys. Comm., **11(2–3)**, 151 (1992).
17. G.I. Stegeman and A. Miller, in *Photonics and Switching*, edited by J.E. Midwinter (Academic Press, London, 1994), pp. 81–145.
18. K.K. Kam, M.K. Debe, R.J. Poirier, and A.R. Drube, J. Vac. Sci. Technol., **A5(4)**, 1914 (1987).
19. M.K. Debe, K.K. Kam, C.J. Liu, and R.J. Poirier, J. Vac. Sci. Technol., **A6**, 1907 (1988).
20. M.K. Debe, J. Appl. Phys., **55**, 3354 (1984); M.K. Debe and T.N. Tommet, J. Appl. Phys., **62**, 1546 (1987).
21. M.K. Debe, R.J. Poirier, and K.K. Kam, Thin Solid Films, **197**, 335 (1991).
22. M.K. Debe and D.R. Field, J. Vac. Sci. Technol., **A9**, 1265 (1991).
23. M.K. Debe, J. Vac. Sci. Technol., **A10(4)**, 2816 (1992).
24. M.K. Debe, J. Vac. Sci. Technol., **21**, 74 (1992).
25. M.K. Debe, R.J. Poirier, D.D. Erickson, T.N. Tommet, D.R. Field, and K.M. White, Thin Solid Films, **186**, 257 (1990).
26. M.K. Debe and R.J. Poirier, Thin Solid Films, **186**, 327 (1990).
27. D. Bloor and R.R. Chance, Eds.: *Polydiacetylenes* (Martinus Nijhoff, Dordrecht, The Netherlands, 1985).
28. D.S. Chemla and J. Zyss, Eds., *Nonlinear Optical Properties of Organic Molecules and Crystals,* Vol. 2 (Academic Press, Orlando, FL, 1987).
29. P.N. Prasad and D.J. Williams, *Introduction to Nonlinear Optical Effects in Molecules and Polymers* (John Wiley and Sons, Inc.: NY, 1991), p. 232.
30. G.M. Carter, M.K. Thakur, Y.J. Chen, and J.V. Hryniewicz, Appl. Phys. Lett., **47**, 457 (1985).
31. J.P. Hermann and P.W. Smith, Digest of Technical Papers—11th International Quantum Electronics Conference, 656 (1980).

32. M.S. Paley, D.O. Frazier, S.P. McManus, D.N. Donovan, U.S. Patent No. 5,451,433 (September 19, 1995).

33. M.S. Paley, D.O. Frazier, H.A. Abdeldeyem, S. Armstrong, and S.P. McManus, J. Am. Chem. Soc., **117(17)**, 4775 (1995).

34. E. Pearson, W.K. Witherow, and B.G. Penn (private communication).

35. M. Thakur and D.M. Krol, Appl. Phys. Lett., **56(13)**, 1213 (1990).

36. M. Samoc, The Australian National University, Laser Physics Centre (private communication).

37. G.I. Stegeman and W. Torruellas in *Electrical, Optical, and Magnetic Properties of Organic Solid State Materials*, edited by A.F. Garito, A.K. Jen, C.Y-C. Lee, L. Dalton (MRS Symposium Proceedings, Materials Research Society, Pittsburg, PA, 1994) p. 397.

38. H.U. Walter, Ed., *Fluid Sciences and Materials in Space ESA* (Springer-Verlag, NY, 1987).

39. B. Antar and V.S. Nuotio-Antar, *Fundamentals of Low-Gravity Fluid Dynamics and Heat Transfer* (CRC Press, 1994).

40. D.O. Frazier, R.J. Hung, M.S. Paley, and Y.T. Long, J. Crys. Growth (unpublished).

41. B.N. Antar, Phys. Fluids, **30(2)**, 322 (1987).

42. J.C. Maxwell Garnett, Philosophical Transactions of the Royal Society of London, **203**, 385 (1904), ibid., **205**, 237 (1906).

43. D. Ricard, P. Rousignol, and C. Flytzanis, Opt. Lett., **10**, 511 (1985).

44. J.W. Sipe and R.W. Boyd, Phys. Rev., **A 46**, 1614 (1992).

45. M. Brust, M. Walker, D. Bethell, D.J. Schiffrin, and R. Whyman, J. Chem. Soc. Chem. Commun., 801 (1994).

46. M.S. Paley, S. Armstrong, W.K. Witherow, and D.O. Frazier, Chem. Mater., **8(4)**, 912 (1996).

47. M.S. Paley, D.O. Frazier, H.A. Abdeldeyem, and S.P. McManus, Chem. Mater., **6(12)**, 2213 (1994).

48. J. Tsiboulkis, A.R. Werninck, A.J. Shand, and G.H.W. Milburn, Liquid Crystals, **3(10)**, 1393 (1988).

49. T. Kim, R.M. Crooks, M. Tsen, and L. Sun, J. Am. Chem. Soc., **117**, 3963 (1995).

50. D.W. Cheong, W.H. Kim, L.A. Samuelson, J. Kumar, and S.K. Tripathy, Macromolecules, **29**, 1416 (1996).

51. W.H. Kim, B. Bihari, R. Moody, N.B. Kodali, J. Kumar, and S.K. Tripathy, Macromolecules, **28**, 642 (1995).

52. H. Muller, O. Nuyken, P. Strtohriegl, Makromol. Chem. (Rapid Commun. 1992), 125.

53. D.J. Sandman, Ed., *Solid State Polymerization* (American Chemical Society, Washington, DC, 1987).

54. D.O. Frazier, R.J. Hung, M.S. Paley, B.G. Penn, and Y.T. Long, J. Crys. Growth, **171**, 288 (1997).

55. B.L. Markham, D.W. Greenwell, and F. Rosenberger, J. Cryst. Growth, **51**, 426 (1981).

56. Z.Z. Ho, C.Y. Ju, and W.M. Hetherington III, J. Appl. Phys., **62**, 716 (1987).

57. J.S. Shirk, J.R. Lindle, F.J. Bartoli, C.A. Hoffman, Z.H. Kafafi, and A.W. Snow, Appl. Phys. Lett., **55**, 1287 (1989).

58. J.W. Wu, J.R. Heflin, R.A. Norwood, K.Y. Wong, O. Zamani-Khamiri, A.F. Garito, P. Kalyanaraman, and J. Sounik, J. Opt. Soc. Am., B, **6(4)**, 707 (1989).

59. M. Matsuda, S. Okada, A. Masaki, H. Nakanishi, Y. Suda, K. Shigehara, and Y. Yamada, Proc. SPIE, **1337**, 105 (1990).

60. M. Hosoda, T. Wada, A. Yamada, A.F. Garito, and H. Sasabe, Jpn. J. Appl. Phys., **30**, L1486 (1991).

61. H. Hoshi, N. Nakamura, and Y. Maruyama, J. Appl. Phys., **70**, 7244 (1991).

62. Y. Suda, K. Shiegehara, A. Yamada, H. Matsuda, S. Okada, A. Masaki, and H. Nakanishi, SPIE, **1560**, 75 (1991).

63. M.K. Casstevens, M. Samoc, J. Pfleger, and P.N. Prasad, J. Chem. Phys., **92**, 2019 (1990).

64. P.N. Prasad and D.J. Williams, *Introduction to Nonlinear Optical Effects in Molecules and Polymers* (Wiley Interscience, NY, 1991), p. 205.

65. R.J. Reeves, R.C. Powell, Y.H. Chang, W.T. Ford, andW. Zhu, Opt. Mater., **5**, 43 (1996).

66. R. Dorsinville, L. Yang, R.R. Alfano, R. Zamboni, R. Danieli, G. Ruani, and C. Taliani, Opt. Lett., **14(23)**, 1321 (1989).

67. Z.Z. Ho and N. Peyghambarian, Chem. Phys. Lett., **148**, 107 (1988).

68. V.S. Williams, S. Mazumdar, N.R. Armstrong, Z.Z. Ho, and N. Peyghambarian, J. Phys. Chem., **96**, 4500 (1992).

69. T. Wada, S. Yamanda, Y. Matsuoka, C.H. Grossmn, K. Shigehara, H. Sasbe, A. Yamada, and A.F. Garito, *Nonlinear Optics of Organics and Semiconductors* (T.Kobayashi, Springer Berlin, 1989), p. 292.

70. M. Hosada, T. Wada, A. Yamada, A. Garito, and H. Sasabe, Jpn. J. Appl. Phys., **30(8B)**, L1486 (1991).

71. P.A. Chollet, F. Kajzar, and J. LeMoigne, SPIE, **1273**, 87 (1990).

72. K. Kumagai, G. Mitzutani, H. Tsukioka, T. Yamauchi, and S. Ushioda, Phys. Rev. B, **48(19)**, 14488 (1993).

73. T. Yamada, H. Hoshi, K. Ishikawa, H. Takezoe, and A. Fukuda, Jpn. J. Appl. Phys., **34**, L299 (1995).

74. M.K. Debe, and K.K. Kam, Thin Solid Films, **186**, 289 (1990).

75. M.K. Debe and R.J. Poirier, Thin Solid Films, **186**, 327 (1990).

76. M.K. Debe, J. Vc. Sci. Technol., **A4(3)**, 273 (1986).

77. H. Abdeldayem, D.O. Frazier, B.G. Penn, W.K. Witherow, C. Banks, D.D. Smith, and H. Sunkara, Opt. Comm., in press.

78. M.Y. Antipin, T.V. Timofeeva, R.D. Clark, V.N. Nesterov, M. Sanghadasa, T.A. Barr, B. Penn, L. Romero, and M. Romero, J. Phys. Chem. A, **102**, 7222 (1998).

79. T. Wada, G.H. Grossman, S. Yamada, A. Yamada, A.F. Garito, H. Sasabe, Mater. Res Soc. Symp. Proc., **173**, 519 (1990).

80. M. Yu Antipin, T.A. Barr, B. Cardelino, R.D. Clark, C.E. Moore, T. Myers, B. Penn, M. Romero, M. Sanhadasa, T.V. Timofeeva, J. Phys. Chem., **101**, 2770 (1997).

81. M.I. Zugrav, W.E. Carswell, C.E. Lundquist, F.C. Wessling, T.M. Leslie, *Materials Research in Low Gravity* (SPIE Proceedings, **3123**, San Diego, California, 28–29 July 1997) p. 110.

GRAVITATIONAL EFFECTS ON MERCURIC IODIDE CRYSTAL GROWTH

BRUCE STEINER,[*] LODEWIJK VAN DEN BERG,[**] URI LAOR[***]
[*] NIST, Gaithersburg, MD 20899, bruce.steiner@nist.gov
[**] Constellation Technology Corp., Largo, FL 33777; work carried out while the author was at
  EG& G Energy Measurements, Goleta, CA
[***] Nuclear Research Centre, Be'er Sheva, 84910 Israel

ABSTRACT

Gravity can affect the physical vapor growth of mercuric iodide in two distinct ways. First, gravity will induce convection during growth, which strongly mixes residual impurities and any elementary gases resulting from imperfect stoichiometry, either of which can then form precipitates in the growing crystal. Second, gravity loads the resulting crystal, which is particularly soft while still hot, especially in the absence of precipitates. We have investigated the effects of these processes on the resulting crystalline regularity and the effects of various types of irregularity, in turn, on performance.

High resolution synchrotron x-radiation diffraction imaging of three generations of crystals, grown both in microgravity and in full gravity, provide graphic evidence of the influence of gravity on mercuric iodide crystal growth. These images tie together the results of other characterization studies, identifying the crystallographic sources of the observed property enhancement in microgravity. The first process, convection, is found to be particularly important, both in its influence on observed crystalline regularity and in the resulting electronic performance of detectors made from these crystals.

As a result of these investigations, the crystalline regularity and performance of *terrestrial* crystals has been substantially improved, although the resulting crystals have not yet achieved parity with the performance of crystals grown in microgravity. We propose new experiments in microgravity for property optimization.

INTRODUCTION

Wafers from mercuric iodide crystals grown both on Spacelab III and on the first International Microgravity Laboratory (IML-1) have been found to display an increased hole mobility•lifetime product, which enables fabrication of energy dispersive radiation detectors with superior resolution. A factor of six enhancement has been achieved [1] and repeated [2] in microgravity.

We identify here, through high resolution synchrotron x-ray diffraction imaging, differences in the crystallographic uniformity of these wafers and comparable ones grown on the ground. Correlations of these differences with other types of structural observation and with the electronic performance of each wafer enable us to identify those crystal irregularities that are most significant. In this way, we have been able to modify terrestrial crystal growth procedures in order to produce enhanced performance. However, the best performance to date has not yet achieved the level of crystals grown in microgravity [3].

Two gravity-sensitive aspects of the physical vapor crystal growth process have been viewed as primary candidates for influential changes in the crystals that result: 1) convection around the growing crystal, and 2) gravity loading directly on the growing crystal itself. Convection during growth clearly mixes residual impurities and any elementary gases resulting from imperfect stoichiometry, either of which readily forms precipitates in the growing crystal in a nonuniform manner. Gravity loading of the crystal during growth indeed affects the lattice planarity, especially during growth in the absence of precipitates, which can harden the lattice.

The high resolution diffraction images of these wafers presented here lead to the identification of chemical uniformity as the principal source of the enhanced performance that has been observed. It appears likely that lattice planarity also plays a role, but its importance remains to be determined. Continued enhancement in properties appears likely.

EXPERIMENTS

Physical vapor crystal growth experiments were performed with three generations of mercuric iodide source material. First generation crystals were grown from source material purified to the level attainable prior to 1981, when final preparations for the Spacelab III experiment were performed [1]. Second generation crystals were grown from material synthesized and purified in-house using best available practice between 1981 and 1988, when final preparations for the IML-1 experiment were carried out [2]. The third generation crystal was grown from material synthesized and purified in-house with special attention to stoichiometry in 1992 [3].

The two experiments in microgravity were designed to duplicate the terrestrial thermal growth environment [1][4][5]. Thus each of the crystals grown in space was directly comparable to the crystal grown on the ground from material of the same generation. Wafers from the crystals grown both in microgravity and on the ground were imaged in Bragg geometry in specially prepared highly parallel 8 keV monochromatic synchrotron radiation in order to explore the crystallographic regularity of each with the essential angular resolution of less than an arc second [6].

The initial imaging experiments, carried out on first generation terrestrial crystals, observed impurities in concentration sufficient to form prominent arrays of precipitates. These precipitate arrays were absent in the crystal grown on Spacelab III from identical first generation material. While the performance of the Spacelab III crystal ($\mu\tau = 42 \times 10^{-6}$ cm^2/V) was enhanced a factor of six over that of the terrestrial crystal, the accompanying variation in lattice orientation in the Spacelab III crystal was unexpectedly increased rather than decreased.

Therefore, two new generations of more highly purified materials were developed; similar crystals were grown; and wafers were prepared from them. In the fabrication of second generation material, attention was devoted primarily to purification [2]. Completion of preparations for IML-1 was necessary before stoichiometric adjustments could be finalized. The performance of the resulting second generation terrestrial crystal was improved by a factor of two over that of the first generation terrestrial crystal. However, this is substantially less than the factor of six increase achieved again in the second generation crystal grown on IML-1.

After preparations for IML-1 were completed, attention could be devoted to the establishment of stoichiometry as well as high purity in a third generation terrestrial crystal. Its performance was again enhanced, but not yet equal to that of the space crystals.

High resolution diffraction images of these crystals reveal consequential differences in crystallographic order. Analysis of these images, correlated with other observations, enables us to identify the sources of the differences in performance of these succeeding generations of mercuric iodide, grown both in microgravity and on the ground. These results point to specific directions for future experimentation designed to duplicate and perhaps even surpass in ground-based crystals the performance of the crystals grown in microgravity to date.

RESULTS

The images of these crystals display structural variation of two types. One of these, arrays of inclusions, dominates the images of first generation terrestrial crystals in the manner shown in Figure 1. The small, discrete features in these arrays invariably remain out of diffraction over an entire Bragg peak for α mercuric iodide. Thus, these anomalous regions consist of one or more additional phases, *i.e.* inclusions. Some of the individual inclusions appear as thin {100}-oriented bands a few micrometers wide. Some of these, in turn, are crossed by identical features, a morphology that has been observed optically after etching of some wafers [7]. Another set of features interspersed with the first appear in these images as round, with characteristic dimensions of 1 μm - 60 μm. Similar features have been observed optically without etching as well [4].

Two types of inclusions have been noted optically in crystals grown either from stoichiometric or from mercury-rich material [4]. Non evaporating residues the size of these inclusions also have been observed and found to consist of organic and metallic impurities [5]. Whether these residues are to be identified directly with these

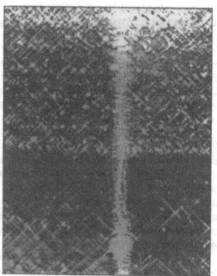

Figure 1. *Asymmetric (1 1 12) image of the central region of first generation terrestrial wafer displaying inclusion arrays formed during growth and vertical lattice twist through seven minutes of arc in the middle.*

two sets of inclusions was not determined definitively. However, after prolonged purification of the starting materials by sublimation, second generation mercuric iodide crystals without inclusions were grown [8], which supports the thesis that both types of inclusions found in first generation crystals are impurities.

In first generation wafers examined by scanning cathodoluminescence microscopy [9], the spectroscopic signatures appear to be concentrated in small (≈5 μm) cells. In Figure 1, similar concentration of diffraction can be observed. The small diffracting regions are delineated by the inclusion morphology.

The white vertical stripe in the center of Figure 1 is an artifact of another aspect of the morphology. The left and the right hand parts of the crystal lattice are rotated with respect to one another by a sharp twist by seven minutes of arc around this <110> axis. The observed seven arc-minutes of folding is comparable to the 20" of arc observed by etching terrestrial wafers several years before the Spacelab III flight [7], and the 6"-7" of arc subsequently determined for other terrestrial wafers by γ-ray diffraction [1]. Less quantitative observation of similar structure in wafers from terrestrial crystals grown about the same time by white beam synchrotron diffraction imaging [10], by high resolution synchrotron diffraction imaging [6], and by laboratory topography and rocking curve observation of virgin wafers [11] as well as in completed detectors [12], all depict <110>-oriented subgrain structure consisting of adjacent grains, tilted with respect to one another by a few minutes of arc. This folding of the lattice thus is characteristic of first generation terrestrially grown mercuric iodide crystals.

The layering of the inclusion distribution visible in Figure 1 may reflect *temporally periodic* variation in the impurity level around the growing crystal. However, alternatively, it may reflect the important role played by even small changes in flow over a vicinal crystal surface steadily growing from solution, as described and analyzed by Chernov [13]. In this instance, the deposition of impurities, particularly those that inhibit crystal growth, may follow closely the local step density, which becomes *spatially periodic*, leading to the deposition of linear inclusion arrays on such growing surfaces.

A comparable image of the first generation Spacelab III crystal in Figure 2, is quite free of inclusions with the result that only small sections appear in diffraction in any one orientation. The six fold increase in hole mobility·lifetime product in this crystal indicates that removal of the inclusions is far more important than the associated increased variation in lattice orientation over the entire crystal to one and a half degrees.

At the time of final preparations for the IML-1 flight in 1992, second generation of mercuric iodide was available through in-house

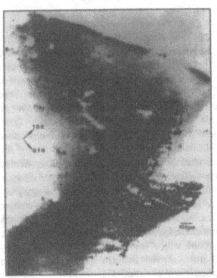

Figure 2. Asymmetric (1 1 10) image of characteristic region of a wafer from comparable first generation crystal grown on Spacelab III. The relatively small area in diffraction displays variation in lattice orientation, associated with the absence of the inclusions prominent in the terrestrial wafer.

synthesis. New, more gentle crystal cutting and polishing techniques were developed for these wafers and reported at a preceding MRS meeting [14].

An image of the IML-1 crystal is shown in Figure 3. The level of inclusions again is low, although a few vertical rows of orientational irregularity can be seen. In this case, however, any inclusions are iodine inclusions, which are characteristic of this second generation material [8]. The entire variation in orientation of this crystal has been reduced to 0.06°. Nevertheless, the electronic performance of this crystal, is comparable to that of the Spacelab III crystal.

A comparable image of a terrestrial crystal grown from second generation material appears in Figure 4. The density of inclusions, which as noted are iodine inclusions, is higher than in the IML-1 crystal. In spite of the higher inclusion level, the lattice orientation of this crystal varies over 0.7°, an order of magnitude greater than the variation in orientation of the IML-1 crystal lattice. This

Figure 3. Symmetric (008) diffraction image of characteristic region of a wafer from the second generation crystal grown on IML-1. Small regions that differ in orientation from the surrounding areas, are visible.

intermediate between that of the growth-purified IML-1 crystal made of the same material and that of the impure first generation terrestrial crystal.

Third generation material, nearly optimized for stoichiometry as well as purity, has been grown on the ground but not yet in microgravity. The diffraction image of a terrestrial crystal, Figure 5, indicates near chemical optimization by the absence of inclusions. This is accompanied by wide variation in lattice orientation, two degrees. In the absence of precipitation hardening, a somewhat wider range than that observed for the second generation terrestrial crystal, 0.7°, is expected. However, part of the variation in lattice orientation observed here may be due to wafer fabrication, which was carried out before the full implications of the change in the lattice were identified. Although superior in electronic performance to the previous generations of terrestrial crystals, it is not yet equivalent to either space crystal. Whether this difference is due to greater variation in lattice

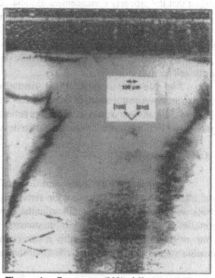

Figure 4. Symmetric (008) diffraction image of characteristic region of second generation terrestrial wafer comparable to the IML-1 wafer. The seed crystal is visible at the top. The restriction of diffraction in the region of new growth to small areas displays short range variation in lattice orientation. The width of the area of new growth that is in diffraction in a given region is correlated with the observed density of iodine inclusions, which thus act to harden the lattice similarly to the impurity inclusions characteristic of Spacelab III generation terrestrial growth.

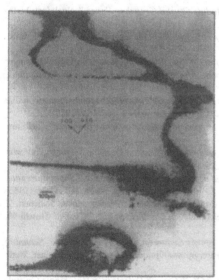

Figure 5. Asymmetric (1 1 10) diffraction image of central region of third generation terrestrial mercuric iodide wafer, displaying variation in lattice orientation associated with increased levels of purification and stoichiometry, which are closer to optimum than in the wafers fabricated from terrestrial crystals of preceding generations.

orientation or to residual chemical effects remains to be determined.

CONCLUSIONS

Six-fold enhancement in the electronic performance of α mercuric iodide crystals grown in microgravity has been traced to reduction in the incorporation of impurities and imperfect stoichiometry. In terrestrial growth, the formation of inclusions is evident both in first generation and in second generation crystals. In the first generation terrestrial crystal, the inclusions are impurities. In the second generation terrestrial crystal, the inclusions are iodine. These observations strongly suggest that the property enhancement associated with suppression in microgravity of both types of suggest that the property enhancement associated with suppression in microgravity of both types of inclusion was achieved by the elimination of convection.

As a result of this knowledge, the superior performance of the space crystals has been approached but not yet equaled in terrestrial growth. The increased performance of purified second generation terrestrial crystals was surpassed by third generation crystals, following attention both to purity and to stoichiometry.

The role of increased lattice orientational uniformity in the achievement of still greater property improvement remains to be observed. Growth in microgravity would address this interesting question. The answer would have both practical and theoretical implications. New generations of space growth of mercuric iodide appear to be called for.

ACKNOWLEDGEMENTS

The Consortium for Commercial Crystal Growth at Clarkson University provided the financial and intellectual support necessary for determination of the state of the art of terrestrial α mercuric iodide wafer uniformity at the time of initial growth in microgravity. This knowledge provided the stimulation of interest essential to the success of this work. Examination of the wafers fabricated from crystals grown subsequently in microgravity and associated terrestrial crystals was made possible by partial support of the NASA Microgravity Sciences and Applications Division.

REFERENCES

[1]   L. van den Berg and W. F. Schnepple, *"Mercuric Iodide Crystal Growth in Space,"* Nucl. Inst. Meth. Phys. Res. **A 283**, 335-338 (1989)

[2]   L. van den Berg, *Growth of Single Crystals of Mercuric Iodide on the Ground and in Space,"* Mater. Res. Soc. Proc. **302**, 73-83 (1993)

[3]   V. M. Gerrish, *" Characterization and Quantification of Detector Performance,"* in T. E. Schlesinger and R. B. James, ed., *Semiconductors for Room Temperature Nuclear Detector Applications*, Semiconductors and Semimetals **43**, 493-530 (1995)

[4]   L. van den Berg, W. Schnepple, C. Ortale, and M. Schieber, *"Vapor Growth of Doped Mercuric Iodide Crystals by the Temperature Oscillation Method,"* J. Cryst. Growth **42**, 160-165 (1977)

[5]   H. A. Lamonds, *"Review of Mercuric Iodide Development Program in Santa Barbara,"* Nucl. Inst. Meth. **213**, 5-12, (1983)

[6]   M. Kuriyama, B. W. Steiner, and R. C. Dobbyn, *"Dynamical Diffraction Imaging (Topography) with X-Ray Synchrotron Radiation,"* Ann. Rev. Mat. Sci. **19**, 183-207 (1989)

[7]   B Milstein, B. Farber, K. Kim, L. van den Berg, and W. F. Schnepple, *"Influence of Temperature upon Dislocation Mobility and Elastic Limit of single Crystal HgI_2,"* Nucl. Inst. Meth. **213**, 65-76 (1983)

[8]   A. Burger, S. Morgan, C. He, E. Silberman, L. van den Berg, C. Ortale, L. Franks, and M. Schieber, *"A Study of Inhomogeneity and Deviations from Stoichiometry in Mercuric Iodide,"* J. Cryst. Growth **99**, 988-993 (1990)

[9]   Y. F. Nicolau and M. Dupuy, *"Study of α HgI_2 Crystals Grown in Space by Differential Scanning Calorimetry, Scanning Cathodoluminescence Microscopy and Optical Microscopy,"* Nucl. Inst. Meth. Phys. Res. **A 283**, 355-362 (1989)

[10]   F. Remy, J. Gastaldi, and G. LeLay, *"Observation of Bulk Growth Defects in α-HgI_2 Single Crystals by Synchrotron X-ray Transmission Topography,"* Nucl. Inst. Meth. Phys. Res. **B 83**, 229-234 (1993)

[11]   S. Gits, *Characterization of Extended Defects in α-HgI_2 Single Crystals,"* Nucl. Inst. Meth. **213**, 43-50 (1983)

[12] M. Schieber, C. Ortale, L. van den Berg, W. Schnepple, L. Keller, C.N.J. Wagner, W. Yelon, F. Ross, G. Georgeson, and F. Milstein, *"Correlation between Mercuric Iodide Detector Performance and Crystalline Perfection,"* Nucl. Inst. Meth. Phys. Res. **A 283**, 172-187 (1989)

[13] A. A. Chernov, *"How does the flow within the boundary layer influence morphological stability of a vicinal face?,"* J. Cryst. Growth **118**, 333-347 (1992)

[14] Bruce Steiner, Lodewijk van den Berg, and Uri Laor, *"High resolution Diffraction Imaging of Mercuric Iodide: Demonstration of the Necessity for Alternate Crystal Processing Techniques for Highly Purified Material."* Mat. Res. Soc. Symp. **375**, 259-264 (1995)

MAGNETIC LEVITATION OF LIQUID DROPLETS

E. BEAUGNON*, R. TOURNIER
Laboratoire de Cristallographie, Matformag, CNRS
BP 166, 38042 Grenoble Cedex 9, France
* beaugnon@labs.polycnrs-gre.fr

ABSTRACT

The magnetic levitation of diamagnetic materials in a high field superconducting magnet is presented. The magnetic force is determined from numerical calculations of the magnetic field components and the stability of levitation is studied using energy calculations. Amongst the numerous materials that can be levitated, experiments are shown for ethanol droplets.

INTRODUCTION

High magnetic field gradients produced by superconducting magnets are able to create forces greater than gravity on almost all materials. The magnetic force is acting on the whole volume of the material and can be adjusted so that it locally compensates the gravity. Although gravity is not suppressed, simulation of microgravity can then be obtained using a homogeneous magnetic force field. The levitation of water and organic materials has already been demonstrated using such techniques [1] and experiments have been even performed on living bodies [2][3].

In order to study the levitation conditions of a liquid droplet in a superconducting magnet, we have calculated the residual force field acting on the liquid as a function of current coil intensity and we have observed experimentally contact-free levitation using video macrophotography.

MAGNETIC FORCE AND FIELD COMPONENTS

The magnetic field $\vec{B}$ created by a current distribution $\vec{j}$ can be numerically calculated from the Biot-Savart law :

$$\vec{B} = \frac{\mu_0}{4\pi} \iiint \vec{j} \wedge \frac{\vec{u}}{r^2} d\tau \qquad (1)$$

An object with a volume V and a magnetic susceptibility χ, placed in a magnetic field gradient, will be submitted to a magnetic force $\vec{F}$. In axisymetric coordinates (r,z), the force vector is given by :

$$\vec{F} = \frac{\chi}{2\mu_0} V \overrightarrow{\text{grad}}(B^2) = \frac{\chi}{\mu_0} V \left[\left(B_z \frac{dB_z}{dz} + B_r \frac{dB_r}{dz} \right) \vec{e_z} + \left(B_z \frac{dB_z}{dr} + B_r \frac{dB_r}{dr} \right) \vec{e_r} \right] \qquad (2)$$

For a solenoid with n turns, an inner radius r_1, an outer radius r_2, and a vertical extent from z_1 to z_2, equation (1) can be integrated twice along r and z. The field and field derivatives calculated at a point (ρ, ζ) are then given for a current I by the following equations :

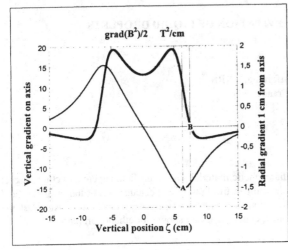

Fig 1 : The vertical gradient along the coil axis and the radial gradient 1 cm beside the axis are plotted as a function of vertical position. Two critical positions are evidenced : vertical stability is obtained when diamagnetic objects are repelled above the maximum force point «A», radial stability is obtained when diamagnetic objects are centered on the axis below the vertical position «B».

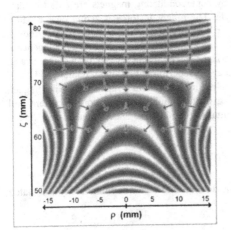 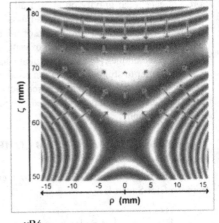

Fig 2a (left) and 2b (right) : The total energy $E = \frac{1}{\rho} \frac{\chi B^2_{(\rho,\zeta,I)}}{2\mu_0} + g\zeta$ is plotted as a function of radial position (horizontal axis, from -16 to 16 mm) and vertical position (vertical axis, from 49 to 81 mm above the coil center). Two fringes are separated by an energy value of 20 erg/gram. In figure 2a (I coil = 94 A), the maximum force is lower than gravity ; no local energy minimum exists : the object falls. In figure 2b (I coil = 97 A) , the levitation is contact-free and stable. A second local minimum also exists above, touching the magnet cryostat walls.

$$B_z = \frac{\mu_0}{4\pi} \frac{nl}{(z_2-z_1)(r_2-r_1)} \int_0^{2\pi} \left[\left[\begin{array}{c} -(\zeta-z)\ln\left(A+r-\rho\cos(\theta)\right) - \frac{\rho\cos(\theta)}{2}\ln\left(\frac{A-(\zeta-z)}{A+(\zeta-z)}\right) \\ +\rho\sin(\theta)\arctan\left(\frac{(\zeta-z)(r-\rho\cos(\theta))}{\rho\sin(\theta)A}\right) \end{array} \right]_{r_1}^{r_2} \right]_{z_1}^{z_2} d\theta \quad (3)$$

$$B_r = \frac{\mu_0}{4\pi} \frac{nl}{(z_2-z_1)(r_2-r_1)} \int_0^{2\pi} \left[\left[\cos(\theta)A + \rho\cos(\theta)\ln\left(A+r-\rho\cos(\theta)\right) \right]_{r_1}^{r_2} \right]_{z_1}^{z_2} d\theta \quad (4)$$

$$\frac{dB_z}{dz} = \frac{\mu_0}{4\pi} \frac{nl}{(z_2-z_1)(r_2-r_1)} \int_0^{2\pi} \left[\left[\frac{r}{A} - \ln\left(A+r-\rho\cos(\theta)\right) \right]_{r_1}^{r_2} \right]_{z_1}^{z_2} d\theta \quad (5)$$

$$\frac{dB_r}{dz} = \frac{dB_z}{dr} = \frac{\mu_0}{4\pi} \frac{nl}{(z_2-z_1)(r_2-r_1)} \int_0^{2\pi} \left[\left[\frac{(\zeta-z)\cos(\theta)}{(\zeta-z)^2+\rho^2\sin^2(\theta)} \frac{(\zeta-z)^2+\rho^2-\rho r\cos(\theta)}{A} \right]_{r_1}^{r_2} \right]_{z_1}^{z_2} d\theta \quad (6)$$

and $\quad \dfrac{dB_r}{dr} = -\dfrac{B_r}{r} - \dfrac{dB_z}{dz} \quad (7)$

where $A = \sqrt{(\zeta-z)^2 + \rho^2 + r^2 - 2\rho r\cos(\theta)}$

LEVITATION STABILITY

It has been demonstrated that stable magnetic levitation can only be obtained for objects with a magnetic permeability lower than the surrounding medium permeability [4]. Diamagnetic materials are obviously good candidates for magnetic levitation, but paramagnetic materials can also be levitated in a more magnetic medium such as over pressurized oxygen [5]. However, magnetic levitation requires also an adequate field geometry and field strength.

Knowing the winding data of a superconducting high field Oxford Instrument magnet, we have calculated the numerical values of equations (3) to (7) in order to estimate the force field available in the vertical room temperature bore of this magnet. The vertical and radial square field gradients in equation (2) are plotted as a function of the vertical position inside the magnet (fig. 1). For a diamagnetic material, levitation can be contact-free, centered on the vertical axis, above the maximum repelling force (point A) and below the point where radial forces drive away the object towards the magnet cryostat walls (point B). The magnetic field intensity needs to be adjusted so that the maximum force (A) is stronger than gravity, but does not raise the object above B. This narrow stable levitation zone is also evidenced when plotting the total energy (magnetic + gravity) as a function of position for different coil current intensities. For a diamagnetic susceptibility $\chi = -0.7 \times 10^{-6}$ emu/g, we have plotted the total energy using a gray density scale for three cases. Figure 2a represents the energy profile for a field intensity below the critical value required for levitation, Figure 2b represents the case of stable levitation, and figure 2c shows that the magnetic forces expel the diamagnetic material towards the magnet bore. In each case, the force intensity and direction around the zero force point are also represented.

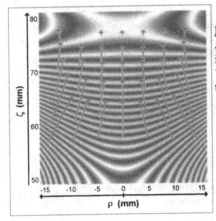

Fig. 2c : For higher values of the field (I = 103 A), the object is raised above the radial force inversion point «B» drawn on figure 1 near ζ = 75 mm. The vertical stability is maintained but the object touches the magnet cryostat walls.

NON WETTING LIQUID DROPLETS

Using video macrophotography, we have recorded the motion of several droplets injected near the equilibrium position presented by figure 2b. Before the droplets merge into a single one, they repeatedly collide but do not coalesce, despite the expected wetting of the liquid on itself. When the liquid droplets finally merge, a small droplet, ejected with a large velocity, is also observed. An example of bouncing is presented on the digitized frames of figure 3.

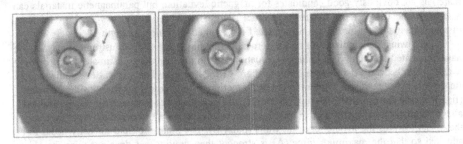

Fig. 3 : Levitated ethanol droplets bouncing in the 32 mm room temperature bore of the superconducting high field coil. The time interval between each frame is 1/5 second. Motion is shown with arrows.

Such a phenomenon seems obvious and frequent (droplets may bounce more than 10 times on each other), but the mechanism for non-wetting still remains unclear and different explanations have already been proposed by authors. A similar effect was observed with levitated liquid helium droplets at low temperature where it was suggested [6] that liquid evaporation creating a gas film avoided contact between the droplets. Thermocapillary convection, driving an air gap between the two droplets, has also been proposed [7], but in our case no droplet temperature difference creating the convection is imposed. It has recently been observed [8] that small temperature differences between the droplets may account for this mechanism ; in our experiment the droplets may bounce on each other until the temperature difference reaches a value below a critical value of a few degrees. Also, isothermal shear flows are able to prevent the coalescence of liquid bodies [9] and this may also explain the behavior that we have observed.

CONCLUSIONS

We have experimentally observed and numerically studied the contact-free levitation of diamagnetic material in a strong magnetic field gradient. Magnetic levitation is not microgravity since residual forces exist near the zero force equilibrium point, but these forces, while centering the diamagnetic material on that point, avoid contact with the walls of the experiment cell. Non wetting effects are observed when levitated ethanol droplets touch each other. The mechanism still remains unclear and needs to be further investigated in order to compare our results with the different mechanisms already proposed by other authors.

REFERENCES

1. E. Beaugnon and R. Tournier, Nature **349**, 470 (1991) ; J. Phys. III (France) **1**, 1423 (1991)
2. J. M. Valles et al., Biophys. J. **73**, 1130 (1997)
3. A. Geim, Physics Today, 36 (Sept 1998)
4. I. J. Lin and T. B. Jones, J. of Electrostatics **15**, 53 (1984)
5. Y. Ikezoe et al. Nature **393**, 749 (1998)
6. M. A. Weilert et al., Phys. Rev. Lett.**77**, 4840 (1996)
7. P. Dell'Aversana and G. P. Neitzel, Physics Today, 38 (January 1998)
8. R. Monti et al., Physics of Fluids **10**, 2786 (1998)
9. P. Dell'Aversana et al., Physics of Fluids **8**, 15 (1996)

Part VI

Microgravity Materials Processing:
Flight Experiments

AC Modulation Calorimetry of Undercooled Liquid Ti$_{34}$Zr$_{11}$Cu$_{47}$Ni$_8$ and Zr$_{57}$Nb$_5$Ni$_{12.6}$Al$_{10}$Cu$_{15.4}$: An MSL–1 Experiment Using TEMPUS

S.C. Glade*, D.S. Lee**, R. Wunderlich[†] and W.L. Johnson*
*Materials Science Dept., 138-78, California Institute of Technology. Pasadena, CA 91125
**Amorphous Technologies International, 27722 El Lazo Road. Laguna Niguel, CA 29677
[†] Technische Universität Berlin, Institut für Metallforschung, Hardenbergstr.36, D-10623 Berlin

Abstract

An AC modulation calorimetry (ACMC) experiment was performed on two bulk metallic glass forming alloys, Ti$_{34}$Zr$_{11}$Cu$_{47}$Ni$_8$ (VIT 101) and Zr$_{57}$Nb$_5$Ni$_{12.6}$Al$_{10}$Cu$_{15.4}$ (VIT 106) on the recent MSL-1 shuttle flight using the TEMPUS hardware. VIT 106 exhibited a maximum undercooling of 140 K in free radiative cooling, while VIT 101 exhibited a smaller undercooling of 50 K. Specific heat measurements were done in both the stable and undercooled liquid regions. These results will be combined with further ground-based measurements in an electrostatic levitator for modeling the nucleation kinetics of these alloys and for calculation of the free energies and entropies of the undercooled liquids in these systems.

Objectives

The primary scientific objective of the experiments conducted in the MSL-1 spacelab mission was the measurement of the specific heat capacity in the stable and undercooled liquid of the two bulk metallic glass forming alloys:

Ti$_{34}$Zr$_{11}$Cu$_{47}$Ni$_8$ VIT 101

Zr$_{57}$Nb$_5$Ni$_{12.6}$Al$_{10}$Cu$_{15.4}$ VIT 106

From the free temperature decay of the specimen between bias temperatures T_1 and T_2, the rate of radiative heat loss was measured. The total hemispherical emmisivity of the specimen as function of temperature can be inferred if the specific heat is known. Furthermore, the method of AC modulation calorimetry employed for the measurement of the specific heat allows the determination of the effective thermal conductivity of the liquid alloy. In cooperation with different scientific groups participating in the TEMPUS experiment on MSL-1, it was also planned to share these specimens for the measurement of the temperature dependencies of specific volume, surface tension, and viscosity.

Background

In recent years, bulk metallic glass forming alloys, which can be produced with dimensions relevant for their use as commercial materials, have attracted considerable attention [1,2]. The exceptional stability of these materials in the metastable undercooled melt makes the undercooled liquid amenable to physical investigations not possible before. Formation of metallic glasses by cooling the liquid alloy from temperatures well above the equilibrium melting point depends critically on the cooling rate, or equivalently on the stability of undercooled melt with regard to crystallization. Following classical nucleation theory, the nucleation rate is determined by kinetic factors such as the viscosity, which expresses the mobility of atoms or groups of atoms, and by thermodynamic factors, which expresses the driving force for the formation of a crystalline nucleus in the melt [3]. The latter term is composed of a surface energy term and the difference in the Gibbs free energy between the undercooled liquid and the competing crystalline phase. The Gibbs free energy difference can be obtained from measurements of the specific heat capacity in the metastable liquid and crystalline phase, and the heat of fusion. Combining the specific heat and viscosity data with measurements of the nucleation kinetics, such as T-T-T curves, can serve to model nucleation kinetics. Thus, differences in the nucleation kinetics and glass forming ability of different alloy compositions may be systematically assessed.

Methods of data acquisition and analysis

Measurements of the specific heat capacity were performed on 8 mm diameter specimens with the containerless electromagnetic processing device TEMPUS, which has been previously described [4,5]. Specimens are positioned by a radio-frequency electromagnetic quadrapole field and heated by a dipole field. Containerless conditions are required to avoid heterogeneous nucleation of the undercooled melt through contact of the specimen with container walls. Furthermore, heating due to the strong levitation forces required for containerless electromagnetic processing would not allow one to perform these experiments under clean ultrahigh vacuum conditions in a 1-g environment.

The heat capacity of the specimen is determined by AC modulation calorimetry. The heating power input to the specimen at bias temperature T_0 is sinusoidally modulated at frequency ω. This results in a temperature response at ω with amplitude ΔT_{mod} such that $\Delta T_{mod}/T_0 \ll 1$, with c_p given by:

$$c_P = \frac{P_{mod}}{\omega \, \Delta T_{mod}} \, f(\omega, \tau_1, \tau_2)$$

P_{mod} is the amplitude of modulated power component. $f(\omega, \tau_1, \tau_2)$ is a correction function accounting for the finite thermal conductivity, with internal relaxation time τ_2, and heat loss, with external relaxation time τ_1 [6]. Due to the typically large difference in the external and internal relaxation times for metallic specimens, there is a frequency where $\Delta T_{mod}(\omega)$ is frequency independent, characterizing the isothermal regime with $[1 - f(\omega, \tau_1, \tau_2)] < 10^{-2}$. Experimentally, this regime can be identified by a phase shift of 90° between the heater current modulation and the temperature response. Calibration of P_{mod} is performed by application AC modulation calorimetry to the crystalline phase with known specific heat. The temperature dependence of the calorimeter constant can be determined by measuring the specific volume and the resistivity, which are deconvolved from current and voltage data of the rf-generator. Figure 1 shows a typical experimental run with the VIT 106 alloy depicting melting and overheating (1), rapid cooling into the undercooled melt (2), and performing AC modulation calorimetry (3). The sample subsequently recalesces (4), possibly into a metastable crystalline phase.

A baseline drift correction is applied to determine ΔT_{mod} with external relaxation time τ_1, and subsequent Fourier analysis (Figure 2). The frequency components at 1.2 Hz and 1.5 Hz are the harmonic motions of the specimen in the potential well of the positioning field and do not represent true temperature variation. Alternatively, a FFT low pass filter is applied to the raw signal and a running average over modulation cycles is performed. The two methods agree within the statistical error of the running average, typically 0.1 K. As such, with $\Delta T_{mod} \approx 3$ K, a relative error in c_p determination of < 4% is obtained. From a step function change in the heating power input, τ_1 can also be determined from the transient behavior of the modulation signal. The data shown in Figure 3 can be very well represented by a single exponential fit as expected for transients with a small temperature difference.

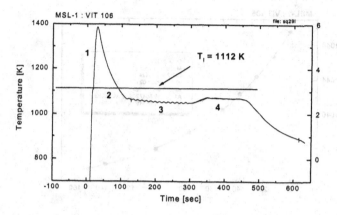

Fig. 1. Typical MSL-1 experiment with VIT 106. .

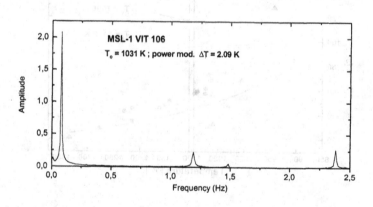

Fig. 2. Fourier transform of temperature signal (3) from Figure1 after correcting for signal drift. Note the frequency components at 1.2Hz and 1.5Hz which are correlated by time-sync video to periodic sample motion in the potential well of the quadrupole positioning fields.

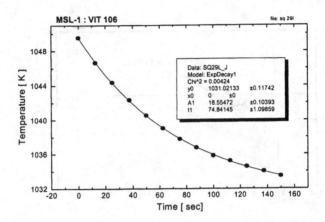

Fig. 3. Position of modulation maxima as function of time for τ_1 determination.
Legend. ● : data points, — : single exponential fit.

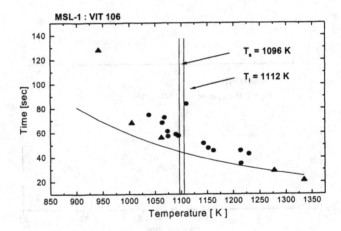

Fig. 4. External relaxation time as a function of temperature for VIT 106
Legend. ● : MSL-1, Δ : ESL- ground based results [7].

Flight results compared with ground results

Figure 4 shows results for the external relaxation time τ_l obtained in the flight experiment for VIT 106 together with similar values obtained in the ESL. The good agreement between these values demonstrates the consistency of the experimental approach. The solid line represents a T^{-3} scaling of τ_l for constant specific heat capacity. The deviation of the experimental data from this scaling reflects the strong increase of the specific heat with decreasing temperature for the undercooled melt, since the total emissivity has only a small $T^{-1/2}$ dependence from the free electron model [8].

In VIT 106 (Figure 5), a singular point near T_l is observed suggesting an unusual behavior in c_p [9]. This was the first melt cycle performed and the anamoly is attributed to the equilibrium melting and resolidification of a minor phase which served as a heterogeneous nucleant during the first melt cycle. It was shown that heterogenous nucleants present initially in the liquid alloy could either be dissolved or passivated by heating the liquid alloy above a well defined temperature limit [10]. In subsequent melt cycles, this phase had to nucleate homogeneously and thus far greater undercooling was achieved. In order to test this hypothesis further, laboratory experiments to analyze the solidification microstructure after rapidly quenching the liquid from different temperatures are under way. AC modulation calorimetry could be performed for this alloy up to an undercooling of 70 K.

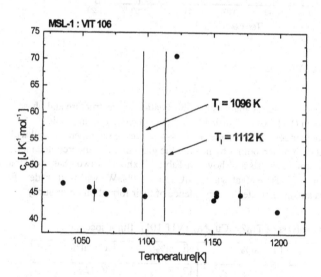

Fig 5. Specific heat capacity of VIT 106.

As compared to ESL ground based experiments where undercooling of over 200 K could be obtained, the VIT 101 flight specimen showed only an undercooling of 50 K, which is attributed to the presence of minor impurities. Specific heat capacity data in the stable melt and the slightly undercooled melt were obtained and are displayed in Figure 5. Specific heat capacity results for VIT 106 are displayed in Figure 6. It should be pointed out that for VIT 101, no anomaly in τ_l or c_p similar to VIT 106 could be observed. The same holds for the measurements performed by the Fecht group on two bulk glass forming alloys of different composition [7].

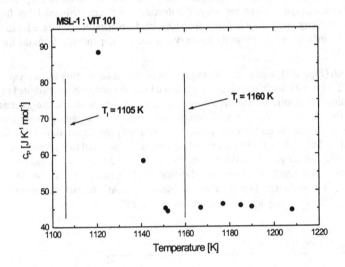

Fig 6. Specific heat capacity of VIT 101.

Other ground based work includes analysis of solidification microstructure and phase selection in the flight specimen (VIT 101) solidified from the undercooled liquid under microgravity conditions. Comparison will be made with a 1-g reference specimen, as well with those phases crystallizing at high undercooling by heating the glass into the undercooled melt. Results of an electron microprobe analysis are shown in Table 1, exhibiting two phases of almost complementary composition in (ZrTi) ≈ 50 at% and (NiCu) ≈ 50 at%. We will further identify the structure of these phases and investigate the dependence of their formation on impurity levels.

Table. 1   EDX spot analysis of $Ti_{34}Zr_{11}Cu_{47}Ni_8$ (VIT 101) flight specimen

Phase	Zr at%	Ti at%	Ni at%	Cu at%
1	7.45	36	12.97	36.42
2	18.99	22.44	8.12	50.44
3	3.27	47.44	3.05	46.24

Conclusions

The specific heat capacity of two glass forming metallic alloys was measured by AC modulation calorimetry in the stable and undercooled (for VIT 106) melt. The temperature dependence of the specific heat as well of the external relaxation time and the specific volume appear were measured. These data will be combined with other thermophysical property data and undercooling experiments to model the nucleation kinetics in these alloys.

Acknowledgements

The authors would like to all the members of the TEMPUS team. This work was supported by NASA under grant number NAG8-954 and the German space agency DARA (Grant No. 50 WM 94-31-4).

References

1. A. Peker and W. Johnson, Appl. Phys. Lett. 63, 2342 (1993)
2. A. Inoue, Y. Ykohama, Y. Shinohara and T. Masumoto, Mat. Trans. JIM, 35, 923 (1994)
3. R. Wunderlich, D. Lee, W. Johnson, and H. Fecht, Phys. Rev. **B55**, 26 (1997)
4. D. Turnbull, J. Chem. Phys. **20**, 411 (1952); F. Spaepen, Acta Met. **23**, 729 (1975)
5. Team Tempus, Containerless Processing in Space, Recent Results (Lecture Notes in Physics, Springer Verlag, Berlin 1995)
6. H. Fecht and W. Johnson, Rev. Sci. Instr., **62**, 1299 (1991)
7. W. Johnson, D. Lee, C. Hays and J. Schroers, "Physical Properties and Processing of Undercooled Metallic Glass Forming Liquids," to be published.
8. A. Sievers, J. Opt. Soc. Am.68, 1505 (1978)
9. R. Wunderlich, R. Sagel, C. Ettl, H. Fecht, D. Lee, S. Glade and W. Johnson, to be published.
10. X. Lin and W. Johnson, J. Appl. Phys. 78, (1995) 6514

GROWTH COMPETITION DURING DOUBLE RECALESCENCE IN FE-CR-NI ALLOYS

D. M. MATSON
Department of Materials Science and Engineering, Massachusetts Institute of Technology, Cambridge, MA 02139, matson@mit.edu

ABSTRACT

The rapid solidification of a Fe-12wt%Cr-16wt%Ni alloy was investigated under containerless processing conditions using both ground-based electromagnetic levitation equipment and aboard the shuttle Columbia using the TEMPUS facility. A high-speed digital video technique was used to image growth of the metastable ferritic phase and the stable austenitic phase into the undercooled melt. Above a critical undercooling, the metastable phase nucleates first. After a delay, a second thermal rise is observed during transformation to the stable phase. Double recalescence events were observed at temperatures consistent with the T_0 temperature of the bcc phase thus defining a value of the critical undercooling for metastable nucleation which is significantly lower than previously predicted. For a given liquid temperature the velocity of the stable fcc phase is greater than that of the metastable bcc phase. The velocity for growth of the stable phase into the semi-solid which forms during primary metastable recalescence was also measured and found to be independent of the initial undercooling. A model based on competitive growth of the two phases successfully predicts the limit where double recalescence events may be detected.

INTRODUCTION

Double recalescence can be observed in the commercially important Fe-Cr-Ni ternary system when the undercooling is sufficient to access the metastable regions of the equilibrium phase diagram. Under these conditions, the primary phase which forms during recalescence from the undercooled melt is the metastable ferritic δ-phase. Subsequent growth of the stable austenitic γ-phase is seen as a delayed secondary rise in temperature. This transformation occurs very rapidly with delay times a strong function of composition. In the range of 69 at% $< X_{Fe} <$ 72 at%, the delay is on the order of microseconds for a (Ni/Cr) atomic ratio of 2.4 while it increases to delay times of several seconds at (Ni/Cr) = 0.6 approaching the pseudobinary eutectic composition.

The growth velocity of the primary phase into the undercooled liquid can be measured during recalescence by examining the rise time across the pyrometer target area. Nucleation is typically initiated at one of the poles of the sample while the pyrometer is focused at the equator. Thus, the occurrence of a delayed double nucleation event and the growth velocity during primary recalescence may be evaluated simultaneously.

Koseki [1] showed microstructural evidence, in samples quenched just after the initiation of the second recalescence event, that growth of the stable phase proceeds around pre-existing metastable dendrites which formed during the primary recalescence event. Using a pyrometry-based technique, Koseki [2], Löser [3], Moir [4], and Volkmann [5] showed that the undercooling required to achieve this phenomena is a strong function of composition and that the undercooling required to observe a double recalescence is depressed far below the equilibrium solidus temperature. Moir [4] and Volkmann [5-7] attribute this depression in temperature required to access formation of the metastable phase to nucleation kinetics and they conclude that phase selection is best described using the Diffuse Interface Theory (DIT) of nucleation [8] with an optimized nucleii composition.

Experimental evidence contrary to these results was obtained by the author [9] using a high-speed video imaging technique. The enhanced spatial resolution of the technique allows for analysis of small (less than 180 micron square) discrete surface elements across the entire visual surface of the droplet instead of an average signal obtained over a wide target region (4 mm). The video results showed that double recalescence events could be detected at temperatures approaching the T_o line in the equilibrium phase diagram. By tracking the location of where the interface intersects the surface of the sample, the velocity may be calculated readily based on knowledge of where nucleation initiated [10]. As before, the occurrence of a delayed double nucleation event and the growth velocity during primary recalescence may be evaluated simultaneously.

A new theory for phase selection based on growth competition, instead of competitive nucleation, was proposed by workers at MIT to explain these results [11]. In this model, growth of the primary and secondary phases is initiated from a common nucleation site. Growth of the stable phase, into the mixture of liquid and metastable solid, proceeds after a known delay time specific to the original liquid composition. The growth rate of each solid phase is determined from experimental measurement. For a specific composition, the growth rate of the metastable phase into the undercooled liquid, V_d, is a strong function of undercooling. The growth rate of the stable phase into the primary metastable dendritic array, or semi-solid mixture, V_{ss}, is independent of the original liquid undercooling. This is because the undercooling that the stable phase experiences following primary recalescence is nearly constant and is consistent with the difference, ΔT_o, between the stable phase and metastable phase T_o temperatures. Should the stable phase break out into the undercooled liquid and overwhelm the metastable phase, the stable phase growth into the liquid, V_g, is always greater than that of the metastable phase. $V_g > V_d$ at any given liquid temperature.

The purpose of this paper is to present the results of microgravity and ground-based experiments in support of the new theory and to attempt to re-examine the results of previous authors to obtain an interpretation consistent with the growth competition theory. The approach is based on analysis of the velocity for growth of the stable phase into the mixture of liquid and pre-existing metastable dendritic array which formed during primary recalescence.

EXPERIMENT

Microgravity Investigations

The TEMPUS containerless electromagnetic levitation facility was used during the first Microgravity Sciences Laboratory (MSL-1R) mission aboard Columbia during STS-94 to investigate the solidification behavior of ternary steel alloys. Experiments were conducted on two flight compositions corresponding to a (Ni/Cr) ratio of 1.18 for a Fe-12wt%Cr-16wt%Ni alloy and an atomic ratio of (Ni/Cr) of 0.66 for a Fe-16wt%Cr-12wt%Ni alloy. Using a set of dual axial and radial 1 MHz pyrometers, the velocity for metastable growth into the liquid and stable growth into the semi-solid were measured. The sample thermal profile was also used to define when a double recalescence event occurred as a function of sample undercooling.

The results of MSL-1R [12] show that the delay time between nucleation of the metastable phase and the stable phase differs significantly between ground-based and microgravity experiments. The authors attribute this difference to a reduction in melt convection in microgravity. Double recalescence events were seen to occur at undercoolings approaching the metastable T_o limit in agreement with the results obtained using the high-speed video imaging technique [9, 12].

Growth velocities for metastable phase growth into the liquid and stable phase growth into the semi-solid were also measured and results do not significantly differ between ground-based and microgravity experiments for the Fe-12wt%Cr-16wt%Ni alloy. The growth velocity of the stable phase into the semi-solid was also measured for the Fe-16wt%Cr-12wt%Ni alloy in microgravity but as this quantity has not been directly measured successfully in ground-based experiments, no comparison is available. V_{ss} for both compositions was seen to be independent of the original liquid undercooling in agreement with the results obtained using the high-speed video imaging technique[9, 12].

Ground-based Investigations

In support of flight operations, Bui [11] and Matson [12] found that double recalescence events could be observed at undercoolings corresponding to the metastable T_o temperature in ground-based containerless electromagnetic levitation experiments utilizing the high-speed video imaging technique. Figure 1 shows a plot of experimental growth velocity as a function of the original liquid undercooling relative to the primary phase solidus. As previously mentioned, note that at any given undercooling, the stable phase grows faster into the liquid than the metastable phase. Under conditions of competitive growth, if the stable phase breaks out it will quickly overwhelm any metastable phase growth into the same liquid. From the figure, if both phases nucleate from the same site, this break out condition may only occur for undercoolings below 135 degrees where $V_{ss} = V_d$. At higher undercoolings, growth of the metastable phase into the liquid is always faster than growth of the stable phase into the semi-solid and therefore a double recalescence event, when it occurs, will always be detected. At lower undercoolings, detection of a double recalescence event, when it occurs, is not certain since the stable phase is growing faster into the semi-solid than the metastable phase can grow into the liquid . The delay between events and the relative distance between the nucleation site and the location where the detection sensor target is aimed now become important.

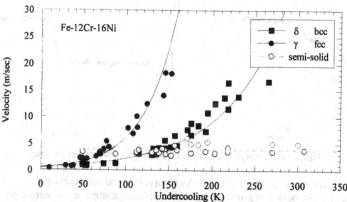

Figure 1 - Growth rate of each phase as measured in ground-based EM levitation testing
Solid symbols indicate growth into undercooled liquid
Open symbols indicate growth into the semi-solid following primary recalescence

Figure 2 shows two digital images of the same droplet in a ground-based undercooling experiment. The first image shows growth competition between the metastable phase and the stable phase. The second image shows a later time when the stable phase has overwhelmed the metastable phase and is growing freely into undercooled liquid. On each image, single pixels are selected and marked as point A near the common polar nucleation site, and point B arbitrarily chosen near the droplet equator. The corresponding intensity profile for each point is shown in the latter portion of the figure. At point A, the profile clearly shows a double recalescence event. The intensity rises in the 18th frame when the metastable phase nucleates at this location and then a second rise is seen in the 74th frame when the stable phase nucleates. At point B, only a single rise at 111th frame can be seen in the intensity profile corresponding to passage of the stable phase as growth occurs into the liquid. Thus, a sensor located at the second location would miss detection of a double recalescence event.

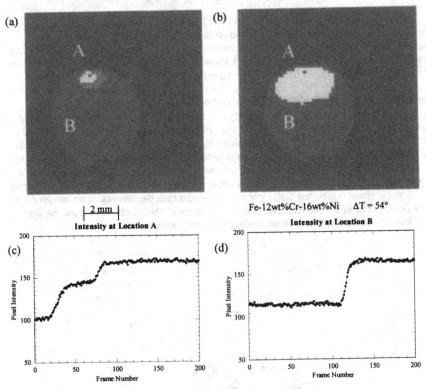

digital video images taken at a rate of 40500 frames per second

Figure 2 - Sensitivity to location of sensor location relative to nucleation site
  a) image of initial period of growth competition between austenitic and ferritic phases
  b) image taken at a time after the austenitic phase has overwhelmed the ferritic phase
  c) pixel intensity profile at a point within the growth competition region
  d) pixel intensity profile at a point outside the growth competition region
      liquid appears as dark grey, austenitic phase as grey, and ferritic phase as white

RESULTS

For the Fe-12wt%Cr-16wt%Ni alloy, measurements of the velocity for growth of the stable phase into the semi-solid were consistent using the two measurement techniques. From flight pyrometry data V_{ss} = 3.83 m/sec and from ground-based high-speed digital video analysis, V_{ss} = 3.63 m/sec. Only flight results have been obtained for the near eutectic composition of Fe-16wt%Cr-12wt%Ni with a measured value of V_{ss} = 0.59 m/sec in microgravity. Interpretation of ground-based results at this composition has proven difficult due to two effects which may be linked with the higher convective flows associated with ground-based tests.

In a majority of tests, high-speed digital imaging showed that nucleation of the second phase occurs at the same site as that of the primary phase. In some instances, ground-based images showed additional nucleation sites within the semi-solid region. At high undercoolings, this effect is not significant to phase selection in that the metastable phase growth into the liquid is fast enough to outpace nucleation site migration. It is significant in its effect on growth velocity measurement as the multiple expanding regions tend to blur together and the growth of individual clusters can not be tracked adequately. At low undercoolings, formation of a secondary nucleation site closer to the edge of the growing metastable dendritic array may cause premature break out thus affecting both phase selection and velocity measurement.

The second effect is seen as a change from nucleation of the stable phase within the metastable array to nucleation along the interface between the growing array and the undercooled liquid. This mechanism shift results in dramatic changes in phase selection as the stable phase quickly overwhelms the slower growing metastable phase, especially at low undercoolings.

With both of these effects active during ground-based experimentation, a digital video technique allows for higher confidence in the results of velocity measurement and phase selection analyses. Because the entire exposed surface is monitored, the delay time, τ_o, at the common nucleation site may be independently verified and is seen to be a slight function of the original undercooling [12]. By using these measured values for τ_o, V_d, and V_{ss}, we are able to predict when double recalescence events may be observed as a function of the distance between the nucleation site and the sensor using the growth competition model. Model predictions for a sensor located at the equator with nucleation at one of the poles is shown in Figure 3. The sharp dividing curve down the center separates the two regions of interest. In the upper region a double recalescence event that occurs will never be observed by a sensor located some critical length from the nucleation site while in the lower region it will always be detected. At the left edge of the figure, no double recalescence events will be observed as the undercooling is insufficient to access the metastable portion of the equilibrium phase diagram. To the right, the model predicts that a double recalescence event will always be observed as above this point $V_d > V_{ss}$ and the stable phase can never overwhelm the growing metastable phase.

This figure can be used to predict the undercooling limit below which an equitoral sensor will erroneously predict that a single recalescence event has occurred when, in fact, double recalescence has taken place. For a seven-millimeter sample the model predicts that this undercooling is 123 degrees. Referring back to Figure 1, this corresponds to a metastable growth velocity of V_d = 3.1 m/sec which does not significantly vary from the measured value of V_{ss} = 3.63 m/sec for this composition. From this analysis it is evident that we can indirectly measure, or at least provide a lower bound for, the growth rate of the stable phase into the metastable dendritic array.

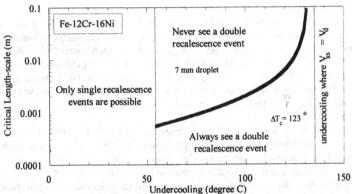

Figure 3 - Growth competition model predictions for when a double recalescence event may be observed as a function of sample undercooling and the distance between the nucleation site and where a sensor target monitors the event

In light of these observations, the data presented by previous investigators may be re-evaluated using the growth competition model to explain apparent phase selection. This work relied on evaluating the limit for when double recalescence events are observed, to predict a critical undercooling required for metastable phase nucleation. If the growth model predictions are accurate, we can re-interpret this limit as a measurement of a critical velocity as a function of system geometry. For short delay times, this velocity for growth of the metastable phase into the liquid will be equal to the growth rate of the stable phase into the semi-solid. We cannot numerically predict the exact behavior of other alloy compositions without experimental verification of the delay times since the driving force for both nucleation and growth within the semi-solid are strong functions of the difference ΔT_o. Composition changes will thus have a large effect on both V_{ss} and delay time.

Since the critical undercooling relates to a critical velocity, we may evaluate the growth rate of the metastable phase into the liquid from previous experimental results at this undercooling. Assuming that this velocity is of the same order as the growth rate of the stable phase into the metastable array we may equate these two values. As seen above the error involved in this treatment is on the order of negative fifteen percent at the alloy composition investigated. Using this approach, it should be possible to obtain an indirect measurement of V_{ss} as a function of composition from the work of previous investigators. This in turn may be used to verify predictions by a new theory (to be developed) to explain growth into a pre-existing dendritic array.

As an example, the critical undercooling data presented by Koseki [2] and Moir [4] using the pyrometry-based data collection technique were used to predict a critical velocity. These velocities are plotted in Figure 4 as a function of the alloy ΔT_o along with the measurements of V_{ss} obtained using the digital video technique[9, 12] and microgravity measurements obtained using the pyrometry-based technique [12]. It is clear that the indirect measurement technique and the direct measurement technique compare favorably.

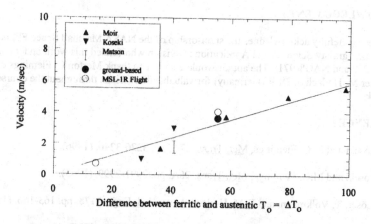

Figure 4 - Comparison between the direct measurement of the growth rate of the stable phase into the metastable dendritic array and indirect calculation of this rate from previously published sources over a wide composition range in the Fe-Cr-Ni system

At higher values for ΔT_o, corresponding to lower (Ni/Cr) ratios, we have more confidence that the values obtained are representative as the driving force for nucleation of the stable phase is higher and delay times are short. The data points in the range $25 < \Delta T_o < 45$ are therefore of reduced confidence and experimental verification using a combination of high-speed digital imaging and pyrometry in a microgravity environment is desirable.

CONCLUSIONS

The use of the low convective conditions attainable with the TEMPUS electromagnetic levitation facility in microgravity enhances the ability to investigate growth of the stable phase into the metastable array which forms during primary recalescence in Fe-Cr-Ni alloys. Longer delay times and a reduced tendency to nucleate the stable phase near the liquid-metastable solid interface have been previously shown to increase the ability to detect double recalescence events and new data supports this observation. Direct measurement of the growth rate into the semi-solid should be conducted in microgravity using both pyrometry and digital imaging techniques.

Ground-based results support the application of a new model based on growth competition, instead of on nucleation kinetics, to define apparent phase selection as a function of composition, undercooling and sensor position relative to the nucleation site. Previous results may be successfully re-evaluated within the framework of the new model and predictions for the growth of the stable phase into the semi-solid may in the future be used as an experimental verification of a model describing this new growth regime. These predictions should be corroborated in microgravity.

ACKNOWLEDGMENTS

The author gratefully acknowledges the sponsorship of the NASA Marshall Space Flight Center Microgravity Sciences and Applications Division who funded this work under contract NAG8-1230 and NAG8-971. The author would also like to thank Merton C. Flemings of MIT and Dieter M. Herlach of DLR (Germany) for valuable discussions throughout the course of this work.

REFERENCES

1. T. Koseki and M.C. Flemings, Met. Trans. **27**A, pp. 3226-3240, (1996).

2. T. Koseki and M.C. Flemings, Met. Trans. **26**A, pp. 2991-2999, (1995).

3. W. Löser, T. Volkmann, and D.M. Herlach, Mater. Sci. Eng.. **A178**, pp. 163-166, (1994).

4. S. Moir and D.M. Herlach, Acta Metall. **45**(7), pp. 2827-2837, (1997).

5. T. Volkmann, W. Löser, and D.M. Herlach, Int. J. Thermo. **17**(5), pp. 1217-1226, (1996).

6. T. Volkmann, W. Löser, and D.M. Herlach, Met. Trans. **28**A, pp. 453-460, (1997).

7. T. Volkmann, W. Löser, and D.M. Herlach, Met. Trans. **28**A, pp. 461-469, (1997).

8. L. Gránásy, J. Non-Cryst. Solids, **162**, pp. 301-303, (1993).

9. D.M. Matson, A. Shokuhfar, J.W. Lum, and M.C. Flemings, in *Solidification Science and Processing*, edited by I. Ohnaka and D.M. Stefanescu, (TMS Warrendale, PA 1996), pp. 19-26.

10. D.M. Matson, in *Solidification 1998*, edited by S.P. Marsh *et al.*, (TMS Warrendale, PA 1998), pp. 233-244.

11. Q.C. Bui, MS Thesis, Massachusetts Institute of Technology, 1998.

12. D.M. Matson, W. Loser, and M.C. Flemings, in *Solidification 1999*, edited by S.P. Marsh *et al.*, (TMS Warrendale, PA 1999).

The Isothermal Dendritic Growth Experiment: Evolution of Teleoperational Control of Materials Research in Microgravity

J.C. LaCombe, M.B. Koss, A.O. Lupulescu, J.E. Frei, and M.E. Glicksman

Materials Science and Engineering Department
Rensselaer Polytechnic Institute
Troy, NY, 12180-3590

ABSTRACT

Exactly one year ago, the Isothermal Dendritic Growth Experiment (IDGE) completed its third and final orbital space flight aboard the United States Microgravity Payload (USMP) on STS-87. The IDGE conducted 180 experiments on dendritic growth in 5-9's succinonitrile (SCN), a BCC material used on USMP-2 and USMP-3, and over 100 experiments on 4-9's pivalic acid (PVA), an FCC material used on USMP-4. IDGE film and telemetry data provide benchmark tip velocity and radii versus supercooling for critically testing transport theory and the interfacial physics of diffusion-limited dendritic growth. Post-flight application of optical tomography is providing the first tip shape data allowing quantitative tests of three-dimensional phase field calculations. Several new discoveries were made during each flight concerning the behavior of dendrites at low driving forces, and the influences of time-dependent pattern features and noise. A summary of these scientific highlights will be provided.

The IDGE instrument was upgraded on each successive flight, improving its optics and electronics, especially the capability for teleoperational control. Near real-time, full gray-scale video was accommodated on USMP-4, allowing investigation of non-steady-state features and time-dependent growth dynamics. A short example of video from space will be shown. USMP-4 science was teleoperated by a student cadre for 16 days from a remote site established by NASA at RPI. This operational experience provides valuable insights, which will be drawn upon for future microgravity experiments to be conducted on the International Space Station.

INTRODUCTION

Dendrites describe the tree-like crystal morphology commonly observed in many material systems—particularly in metals and alloys that freeze from supercooled or supersaturated melts. There is a high level of engineering interest in dendritic solidification, because of the role that dendrites play as components of the microstructures of cast alloys. Microstructure plays a key role in determining the physical properties of cast or welded products. In addition, dendritic solidification constitutes an example of non-equilibrium physics as well as dynamic pattern formation, where an amorphous melt, under simple starting conditions, evolves into a complex ramified microstructure [1].

Although it is well known that dendritic growth is controlled by the transport of latent heat from the moving solid-melt interface as the dendrite advances into a supercooled melt an accurate and predictive model of dendritic solidification has not been developed. Current theories focus on the transfer of heat or solute from the solid-liquid interface into the melt, as well as the interfacial stability and its related selection physics. However, the effects of gravity-induced convection hamper the application of these models, preventing. the basic theories from being adequately tested solely under terrestrial conditions [2].

BACKGROUND

The Isothermal Dendritic Growth Experiment (IDGE) was comprised of a series of three NASA-supported microgravity experiments, all of which flew aboard the space shuttle, *Columbia*. This experimental space flight series was designed and operated to grow and record dendrite solidification in the absence of gravity-induced convective heat transfer, and thereby produce a wealth of benchmark-quality data for testing solidification scaling laws [3, 4].

The first flight of the IDGE took place aboard STS-62, as part of the second United States Microgravity Payload (USMP-2) in March of 1994 [5], while the second flight (USMP-3) flew aboard STS-75, in February/March of 1996. Both flights used ultra-pure succinonitrile (SCN) as the test material. SCN is an organic crystal that forms dendrites similar to the BCC metals when it solidifies, making it an excellent analog material for studying many aspects of the solidification process in ferrous metals. The third and final IDGE flight (USMP-4) launched on STS-87, in December of 1997, employed a different test material, pivalic acid (PVA). PVA is an FCC organic crystal that solidifies like many non-ferrous metals. PVA, like SCN, has convenient properties for conducting benchmark experiments. However, unlike succinonitrile, PVA exhibits a large anisotropy of its crystal-melt interfacial energy, which is a key parameter in the selection of dendritic operating states.

THE EXPERIMENTS

The major results and conclusions of the three IDGE microgravity experiment sets are discussed here along with a modeling effort that was undertaken in order to better understand certain aspects of the experimental data.

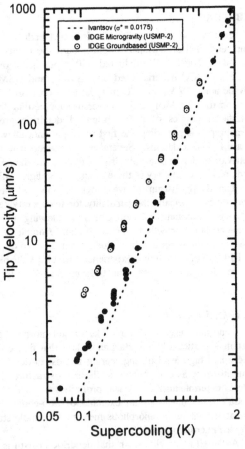

Figure 1: Velocity vs. supercooling for succinonitrile dendrites.

USMP-2 (SCN)

The main conclusions drawn from comparing the on-orbit data to terrestrial dendritic growth data (obtained using the same apparatus and techniques) are as follows. 1) Convective effects under terrestrial conditions cause growth speed increases (Figure 1) up to a factor of 2 at the lower supercoolings ($\Delta T < 0.5$ K), with convection effects remaining discernible under terrestrial conditions up to supercoolings as high as 1.7 K—far beyond what was thought. 2) In the supercooling range above 0.47 K, microgravity data remain virtually free of convective or chamber-wall effects, and may be used reliably for examining diffusion-limited dendritic growth theories. 3) The diffusion solution to the dendrite problem, combined with a unique scaling constant, σ^*, will not provide accurate prediction of the growth velocity and dendritic tip radii (Figure 2). 4) Growth Péclet numbers (Figure 3) calculated from Ivantsov's solution deviate systematically from the IDGE data observed under diffusion-limited conditions. 5) The scaling parameter σ^* (Figure 4) does not appear to be a constant, independent of supercooling. Finally, 6), the σ^* measurements from the terrestrial and microgravity data are in good agreement with each other, despite a difference of over six orders of magnitude in the quasi-static acceleration environment of low-earth orbit and terrestrial conditions [6, 7].

USMP-3 (SCN)

The second IDGE flight on USMP-3/STS-75 mostly supported the above conclusions. However, at present at a still non-final stage of the analysis, there are some important modifications. With sufficient repeated observations [8], it now appears that the terrestrial and microgravity σ^* are distinguishable, with the microgravity σ^* larger than those measured under terrestrial conditions. However, even with the built up statistic of repeated experiment cycles, the

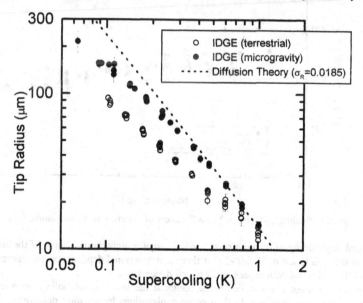

Figure 2: SCN dendrite tip radius of curvature.

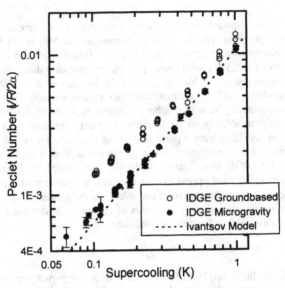

Figure 3: Péclet number as a function of supercooling for succinonitrile dendrites.

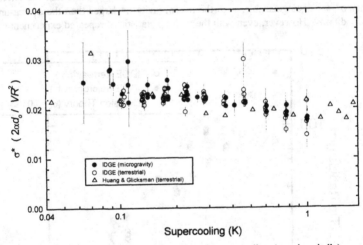

Figure 4: Scaling constant, σ* as a function of supercooling (succinonitrile).

functional dependence of σ* with supercooling remains ambiguous. Some of the additional data supports the conclusion of USMP-2 that there is a functional dependence on supercooling, while some of the additional data argues against such dependence.

The second flight also clarified some issues at the lower supercoolings such as the role of convection [9], wall proximity [10] or other explanations by showing definitively that the low-supercooling deviation from the transport solution is not due to convection [11, 12]. This data

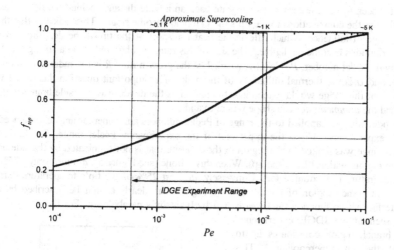

Figure 5: Fractional contribution to the thermal conditions at the tip, due to latent heat released locally in the tip region.

also led to an improved understanding of the extent to which the low temperature effects are due to wall proximity effects [13]. Finally, the second flight yielded sufficient data to make a three-dimensional reconstruction of the non-parabolic, non-body-of-revolution dendritic tip shape [14-16].

Moving Heat Source Analysis

The method of moving heat sources is applied to the problem of dendritic growth in order to examine how the actual, non-paraboloidal shape of the dendrite tip and its trailing side branch structure affects the transport process. The model describes the diffusive thermal transport processes around a body of revolution (representing the dendrite) advancing into a quiescent melt at a constant rate. The latent heat produced at each of the points along an advancing solidification front is superposed to determine the net change in temperature at an arbitrary point in the surrounding melt. The calculation is performed by prescribing the interface shape used in the model to be a shape-preserving body of revolution, which advances into the melt at a constant velocity. Once the shape and the growth velocity are specified, we solve the heat equation (formulated by the moving heat source method) to determine the net thermal field experienced by the tip of the interface which corresponds to the specified growth conditions.

Results of this work indicate that when corrections to Ivantsov's classical, infinite, parabolic tip shape are incorporated, enhanced agreement with experimental data is possible relative to the Ivantsov solution. Changes in the interface shape from Ivantsov's paraboloid essentially "shift" the predicted Pe vs. Δ (Δ is the dimensionless supercooling) relationship away from the Ivantsov result. For interface shapes that are wider than a parabola, Pe will be lower than Ivantsov's result (for a given Δ). And for an interface shape that is narrower than a parabola will raise Pe.

The superposition of the latent heat sources (mentioned above) is accomplished by integration starting at the tip, and extending back to the interface areas behind the tip. When done in this manner, it is seen to be unnecessary to integrate back an infinite distance behind the tip. Instead, far from the tip, the contributions to the tip's thermal field become small. This indicates that there is some "range of influence" which affects the heat transport at the tip of the dendrite, beyond which contributions become negligible. The size of this range is observed to be a strong function of the growth Péclet number, $Pe=VR/2\alpha$, where V is the growth rate, R is the radius of curvature of the tip, and α is the thermal diffusivity of the melt. The important question this raised was whether or not this range would extend into the region of the dendrite where side branches were present, and the *assumed* interface shape was not valid.

The model was next applied to the range of Péclet numbers and supercoolings experienced in the IDGE experiments. Using the upper end of the experimental Péclet range, $Pe \sim 0.01$, the integration range was limited to the region of the dendrite that is not dominated by the side arms (i.e. the tip region, within $12R$ of the tip). When this is done (see Figure 5), *at most*, only ~75% of the tip temperature is accounted for. Conversely, *at least* 25% of the tip's temperature derives from the side branch region of the dendrite—a region that clearly cannot be described by any simple interface shape function (and clearly, not by Ivantsov's paraboloid). For the lower Péclet numbers seen in the IDGE experiments, the side branch region contributes up to ~60% of the tip supercooling. This observation joins earlier work by Schaefer [17] in suggesting that under commonly observed growth conditions, the side branch region of a dendrite contributes significantly to the thermal conditions at the tip itself.

The importance of the side branch region has a notable implication to another aspect of the experimental data. Since there is a considerable stochastic aspect to the side branch structure, it follows that the scatter in the IDGE experimental data seen in the supercooling range above ~0.44 K (Figure 6) may be explained by the stochastic variations in the side branch structure. The significance of the side branch region (demonstrated above), combined with the random nature of the interface structure make it a reasonable explanation.

<u>USMP-4</u>

The data and subsequent analysis from the final flight experiment are currently at a preliminary stage, based on images received using telemetry from space. We compared the dendritic growth speed of PVA as a function of the supercooling to both terrestrially measured PVA data, and an estimate scaled from prior SCN

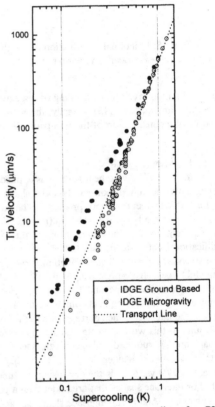

Figure 6: Velocity vs. supercooling for PVA dendrites grown under microgravity conditions.

microgravity data. The preliminary results of these tests indicate that the PVA data are in good agreement with the SCN data (Figure 6). This implies that dendritic growth in PVA is, like SCN, diffusion-limited, with little, if any, kinetic response. This observation conflicts with the conclusion reached by other investigators that there are large interfacial kinetic effects in PVA.

Figure 7 shows some preliminary data assessing the nature of boundary layer interactions during growth of PVA dendrites. We see that when the nearest neighbor distance exceeds about two thermal diffusion distances, λ, where the diffusion distance $\lambda = \alpha/V$ (α is the diffusivity and V is the tip speed) the velocity levels off at its maximum steady–state rate. When nearest neighbor spacings fall below about 2λ, the velocity is reduced through thermal interactions of the boundary layers. This phenomenon of neighbor interactions has never been observed before, because microgravity conditions are needed to insure growth limited by thermal diffusion from the solid–melt interface.

Currently we are extracting more accurate velocity and tip radius, shape, and side-branching measurements from post-flight 35mm film and videos. The video data, obtained at 30 frames per second is revealing transient aspects of the experimental process that have not been identified during the first two USMP missions. Figure 8a illustrates the displacement vs. time of a PVA dendrite grown under ground-based conditions at a supercooling of 0.5 K. The plot outwardly appears to exhibit a constant growth rate, though a detailed examination of these ~1900 data points suggests otherwise. Figure 8b shows the residuals resulting from a straight-line regression through the displacement data. The spread in the residuals over any small window in time is indicative of the 20μm pixel size of the video camera. However, the trend of the residuals indicates that a constant velocity does not properly describe the displacement behavior, and in fact, the growth rate is super-linear in time. This transient behavior is observed over nearly the entire range of supercoolings, and in microgravity experiments as well. The mechanism for this

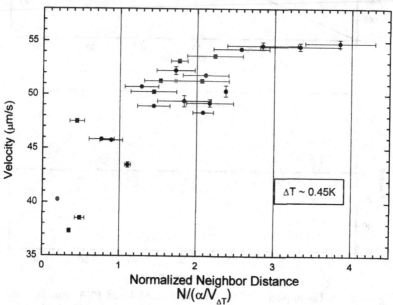

Figure 7: PVA dendrite growth velocity versus distance to nearest neighbor.

behavior is not clear at the present time. Possible causes include container interactions and interactions with other equiaxed arms of the dendrite.

Telescience
    In addition to our investigation of dendritic solidification kinetics and morphology, the IDGE has participated in the development of remote, university-based teleoperations. These trials and tests point the way to the future of microgravity science operations on the International Space Station (ISS). NASA headquarters and the Telescience Support Center (TSC) at LeRC, set a goal for developing the experience and expertise to set up remote, non-NASA locations from which to control space station experiments. Recent IDGE space shuttle flights provide proof-of-concept and tests of remote space flight teleoperations [18].

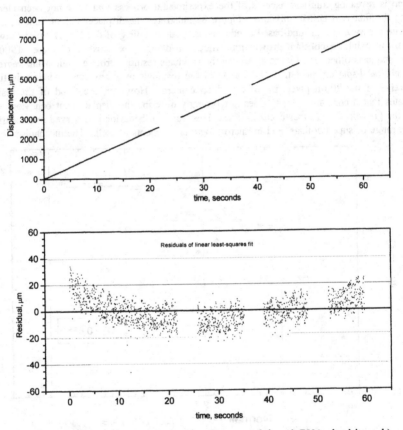

Figure 8: a) Tip displacement vs. time for ground based PVA dendrite.   b) Residual plot from a linear regression contains a trend, indicating that the dendrite is accelerating over time.

242

SUMMARY AND CONCLUSIONS

The data and analysis performed on the dendritic growth speed and tip size in SCN demonstrates that although the theory yields predictions that are reasonably in agreement with experiment, there are significant discrepancies. However, some of these discrepancies can be explained by accurately describing the diffusion of heat. One key finding involves the recognition that in order to satisfactorily characterize the transport process, it is necessary to represent the dendrite's interfacial morphology ranging from the tip, and extending back, well into the side branching region. When such corrections are incorporated into our understanding, the role of heat transfer in dendritic growth is validated, with the caveat that a more realistic model of the dendrite than a paraboloid is needed to account for heat flow in experimentally observed dendrites. We are currently conducting additional analysis to further confirm and demonstrate these conclusions.

The data and analyses for the growth selection physics remain much less definitive. From the first flight, the data indicated that the selection parameter, σ^*, is not exactly a constant, but exhibits a slight dependence on the supercooling. Additional data from the second flight are being examined to investigate the selection of a unique dendrite speed, tip size and shape.

The IDGE flight series is now complete. We are currently completing analyses and moving towards final data archiving. It is gratifying to see that the IDGE published results and archived data sets are being used actively by other scientists and engineers. In addition, we are also pleased to report that the techniques and IDGE hardware system that the authors developed with NASA, are being currently employed on both designated flight experiments, like EDSE, and on flight definition experiments, like TDSE.

ACKNOWLEDGEMENTS

The authors would like to thank NASA's Life and Microgravity Sciences and Application Division, (Code U), Washington, DC, for support under NASA Contract No. NAS3-25368 as well as the NASA Graduate Student Researcher Program.

REFERENCES

1. M.E. Glicksman and S.P. Marsh, The Dendrite, in Handbook of Crystal Growth, D.T.J. Hurle, ed., (Amsterdam: Elsevier Science Publishers, 1993), 1077-1122.

2. M.E. Glicksman and S.C. Huang, Convective heat Transfer During Dendritic Growth, in Convective Transport and Instability Phenomena, Z.a.O. Karlsruhe, ed., : , 1982), 557.

3. M.E. Glicksman, M.B. Koss, and E.A. Winsa, "The Chronology of a Microgravity Spaceflight Experiment: IDGE," Journal Of Metals, 47 (8) (1995), 51-54.

4. M.E. Glicksman, et al., Metallurgical Transactions A, 19A (1988), 1945.

5. M.E. Glicksman, M.B. Koss, and E.A. Winsa, "Dendritic Growth Velocities in Microgravity," Physical Review Letters, 73 (4) (1994), 573-576.

6.   M.E. Glicksman, et al., "Dendritic Growth of Succinonitrile in Terrestrial and Microgravity Conditions as a Test of Theory," Iron and Steel Institute of Japan International, 35 (6) (1995), 604-610.

7.   M.B. Koss, et al., "Manuscript submitted to Metallurgical and Materials Transactions," Unpublished Research, (1998),

8.   A.O. Lupulescu, et al., *Unpublished research*, 1998.

9.   R.F. Sekerka, S.R. Coriell, and G.B. McFadden, "Stagnant Film Model of the Effect of Natural Convection on the Dendrite Operating State," Journal of Crystal Growth, 154 (1995), 370-376.

10.   V. Pines, A. Chait, and M. Zlatkowski, "Thermal Diffusion Dominated Dendritic Growth - An Analysis of the Wall Proximity Effect," Journal of Crystal Growth, 167 (1996), 383-386.

11.   M.B. Koss, et al., "The Effect of Convection on Dendritic Growth Under Microgravity Conditions," Chemical Engineering Communications, 152-153 (1996), 351-363.

12.   L.A. Tennenhouse, et al., "Use of Microgravity to Interpret Dendritic Growth Kinetics at Small Supercoolings," Journal of Crystal Growth, 174 (1997), 82-89.

13.   L.A. Tennenhouse, et al., *Unpublished research*, 1998.

14.   J.C. LaCombe, et al., "Three-Dimensional Dendrite-Tip Morphology," Physical Review E, 52 (3) (1995), 2778-2786.

15.   J.C. LaCombe, et al., Dendrite Tip-Shape Characteristics, *proceedings of* Materials Research Society Fall Meeting. 1995. Boston, MA: MRS.

16.   J.C. LaCombe, et al., *Unpublished research*, 1998.

17.   R.J. Schaefer, "The Validity of Steady-State Dendrite Growth Models," Journal of Crystal Growth, 43 (1978), 17-20.

18.   M.B. Koss, et al., Development of a University Based Remote Teleoperations Site for the Performance of Experiments in Microgravity, *proceedings of* 8th International Symposium on Experimental Methods for Microgravity Materials Science. 1996: The Minerals, Metals & Materials Society.

ZEOLITE CRYSTAL GROWTH IN SPACE

A. Sacco, Jr., N. Bac, J. Warzywoda, G. Rossetti, Jr., and M. Valcheva-Traykova
The Center for Advanced Microgravity Materials Processing,
Northeastern University, Boston, MA 02115, asacco@coe.neu.edu

ABSTRACT

The extensive use of zeolites and their impact on the world's economy has resulted in many efforts to characterize their structure and improve the knowledge base for nucleation and growth of these crystals. Zeolite crystal growth (ZCG) experiments have been conducted in space by many researchers. The results have been varied with little overall constancy with respect to the types of zeolites grown, the procedures used, and methodologies employed for comparison of terrestrial and microgravity processed crystals. Results from this laboratory have indicated that solutions must be mixed in space, and that the nucleation mechanism as well as the growth mechanism is affected by the "suppression" of fluid motion. Earlier flight experiments (STS-43, STS-50 and STS-57) indicated that by controlling the nucleation event, the size of the crystals could be increased with improved structural quality compared to the ground-based controls. This was hypothesized to be the result of hindrance of the solution/dissolution mechanism caused by reduced fluid motion. USML-2 (STS-73) zeolite experiments were designed to further enhance the understanding of nucleation and growth of zeolite crystals, while attempting to provide a means of controlling the defect concentration in microgravity. Zeolites A, X, and Silicalte were grown during the 16-day United States Microgravity Laboratory number 2 (USML-2) mission. All zeolites were successfully grown in the 16 day mission. The zeolite A and X crystals were in general larger in dimension (10-50%), but not substantially different in crystal perfection as measured by their lattice parameters. However, They were significantly smoother as indicated by AFM. Zeolite β as well as Silicalite were only slightly larger in size but, appear to have different distributions of Aluminum atoms as determined by selective catalytic reactions.

INTRODUCTION

Zeolites are crystalline aluminosilicates with cations to balance the framework charge. Zeolites have an extensive three-dimensional network of oxygen ions. Situated within the tetrahedral sites formed by the oxygen atoms can be either an Al^{+3} or Si^{+4} ion. This unique structure has proven to be extremely effective in processing materials at the molecular level and has led to zeolites becoming the backbone of chemical and petrochemical process industries. The extensive use of zeolites in refining (Fluidized Catalytic Cracking [FCC]), petrochemical processing, laundry detergents, agriculture, and environmental cleanup as catalysts, adsorbents and ion exchangers caused the market to mushroom in the mid-1980's, reaching a worldwide sales of about $2 billion. The existing markets for zeolites are expected to grow at a rate of 3 to 4 percent through the year 2000 [1]. Continued growth for zeolites includes cultivating markets beyond petroleum refining and detergents. New applications for these molecularly selective materials are in areas such as selective membranes, chemical sensors, molecular electronics, quantum-confined semiconductors and zeolite-polymer composites. Because of their ability to "select" individual molecules to chemically and physically process, extensive research is on going to learn how to "customize" these materials to increase their catalytic and adsorption efficiency. The promise of controlled fluid motion and the associated effects on heat and mass transfer made available in low earth orbit provides a unique opportunity to study the nucleation and growth process.

EXPERIMENTAL

Thirty-eight experiments were performed while in low earth orbit. These thirty-eight experiments represented nineteen distinct zeolite formulations: 6 were zeolite A formulations, 6 were X formulations, 4 were zeolite β compositions and 3 were silicalite solutions.

Equipment

Two identical experiments were performed for each of these different formulations. Each experiment was performed in individual reactors ("autoclaves"), and two such units were combined to produce one experiment. These "autoclaves" consisted of two chambers: one held the aluminate solution, the other the silicate solution. They were Teflon lined and were build of aluminum and titanium. The aluminum autoclaves were used in the temperature region 25^0C to 150^0C. Titanium autoclaves were used for the temperature region of 150^0C to 225^0C. Nineteen experiments, one for each of the nineteen different solutions, were performed in transparent autoclaves identical in internal construction to the metal autoclaves. These were used to visually determine the mixing characteristics and protocols for mixing all the experimental solutions. This was necessary not only to maximize the homogeneity of the solutions, but also to minimize the formation of bubbles or foams. The formation of regions of in-homogeneity would result in extraneous zeolite impurity phases. The formation of small bubbles increases surface for heterogeneous nucleation [2], and thus makes nucleation control difficult.

The Zeolite Crystal Growth Furnace is described elsewhere [3] so only a brief description will be given here. The furnace was designed to take up the equivalent of two mid-deck lockers ($\cong$ 4.5 sq. ft.). Maximum heat up power was not to exceed 225 watts for 10 hr and a steady state heat loss of only 150 watts was allowed. The furnace had a circular cross section that contained nineteen separate, redundantly powered and controlled heater tubes. The autoclaves, consisting of two separate autoclave "reactors", were positioned in each of these nineteen tubes. Temperature control was good to 1^0C axially across the solution chamber, and radially to about 0.1^0C. The furnace was configured as three zones radiating from a hot center region: Zone 1($180-225$ ^0C) containing one set of autoclaves; then a middle zone: Zone 2 ($105-150^0C$) which contained six sets of autoclaves; and finally an outer zone: Zone 3 (96^0C) that contained the remaining twelve sets of autoclaves. The furnace was controlled, data was archived and faults monitored and annotated with two independent CPU's. Control temperatures could be re-set, and other corrective actions taken by a payload specialist, if required.

Methodology

Earlier work in this laboratory [4,5] has shown that if large crystals were desired a nucleation control agent was necessary. In the results presented here, 2,3'2"-nitrilotriethanol (TEA) as well as 2,2-Bis(hydroxymethyl)-2,2',2"-nitrilotriethanol (BIS) where used as nucleation control agents. These were used for both the Zeolite A and X formulations. In the results presented here, zeolite β and silicalite formulations were not processed with any attempt to control the nucleation or growth events.

RESULTS

Four different zeolite types were crystallized during the USML-2 mission. The results for zeolites A, X, Beta and Silicalite are reported here. The zeolite crystals grown in microgravity and their terrestrial/controls are compared with techniques including Scanning Electron Microscopy (SEM), Particle Size Distribution (PSD), X-ray Diffraction (XRD), Transmission Electron Microscopy (TEM), and Atomic Force Microscopy (AFM).

Zeolite A

Figure 1 illustrates the kind of results observed for zeolite A. Typically, the results were and increase in linear dimension of 10-50% compared to the ground based controls. The largest zeolite A crystals formed were approximately 85 μm or about 42 times greater than the commercial crystals (~2 μm). This corresponds to a volume increase of about 77,000 times for the flight crystals over the commercial crystals. Figure 2 illustrates the PSD's for the samples shown in Figure 1. As indicated microgravity substantially effects the nucleation and growth history. In the case of zeolite A there was typically two crystal populations, one large and another smaller. These PSD's must be evaluated with caution, for often in the case of ground-based controls, there is quite a large amount of agglomeration which results in a false shift in average crystal size. Flight samples show no agglomeration. This type of PSD is typical of many of the A solutions made with TEA. Earlier lattice parameter studies [6, 7] had indicated that, although the difference was small, the crystals grown in microgravity had smaller unit cells (e.g., 1848.97 $Å^3$ versus 1857.10 $Å^3$ terrestrial control). A more detailed analysis has shown that there is little if any significant difference in lattice parameters for zeolite A. This suggests, at least in simple systems such as zeolite A, composition plays a more important role than does the fluid dynamics in lattice perfection.

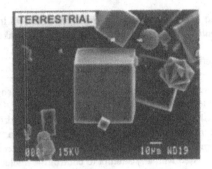

Figure 1 Typical results observed for Zeolite A

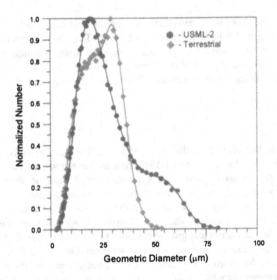

Figure 2 PSD's of Zeolite A Crystals Grown in Orbit Compared to Terrestrial Control (TEA used as nucleation control agent)

Figure 3 illustrates the crystals formed of zeolite A using the nucleation control agent BIS. Here the use of BIS produced a more uniform population as illustrated in Figure 4. For this formulation the larger particle size population dominates the distribution. This suggests that with BIS as the nucleation control agent the nucleation and growth history is different than that observed for terrestrial zeolite A. Again, an analysis of these crystals indicates that the lattice parameters show only slight differences. However, clearly the effect of processing in space is substantially different for these two different nucleation control agents.

Figure 3 Zeolite A Crystals Produced in Space and their Ground Based Controls (BIS used as a nucleation control agent)

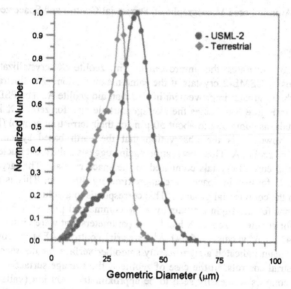

Figure 4 PSD's of Space and Terrestrial Control Crystal (BIS used as a nucleation control agent)

Atomic Force Microscope micrographs (Figure 5) also indicate a much smoother surface for crystals grown in space. The space-grown crystals show regular layered growth for the Zeolite A crystal and exhibit the square growth planes that lead to A's cubic structure. The terrestrial/control counterpart, however, exhibits a rough surface with an unusual "liquid-drop-like" surface structure. At this time it is believed that this "drop-like" structure is the result of un-reacted "gel'. This could not be washed off without changing the surface morphology.

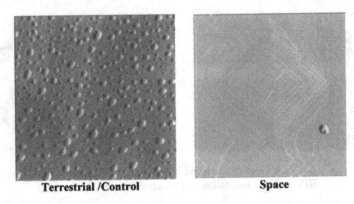

Terrestrial /Control　　　　　　　**Space**

Figure 5 Atomic Force Micrographs of Terrestrial/Control and Space zeolite A Crystals

Zeolite X

Figure 6 illustrates the improvements in zeolite X crystallization with SEM micrographs of USML-2 crystals at the same magnification. As illustrated in Figure 6, zeolite X shows greater improvement in size than did zeolite A. The PSD for the crystals illustrated in Figure 6 indicates the average particle size for zeolite X is 75 μm for the space crystals, as compared to about 50 μm for their terrestrial control (Figure 7). More important, however, is the observation that the earth based system shows a large population of zeolite A. This observation again suggests a different nucleation history for the terrestrial control crystals compared to the space crystals. The largest particle sizes for zeolite X formed in space were approximately 300 μm. This is about 107 times larger than the commercial product. This corresponds to a volume increase of about 1.24 million times for the flight crystals over the commercial product. Similar to what was observed for zeolite A, zeolite X had lattice parameters that were virtually the same (e.g., flight 15343.25 $Å^3$ versus 15354.41 $Å^3$ terrestrial control). AFM micrographs of zeolite X crystals again indicated a significantly smoother surface for the space crystals versus their terrestrial controls. In the case of zeolite X the average surface morphology of the terrestrial controls was easily seen to be approximately 300 nm (valley to peak). The surface morphology of the space grown crystals was less than 20 nm. This appears to be typical for all space grown crystals of zeolite A as well as zeolite X. Smoothness in the surface indicates a more controlled (i.e., slower) growth process.

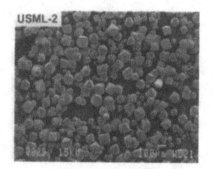 

Figure 6 Improvements from Space Research – Zeolite X

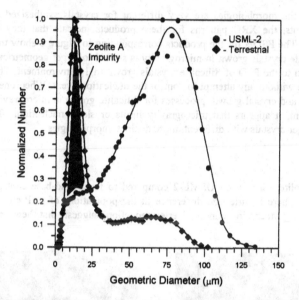

Figure 7 PSD's of Zeolite X Crystals Comparing Space and Grown Controls

Silicalite

Figure 8 illustrates typical results from the Silicalite synthesis using untreated silica gel (without nucleation control). SEM micrographs shown in Figure 8 indicate clearly that the intergrown morphology of Silicalite crystals grown in microgravity is substantially different from the morphology of Silicalite crystals grown in a 1g environment. The terrestrial crystals were in the form of spherulitic agglomerates, while the microgravity products appear to consist of fewer, well-formed crystals.

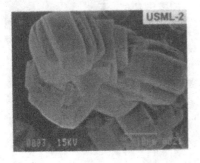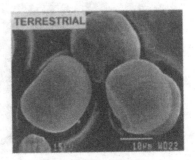

Figure 8. SEM Micrographs of Silicalite from Untreated Silica Gel

Although the morphologies are very different for crystals synthesized in µg and 1g environments, the XRD patterns for these products indicate that they both are pure Silicalite. The PSD's for the products corresponding to Figure 8 show that the PSD of the Silicalite crystals grown in microgravity is shifted to larger geometrical diameters in comparison to the PSD of Silicalite crystals grown in 1g environment. The silicate gels were made without any attempt to control the nucleation event. When one compares the nucleation and crystal growth processes for Silicalite grown in microgravity and in a 1g environment, it appears that microgravity limits or slows nucleation. This results in fewer, larger crystals with different, more distinct morphologies.

Zeolite β

The zeolite β grown on USML-2 compared to its ground-base control is shown in Figure 9. There is little size difference in the populations of β. Preliminary catalytic testing (using Meervein-Ponndorf-Verley reduction) suggests that there are fewer Lewis active

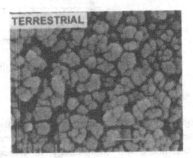

Figure 9 Zeolite β: Flight as Compared to Terrestrial Control Crystals

sites in the space crystals than in the ground-based controls. This implies that the aluminum distribution is different. At present, more catalytic tests are being performed as well as a more detailed XRD (e.g., line profile analysis) is being done to verify this finding.

CONCLUSIONS

Zeolite A and zeolite X samples grown in microgravity show enhanced size of the largest crystals (10-50%), and often in the average crystal size in comparison to their terrestrial/control counterparts. Earlier data [7] suggested that smaller unit cell volumes of the flight samples in comparison to the terrestrial/control samples appeared to be consistent with fewer lattice defects in their structure. However, more sophisticated analysis of the lattice parameters indicate little if any difference in these parameters. However, it must be recognized that the formulations discussed here were different from our earlier flights. From our earlier work, we have learned how to control the solubility of the silica source, in effect simulating the slowing of the growth and nucleation process that occurs naturally in microgravity. Thus, it was expected that the lattice defects in our latest crystal samples should be similar. This effect is clearly seen in more complex system such as silicalite and zeolite β.

Flight Silicalite crystals grown from untreated silica gel (no attempt to control nucleation or growth) yielded larger crystals with an intergrown morphology. This morphology is different in comparison to terrestrial crystals synthesized from untreated silica gel, which grew as spherulitic agglomerates. No agglomeration is observed in the flight samples due to the lack (or reduction) of settling of the silica gel substrate and/or the Silicalite product.

Preliminary results from analysis of zeolite β suggest that the aluminum distribution in the flight crystals may be different from that of the terrestrial controls. This is somewhat consistent with earlier results that indicate the structure of the zeolites produced in space have fewer defects. However, at this time this conclusion needs to be substantiated.

ACKNOWLEDGMENTS

The authors acknowledge NASA for funding. Thanks are extended to the crew of STS-73, especially Ken Bowersox, Kathryn Thornton, Fred Leslie, Michael Lopez-Alegria, Kent Rominger, and Catherine Coleman for a great job in orbit. In addition thanks are extended to Lisa McCauley of Battelle for her support, and Jack Ferraro (WPI), and the ZCG assembly crew at KSC: Ipek Guray, Teran L. Sacco, Michelle Marceau, and Robert Whitmore (Battelle).

REFERENCES

1. Chemical Week, **35**, June 5, 1996.
2. E. N. Coker, A. G. Dixon, R. W. Thompson, and A. Sacco, Jr., Microporous Materials 3 pp.637-646 (1995).

3. M. R. Fiske and R. A. Olsen, (AAIA Paper 92-0785, 30th Aerospace Sciences Meeting, Reno, NV, January 1992.
4. G. Scott, R. W. Thompson, A. G. Dixon, and A. Sacco, Jr. Zeolites 10, pp.44-49 (1990).
5. M. Morris, A. G. Dixon, A. Sacco, Jr., and R. W. Thompson Zeolites 13, pp. 113-121 (1993).
6. A. Sacco, Jr., N. Baç, E. N. Coker, A. G. Dixon, J. Warzywoda, and R.W. Thompson Joint L+1 Year Science Review of USML-1 and USMP-1, NASA Conference Publication 3272, May 1994.
7. A. Sacco, Jr., N. Baç, J. Warzywoda, I. Guray, M. Marceau, T. L. Sacco, and L. M. Whalen (AIP Proc. 420, Space Technology and Applications International Forum, Part 2, Albuquerque NM 1998), pp.544-549.

MICROGRAVITY SOLIDIFICATION OF IMMISCIBLE ALLOYS

J. B. Andrews*, L. J. Hayes*, Y. Arikawa*, S. R. Coriell**,
*Department of Materials and Mechanical Engineering
University of Alabama at Birmingham
Birmingham, AL 35294
**Metallurgical Division, National Institute of Standards and Technology
Gaithersburg, MD 20899

ABSTRACT

This paper covers findings obtained from the microgravity directional solidification of immiscible aluminum-indium alloys during the Life and Microgravity Spacelab Mission in 1996. Three alloys, one of monotectic composition and two alloys containing an excess of indium above the monotectic (i.e., hypermonotectic compositions) were solidified using the Advanced Gradient Heating Facility (AGHF). Samples were processed in specialized ampoule assemblies containing pistons and a high temperature spring in a partially successful attempt to prevent void formation due to thermal contraction of the melt and solidification shrinkage. A comparison of compositional variations between microgravity processed and ground processed samples revealed compositional variations along the length of ground processed samples which were representative of results anticipated due to convective mixing in the melt. Flight samples showed an initial compositional variation indicative of minimal mixing in the melt. However, a discontinuity in the microstructure was observed which coincided with the presence of a void in the flight sample.

INTRODUCTION

Many alloys that exhibit immiscible behavior in the liquid state (e.g. hypermonotectic alloys) show great potential for engineering applications. However, in order to obtain the potentially beneficial characteristics for these alloys it is necessary to produce the proper microstructure. Alloys intended for use as catalysts, electrical contacts and bearings can make use of a microstructure consisting of one phase dispersed in the other. Other possible applications such as superconducters and high performance magnets, require a more difficult to obtain aligned fibrous microstructure. Unfortunately, controlling the solidification process in order to produce desirable microstructures in these alloys is challenging. An improved understanding of solidification processes is needed, but the study of solidification processes on Earth is hindered by inherent flows which occur in these systems and by sedimentation of the more dense of the two liquid phases.

The objective of this ongoing effort is to gain an improved understanding of solidification processes in immiscible alloy systems. A portion of the study involves the development of experimental techniques, which will permit solidification of immiscible alloys to produce aligned microstructures [1]. A parallel effort is underway to develop a model for the solidification process in these alloy systems [2]. Results from experimentation will be compared to those predicted from the model and utilized to improve the model. In order to permit solidification under the conditions necessary to form fibrous structures in these immiscible alloys, experimentation must be carried out under low-gravity conditions.

Mat. Res. Soc. Symp. Proc. Vol. 551 © 1999 Materials Research Society

The first phase of the Coupled Growth in Hypermonotectics (CGH) experiment flew aboard the Life and Microgravity Spacelab (LMS) mission during the summer of 1996. Processing conditions were controlled in an attempt to force the production of an aligned fibrous phase in the microstructure of processed samples. The Advanced Gradient Heating Facility (AGHF) was used to directionally solidify the immiscible alloys. Alloys in the aluminum-indium system were studied and specialized aluminum nitride ampoules were used in order to control several undesirable effects that are sometimes observed during low gravity processing. Three alloy compositions were processed during this mission in order to permit comparison with the model over a composition range. Two ground based control samples were subsequently processed under conditions identical to those of two of the flight samples in order to allow a direct comparison of the results.

BACKGROUND

A phase diagram for the aluminum-indium immiscible alloy system being studied in this investigation is shown in Figure 1 [3]. One of the most prominent features of this phase diagram is the miscibility gap in which two separate liquids co-exist over a temperature and composition range. It is also apparent that a transformation can occur in this system which involves the decomposition of one liquid phase to form a solid and another liquid phase (i.e. the monotectic reaction $(L_1 \rightarrow S_1 + L_2)$.

The monotectic reaction is quite similar to the eutectic reaction $(L \rightarrow S_1 + S_2)$ which occurs in many alloy systems. It is well known that many eutectic alloys can be directionally solidified to produce an aligned fibrous microstructure. Since monotectic systems fit into the "eutectic type" category it is tempting to hypothesize that similar processing conditions could be used to produce aligned fibrous microstuctures in monotectic alloys. However, the fact that one of the product phases in a monotectic is a liquid can lead to substantially different behavior than that observed in eutectic systems where both product phases are solids.

In a "eutectic type" system, fibrous composite structures can be formed during directional solidification as a result of the unmixing of the liquid phase to form the two product phases. In this reaction and as shown schematically in Figure 2, the solute rejected during the formation of one of the product phases is consumed in the formation of the other. This "coupled growth" process has been observed in eutectic systems and in some monotectic systems for alloys of monotectic composition.

It is well known that in eutectic alloy systems, alloys which are off of eutectic composition can be directionally solidified to produce fibrous microstructures [4,5]. In fact, the use of off-eutectic compositions provides a means of controlling the volume fraction of the fibrous phase. However, there has been some controversy whether off-monotectic alloys, especially those with solute contents higher than the monotectic (i.e. hypermonotectic alloys), can be directionally solidified under conditions which would lead to a stable coupled growth process and the development of a fibrous microstructure.

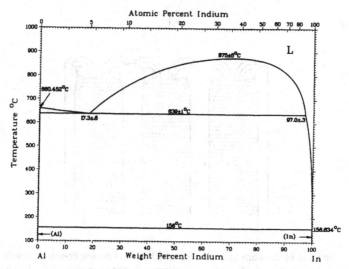

Figure 1. Phase diagram for the aluminum-indium alloy system

Interface Instability

Steady state coupled growth in an off-monotectic alloy should involve two diffusion fields. One of these fields would be associated with the "unmixing" process, which is part of coupled growth. This coupled growth diffusion field should only extend into the liquid a short distance, approximately one half the interfiber spacing. A separate, much larger diffusion field would arise due to the alloy being off of monotectic composition. The composition of the liquid far removed from the solidification front would be equal to the alloy composition, while the composition at the front would be close to the monotectic composition as is shown in Figure 3a. This composition variation forms a solute depleted boundary layer.

The presence of a solute depleted boundary layer in the liquid at the solidification front can cause interfacial instability. The composition variation with position in the liquid results in a variation in the temperature at which the miscibility gap would be entered (See Figure 3b). Just adjacent to the interface, a temperature as low as the monotectic temperature can be tolerated without entering the miscibility gap. However, as the composition increases with distance from the interface, the temperature at which the miscibility gap would be entered increases as well. If the thermal gradient in the sample is insufficient to keep the local temperature above the local miscibility gap temperature, the second phase immiscible liquid can form droplets in advance of the solidification front. This interface instability could disrupt the coupled growth process. Using a simple constitutional supercooling analysis [6], the conditions which must be met in order to avoid interface instability are given by the relationship.

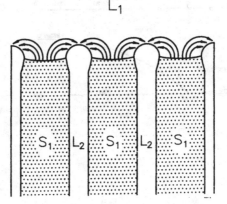

Figure 2. Schematic of the unmixing process which occurs during coupled growth. The solute rejected during formation of the S_1 phase is consumed during formation of the L_2 phase.

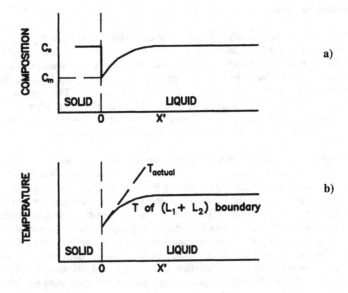

Figure 3. a) The solute depleted boundary layer anticipated in the liquid during directional solidification of a hypermonotectic alloy. b) The variation in the temperature at which the L_2 phase will form as a function of distance. If the temperature gradient in the sample keeps the local temperature above the curve, interface stability is possible.

$$\frac{G_L}{V} > \frac{m_L \left(C_o - C_m \right)}{D_L} \qquad (1)$$

where
G_L = temperature gradient in the liquid
V = solidification front velocity
m_L = slope of the $(L_1 + L_2)$ two-phase boundary
C_o = alloy composition
C_M = monotectic composition
D_L = diffusivity of solute in the liquid

If the above conditions are met, it should be possible to produce an aligned fibrous composite microstructure even in a hypermonotectic alloy. It is important to note that since the formation of the L_2 phase in advance of the solidification front is suppressed, difficulties due to sedimentation of the L_2 phase during directional solidification would be avoided. However, there are other gravity driven phenomena that make ground based experimentation difficult.

Convective Instability

While the use of a high thermal gradient to growth rate ratio should make it possible to achieve steady state coupled growth during directional solidification, the solute depleted boundary layer appears to give rise to undesirable fluid flow in the sample. In almost every known immiscible alloy system the solute has a higher density than the solvent. This implies that for vertical solidification, the solute depleted boundary layer in advance of the solidification front will have a lower density than the liquid above it, as shown in Figure 4. This density variation is expected to result in convective flows which can lead to difficulties during solidification. In most cases the resulting convective flow can produce compositional variations that are sufficient to prevent coupled growth over a substantial portion of the sample [7].

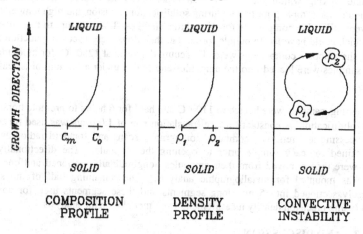

Figure 4. The anticipated composition variation in the liquid in advance of the solidification front and the resulting variation in density. The presence of a low density liquid beneath a heavier liquid can lead to convective instability.

In order to obtain steady state coupled growth conditions in immiscible alloys both interfacial stability and convective stability are required. It should be possible to obtain interfacial stability by directionally solidifying alloys using a high thermal gradient to growth rate ratio. This requirement can best be met by using a furnace that can produce a high thermal gradient in the sample during solidification. Convective stability presents more of a problem. Analysis implies that the only way to reduce convective flows to the level required for this study is by carrying out directional solidification under microgravity conditions.

Microgravity processing has the potential benefit of allowing solidification of hypermonotectic alloys under steady state growth conditions. The steady state conditions obtained will provide the opportunity to prove that interface stability and coupled growth are possible in hypermonotectic alloys and can result in the production of aligned microstructures. By processing samples over a composition and growth rate range, data can be obtained which is imperative for the verification and extension of the solidification modeling work underway. In addition, new data will be made available on interface stability limits in these alloy systems.

PROCESSING AND ANALYSIS

For this investigation aluminum-indium alloys were produced from five nines purity components by vacuum induction melting. The samples were sealed under vacuum in aluminum nitride ampoules which contained several sample segments, pistons and a spring in order to maintain the desired growth conditions during processing. Thermocouples were cemented in grooves along the exterior of the ampoule in order to monitor temperatures and determine thermal gradients and growth rates.

A great deal of care was used in the preparation of the alloys and ampoule assemblies in order to minimize the presence of any free surfaces or voids in the samples during processing. Free surfaces were undesirable because the surface tension driven flow at those surfaces could lead to undesirable mixing within the sample. In order to avoid free surfaces generated due to contraction of the sample prior to and during solidification, a piston and high temperature carbon spring were utilized to compensate for the volume changes. In an attempt to reduce any residual gasses which could have led to bubble formation, the alloys were vacuum induction melted. In addition, the ampoule components were all vacuum degassed at 1250° C for more than 6 hours. Finally, samples were loaded into the ampoules and sealed under a vacuum of at least 1×10^{-4} Torr.

For processing, samples were heated to 1100°C and held for 6 hours to promote homogenization prior to solidification. A constant furnace translation rate of 1.0 µm/s was used for all samples. After processing and retrieval, computed tomography scans and standard radiographic images were obtained for each sample prior to opening the ampoule. The directionally solidified samples were then removed from their ampoules, longitudinally sectioned and one half of the sample was mounted for metallographic analysis. The remaining half of the sample was transversely sectioned into 5 mm long segments and these segments used for compositional analysis (using precision density measurements) and metallographic analysis.

RESULTS AND DISCUSSION

The Coupled Growth in Hypermonotectics project involves the directional solidification of a range of alloy compositions over a range of growth rates and will eventually utilize facilities aboard the International Space Station. For the Life and Microgravity Spacelab (LMS) mission, three alloy compositions in the aluminum-indium immiscible system were directionally solidified. The Advanced Gradient Heating Facility (AGHF) was used to directionally solidify these samples at a furnace translation rate of 1.0 μm/s.

Void Formation

Radiographic analysis of the flight and ground based samples revealed the presence of several voids in two of the flight samples. These voids were undesirable for several reasons. First, the voids provide free surfaces, which, with the high thermal gradients in this experiment, can allow surface tension induced flow and undesirable mixing in the liquid. There is also a potential problem with inconsistency in growth conditions along the sample. The voids will obviously change the local cross sectional area of the sample and result in local variations in both thermal gradient and growth rate. These variations in local conditions complicate analysis of the samples.

One of the questions being addressed in this study concerns the source of the voids. This topic has been covered in a preliminary manner in another paper [8] and will be addressed in detail in a subsequent publication. However, a brief discussion of void formation is warranted here.

There are several factors that could have led to the development of voids in these samples. For example, if the high temperature spring failed or if one of the pistons stuck, the ampoule would not be able to compensate for thermal contraction in the melt or for solidification shrinkage. This type of failure would result in the formation of voids. Another possibility is leakage of gas past the seal at the bottom of the ampoule. Gas that seeped into the ampoule could result in bubble formation in the melt. Each ampoule was helium leak tested after assembly in order to test for this possibility. However, it is conceivable that changes in the integrity of the seal occurred during subsequent handling or during heating for processing.

Another possibility for gas bubble formation involves gas released within the ampoule either due to decomposition of a component at temperature or due to a reaction. A reaction between the aluminum or indium and the aluminum nitride ampoule is not anticipated. In addition, reactions between the ampoule and carbon spring were not observed during testing. The vapor pressures of aluminum and indium are both rather low at the temperatures utilized and should be easily overcome by the spring pressure. Investigations are continuing in this area,

Compositional Analysis

Precision density measurements were utilized to determine compositional variations along the length of the samples processed. The results obtained for a ground processed 18.5 wt % In hypermonotectic sample are shown in Figure 5. The high initial solute content and the decrease in solute content with distance is indicative of a sample processed under conditions of convective mixing. The anticipated variation in solute content with position due to convective mixing in off eutectic alloys has been modeled by Verhoeven and Homer [9]. Verhoeven and Homer's analysis, which does not include the Soret effect, predicts the composition along the sample will vary according to the relationship:

$$\overline{C}_S = C_M - \frac{C_M - C_O}{1 - e^{-p\delta}}(1 - g)^{-\frac{I}{1-e^{p\delta}}} \qquad (2)$$

where
- δ = boundary layer thickness
- p = solidification rate divided by diffusion coefficient
- g = fraction solidified
- C_M = monotectic composition
- C_O = alloy composition

Using this relationship and a boundary layer thickness, δ, of 200 μm produces the solid curve shown in Figure 5. While the degree of fit to the data is only marginal, the composition variations observed can be taken as evidence of convective instability and mixing in the melt during processing.

The results obtained from compositional analysis of the 18.5 wt % In flight sample are shown in Figure 6. It should be noted that in this case the initial composition was quite close to the overall alloy composition of 18.5 %. However, the composition began to increase as solidification progressed. This variation occurred presumably due to the presence of two voids which formed

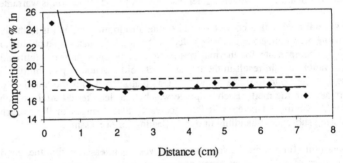

Figure 5. Composition variation along a ground processed hypermonotectic sample containing 18.5 wt % In. The variations shown are representative of those anticipated for convective instability.

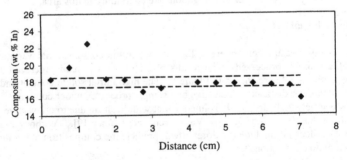

Figure 6. Compositional variation along a flight processed hypermonotectic sample containing 18.5 wt % In.

in the melt during processing. These voids would have provided free surfaces which could lead to surface tension driven flow. The net result would again be mixing in the melt and a change in solidification conditions from those seen initially. Additional evidence of these changes were observed during microstructural analysis.

Microstructural Analysis

Analysis of the ground processed 18.5 wt % In control sample revealed a variation in microstructure close to that anticipated. As discussed in the previous section , the first to freeze portion of the sample had an indium content considerably higher than the overall alloy composition. This region, which contained up to 7.5 % more indium than the monotectic, was too indium rich to permit solidification under interfacially stable conditions. As a result a dispersed microstructure was initially produced. As solidification of this sample proceeded, the indium content dropped (as anticipated) bringing the local composition down to a value relatively close to the monotectic composition. Under these conditions a partially fibrous microstructure began to form. However, this growth process was disrupted when the interface was approximately 30 mm from the bottom of the sample resulting in arrays of spherical particles. The microstructure along the remainder of the sample varied from partially fibrous to aligned spheres and back in a random fashion.

Analysis of the 18.5 wt % In flight sample revealed a totally different variation in microstructure with position. In this sample, the first to freeze region almost immediately began to form a fibrous structure. This finding is consistent with that predicted for substantially reduced mixing in the liquid. As solidification progressed the microstructure remained totally fibrous with the exception of a few irregular indium particles at a location 13mm from first to solidify. A transverse section of this fibrous structure is shown in Figure 7. The formation of these indium

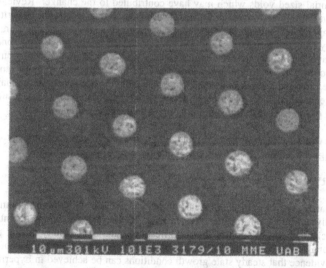

Figure 7. Scanning electron micrograph of a fibrous growth region in a microgravity processed 18.5 wt % In immiscible alloy.

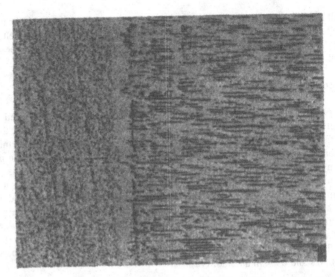

Figure 8. The fibrous to dispersed transition observed in a 18.5 wt % In microgravity processed sample just as the solidification front passed to voids.

particles coincided with the interface passing two voids along the edge of the sample. However, a fibrous microstructure continued to form. At a location 26 mm from the first to freeze the microstructure suddenly changed from completely fibrous to dispersed. This microstructural transition is shown in Figure 8. At this location the solidification front would have just passed two substantial sized voids which may have contributed to the change. Review of acceleration levels during solidification indicated no disturbance at this location. Unfortunately, data on sample temperatures, growth rates and thermal gradients were lost for this sample due to a loss of telemetry from the furnace facility. As a result, details on the processing condition variations that may have caused this change are unavailable. However, there is speculation that this change may have occurred due to the formation of a void in the hot side dummy sample. This void would have provided a break in the heat flow path from the hot zone into the main sample which would have resulted in a significant increase in growth rate. The net result would be the breakdown of interface stability and production of a dispersed structure.

Additional work is underway on the flight and ground based control samples processed as part of this investigation. These results will be reported in a subsequent publication.

SUMMARY

To summarize our findings to date, three immiscible aluminum-indium samples were directionally solidified during the LMS mission. Two of these samples contained voids that were sufficient in size to modify the solidification parameters locally during processing. The compositional and microstructural results obtained from the initial portion of a flight sample provides evidence that steady state growth conditions can be achieved in hypermonotectic alloys through microgravity processing. Work is continuing on the analysis of these samples and on additional samples processed as part of the Coupled Growth in Hypermonotectics experiment.

An extensive amount of metallographic analysis remains to be done on these samples. This work will involve determination of the interfiber spacings and volume fractions of phases as a function of position along the samples. The specific microstructural details will then be used to test the monotectic solidification model developed as part of this investigation.

ACKNOWLEDGEMENTS

The authors wish to thank the National Aeronautics and Space Administration for financial support of this effort through contract NAS8-39717.

REFERENCES

1. Andrews, J. B., O'Dell, J. S., Cheney, A. B., Arikawa, Y., and Hayes, L. J., "Ampoule Design and Testing for Microgravity Experimentation on Coupled Growth in Hypermonotectics," *Proceedings of the 8th International Conference on Experimental Methods for Microgravity Materials Science*, a publication of the Minerals, Metals and Materials Society, R. S. Schiffman editor, 1996.
2. Coriell, S. R., Mitchell, W. F., Murray, B. T., Andrews, J. B., and Arikawa, Y., "Analysis of Monotectic Growth: Infinite Diffusion in L_2 – Phase," *Journal of Crystal Growth* 179 (1997) pp. 647-657.
3. Massalski, T. B., *Binary Alloy Phase Diagrams*, ed. W. W. Scott, ASM International, 1990, pp. 162.
4. Mollard, F. R. and Flemings, M. C., "Growth of Composites from the Melt – Part I," *Transactions of the Metallurgical Society of AIME*, Vol. 239, 1967, pp. 1526-1533.
5. Mollard, F. R. and Flemings, M. C., "Growth of Composites from the Melt – Part II," *Transactions of the Metallurgical Society of AIME*, Vol. 239, 1967, pp.1534-1546.
6. Tiller, W. A., Jackson, K. A., Rutter, J. W., and Chalmers, B., "The Redistribution of Solute Atoms During the Solidification of Metals," *Acta Metall* Vol. 1, 1953, pp. 50-65.
7. Hayes, L. J. and Andrews, J. B., "The Influence of Convection on Composition and Morphology of Directionally Solidified Hypermonotectic Alloys," *Proceedings of the 7th International Conference on Experimental Methods for Microgravity Materials Science*, a publication of the Minerals, Metals and Materials Society, R. S. Schiffman editor, 1995, pp. 87-92.
8. Andrews, J. B., Hayes, L. J., Arikawa, Y., and Corriell, S. R., "Directional Solidification of Immiscible Al-In Alloys Under Microgravity Conditions," *Proceedings of the 10th International Symposium on Experimental Methods for Microgravity Materials Science*, a CD-ROM Publication by TMS, in press.
9. Verhoeven, J. P. and Homer, R. H., "The Growth of Off-Eutectic Composites from Stirred Melts," *Metallurgical Transactions*, Volume 1, 1970, pp. 3437-3441

Part VII

Materials for Space and Other Hostile Environments

RECENT MEASUREMENTS OF EXPERIMENT SENSITIVITY TO G-JITTER AND THEIR SIGNIFICANCE TO ISS FACILITY DEVELOPMENT

R. A. HERRING*, B. TRYGGVASON**
Microgravity Sciences Program, Canadian Space Agency, St. Hubert, PQ, J3Y 8Y9
**Canadian Astronaut Program, Canadian Space Agency, St. Hubert, PQ, J3Y 8Y9

ABSTRACT

Recent experimental measurements of various microgravity experiments have been taken on the Mir and Space Shuttle under different conditions of microgravity using the Microgravity-vibration Isolation Mount (MIM). The results to date show a clear difference when the experimental measurements are taken from g-levels offered by the Mir and the Space Shuttle (non isolated) to g-levels offered by MIM (isolated) which have been reduced by two orders of magnitude. Concern for the International Space Station (ISS) experimental facilities arises when the quality of microgravity on the Mir and Space Shuttle (non isolated), which is believed to be not good enough, has been measured to be better than the ISS Requirement established by NASA for isolated racks, which will be significantly better than those racks not isolated.

INTRODUCTION

There is a need to know the quality of microgravity necessary for taking good measurements of fundamental material properties and enabling some types of materials processing. Experiments sensitive to g-jitter have been performed using Mir and the Space Shuttle as the reference g-jitter microgravity condition [1, 2, 3, 4]. These experiments had their platform either isolated, non-isolated or had induced g-pulses performed using the Microgravity-vibration Isolation Mount (MIM). The Mir experiments involved 1) measurements of intrinsic metal diffusion in metals and semiconductors, 2) measurements of Soret coefficients and the Ostwald ripening phenomenon and, 3) processing of semiconductor materials and glasses. The Space Shuttle measurements were taken on the STS-85 flight and involved 1) the generation of resonance patterns experiments, 2) wave maker experiments, 3) bubble motion experiments, 4) liquid diffusion experiments and, 5) Brownian motion of small particles in density-matched fluids. More recent measurements include the growth of protein crystals on/off MIM which were returned from Mir on STS-89. While not all of these experiments have produced stellar results, many of them have provided clear indications of differences when the measurements were taken in isolated versus non-isolated conditions. As well, some of the results of these experiments are still being analyzed with their interpretations expected in the near future. The results of current analyzed experimental measurements are significant since they predict significantly greater sensitivity to g-jitter than the current ISS vibratory specification for ARIS isolated racks. These results are in agreement with many analytical and numerical studies conducted over the years using single frequency, sinusoidal g-inputs to physical systems in an attempt to predict the g-jitter sensitivity where it is predicted that no differences should exist, although it is now established that these models do not represent the real condition of g-jitter which is comprised of low-level accelerations of a microgravity platform of varying amplitudes (μg to mg) over a large frequency range (~0.01 Hz to ~300 Hz). There is a need to continue taking these experimental measurements, as well as, improve existing theoretical models and simulation methods in order to know the quality of gravity necessary to make good measurements of fundamental material properties and enable some types of materials processing.

Mat. Res. Soc. Symp. Proc. Vol. 551 © 1999 Materials Research Society

It should be noted that if the acceleration environment on Mir and the Space Shuttle are not good enough for experimental measurements then those measurements taken on platforms which just meet the ISS requirements or are non-isolated will also not be good enough. If the results seen thus far are accurate predictors of the sensitivity for experiments planned for the ISS then the various experimental facilities being developed for the ISS and not mounted in an ARIS will need to incorporate provision for isolation systems. These results have raised interest and concerns of facility developers in Canada, Europe and Japan. While any decision to incorporate isolation systems as integral components of facilities implies additional costs, it will be far more cost effective to do this as part of the facility development, rather than to try to fix facilities once they are on orbit in the ISS.

The following will present some recent results obtained from materials science and fluid science experiments flown on Mir and STS-85 which were designed to investigate g-jitter effects on scientific measurements and materials processing in microgravity. These results will be discussed as to their significance of ISS facility development.

RESULTS

G-Jitter Effect on Diffusion in Liquid Metal Systems

The results of Smith in Herring[1, 3, 4] on the measured variation in diffusion coefficient with temperature and g-environment for the lead-gold system (Figure 1) have recently been verified [5, 6] and expanded upon to include the Pb-Ag system. The data points in Figure 1 labeled QUELD-I and QUESTS were conducted on STS-52 and a sounding rocket respectively. Neither of these was on an isolation system. These results and those obtained with QUELD-II on the Mir with the MIM in latched mode all lie on a line that is significantly higher than the comparable coefficients obtained with the MIM operating in isolation mode or in the driven mode described above.

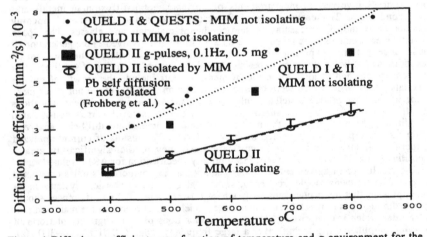

Figure 1 Diffusion coefficients as a function of temperature and g-environment for the Pb-Au and Pb self diffusion. Note isolation decreases diffusivity by 2-3 times and monochromatic g-jitter has no effect [1, 3, 5].

The diffusion coefficient is clearly lower in the isolated mode by ~2-3 times than it is for the non-isolated mode. The diffusion coefficients measured in the isolated mode most likely better reflect the true intrinsic diffusion coefficient. Of particular significance is the result obtained in the 0.1 Hz square wave driven mode. This shows that a relatively simple disturbance does not effect the diffusion coefficient as does the three axis random g-jitter. In the driven mode the MIM flotor was driven in a direction of 45 degrees to the length axis of the sample with a square wave motion having amplitude of 0.5 mg and a period of 0.1 Hz. In terms of the frequency content of this driven mode, the square wave acceleration pattern includes a sinusoidal oscillation at the fundamental frequency and sinusoidal harmonics that decrease rapidly with frequency. On Figure 1 we have added diffusion coefficients obtained by Frohberg et. al. [7] who's results are comparable to the lead-gold coefficient obtained for the non-isolated mode and closely follow a parabolic relationship with temperature whereas the isolated results more closely follow a linear relationship with temperature. Results for a Pb-Ag system which emphasizes an alloy system of larger molecular weight difference than the Pb-Au system behaved in the same manner where g-jitter increases the measured diffusion coefficient and the relatively simple disturbance of a 0.1 Hz square wave g-pulse does not effect the diffusion coefficient.

G-Jitter Effect on the Motion of Encapsulated Bubbles

The motion of a bubble encapsulated in water, having a diameter of ~8 mm and Reynold no. ~50 and high Pr no., was observed [8] to move proportional to the g-vector during g-pulse inputs and to have displacements as a function of frequency (0.1 Hz to 3.25 Hz) of only ~1/2 that predicted by a theoretical model [9]. Of particular significance was the motion of the bubble over a wide range of frequencies for both the isolated and non-isolated cases (Figure 2) which clearly shows that high frequency accelerations drive fluid currents in density mismatched (air-water) systems.

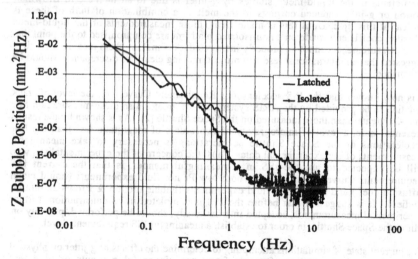

Figure 2 Measured bubble (8.4 mm diameter) displacements for MIM-latched and MIM-isolating cases showing good correlation with the isolation performance of MIM-II during STS-85 [8].

271

Brownian Motion Studies

Measurements of the displacement of small (1 μm and 5 μm dia.), density-matched particles in water [10] as a function of time have shown that the particles in a non-isolated fluid, i.e., experiencing g-jitter, are displaced significantly more than particles in an isolated fluid by almost two times. Enhanced Brownian motion of these particles in water, a high Pr no. fluid, would effectively be the same as an increase in the rate of diffusion.

DISCUSSION

The effects of vibratory accelerations, i.e., g-jitter, on materials processing and experimental measurements are now becoming well known from the studies performed on Mir and Space Shuttle flights. These experimental studies were possible with the development of the Microgravity Vibration Isolation Mount (MIM) which accumulated over 2500 hours of operational time on Mir in support of science experiments with many of these chosen to specifically investigate the effects of g-jitter on the results. The trend in the experimental data clearly shows that the experimental measurements taken in isolation are different than those measurements taken without isolation. The most important results of these experiments show that 1) g-jitter increases the diffusion coefficients by ~2-3 times for liquid-metal studies, i.e., low Pr no. fluids, 2) for the liquid-metal studies, g-jitter has a more severe effect that single frequency g-inputs which had no effect on the measured diffusion coefficients, 3) high frequency g-jitter accelerations create current flows in water, i.e., high Pr no. fluids, as observed by the motion of bubbles as a function of frequency and, 4) the Brownian motion of particles are significantly increased by the effect of g-jitter which in effect has increased the rate of diffusion of the particles in the fluid. Whether the measured increase of the diffusion coefficients in the liquid-metal studies by g-jitter is due to an increase in Brownian motion or g-jitter-induced currents in the melt or a combination of both of these is unclear at this time. Until this is made clear, those measurements of the liquid-metal diffusion coefficients made on a non-isolated platform are compromised to the point that these measurements cannot be properly interpreted, thus making them meaningless. It is suggested that an international research project, making clear this discrepancy, should be established.

It is now known that the ISS accelerations Requirement Curve, i.e., the level of rack isolation to be provided by the ARIS system, is ~2x greater than those measured on Mir and ~3x greater than those measured on the Space Shuttle [2]. As is shown by the results presented here, a level of microgravity of two orders of magnitude less than the Mir accelerations or the Space Shuttle accelerations is necessary to take meaningful measurements of the experimental data. Thus, the concern of the experimentalists, who will be expected to use the ISS in a meaningful manner, is that the current ISS Requirements Curve is too large for both low Pr no. fluid experiments (liquid metal diffusion coef.) and high Pr no. fluid experiments (bubble motion & Brownian motion studies). It is suggested that before the ISS is completed in its construction, further experimental measurements, accepted internationally, should be performed, possibly on Mir or the Space Shuttle, in order to establish a meaningful ISS requirements level.

The current state of simulations attempting to determine the effects of g-jitter on physical systems typically comprise of single frequency sinusoidal g inputs or of a few frequencies matching the resonance peaks of the microgravity platform. These similations usually predict little effect of g-jitter on the scientific measurements, which agrees well with those results presented here for Au (and Ag) diffusion in Pb.

Unfortunately, these simulations do not come close to representing the true g-jitter condition, as varified by the liquid-metal experiments, where at every frequency there are variations in the acceleration. The extent of g-jitter can be seen when taking the Fourier transform of the Space Shuttle's acceleration power spectrum where a non-Gaussian, non-symmetrical distribution of the probability spectrum results from the significant deviations occurred by the large amplitude pulses and variations in the acceleration spectrum with time, respectively [11]. Thus, g-jitter on the microgravity platforms being used for experimental measurements is much more severe than current simulations predict.

Further work concerning the determination of accelerations levels of a microgravity structure over a frequency range, $A(f)$, can be approximated from Newtonian theories of motion which, even though they give highly complex equations, needs to be performed for each microgravity platform and every microgravity facility. From preliminary studies [12], the parameters, needing to be considered, can be expressed as,

$$A(f) = B_j \text{(power/mass)} \qquad (1)$$

where B_j is a constant representing the fraction of the total energy dissipated by mechanical means versus heat loss, sound, etc., and the power to mass ratio is the most important parameter for determining the acceleration levels on microgravity platforms or those acceleration levels created by the experimental facility during its operation.

Currently the only means to study the true effects of g-jitter is by experimentation using an isolation system which can reduce g-jitter to a level where true differences can be measured. The effects of G-jitter on the science being performed can only be measured if it can first be reduced to an insignificant level, i.e., establishing a good reference condition.

ACKNOWLEDGMENTS

The authors are grateful to its PIs and MIM support staff for all of their foresight and success in participating in the projects supported by the Microgravity Sciences Program, Canadian Space Agency.

REFERENCES

1. R. A. Herring, IAF - 97-IAA.12.2.03, Turin (1997).

2. B. Tryggvason, IAF -97-J.2.04, Turin (1997).

3. R. A. Herring, B. Tryggvason, IAF - 98, Melbourne (1998).

4. B. Tryggvason and R. A. Herring, G-jitter Workshop, ESTEC (September, 1998).

5. J. Robert and R. W. Smith, "Diffusion in Liquid Metals" QUELD II Review Meeting, CSA Headquarters (August, 1998).

6. R. W. Smith, G-jitter Workshop, ESTEC (September, 1998).

7. G. Frohberg, K. H. Kraatz and H. Wever, in: Scientific Results of the German Spacelab Mission D1, Norderney Symposium (August 1986) pp. 144-151.

8. S. Farris, K. S. Rezkallah and J. Bugg, "A Study on the Motion of a Bubble Subjected to Simulated Broadband G-jitter: Results from a Recent Shuttle Flight" Proc. 1st Pan Pacific Workshop on Microgravity Science, Tokyo (July 8-10 1998).

9. R. Mei et. al., "A note on the history force on spherical bubble at finite Reynolds number" Physics of Fluids 6(1) 418-420 (1994).

10. M. Ouelette and B. Tryggvason, FLEX Review Meeting, CSA Headquarters (September, 1998).

11. J. R. Thomson, J. Casademunt, F. Drolet and J. Viñals, "Coarsening of solid-liquid mixtures in a random acceleration field" Phys. Fluids 9 (5) 1336-1343 (1997).

12. B. Tryggvason, J. DeCarufel and W. Stewart, "Microgravity Vibration Isolation Systems: General Performance Model Applied to Various Microgravity Platforms". To be published.

BORON CONTAINING POLYIMIDES FOR
AEROSPACE RADIATION SHIELDING

Stephen C. Ko, Christopher S. Pugh, Richard L. Kiefer, Robert A. Orwoll, College of William and Mary, Department of Chemistry, Williamsburg, VA; Sheila A. Thibeault, Glen C. King, NASA Langley Research Center, Materials Division, Hampton, VA.

ABSTRACT

In interplanetary travel and high altitude flight, humans will be exposed to high energy charged particles from solar flares and galactic cosmic rays. These particles lose energy in a material by Coulomb interactions and nuclear collisions. In nuclear collisions, large amounts of energy are transferred and secondary particles are formed from both the projectile and the struck nucleus. A significant portion of these particles are neutrons which can only lose energy by collisions or reactions with a nucleus. Hydrogen-containing materials, such as polymers, are most effective in reducing the neutron energy. When reduced to very low energies, neutrons have a high probability of reacting with a nucleus. Such reactions are dangerous in the human body, and can cause electronic equipment failure. Low energy neutrons react particularly well with a stable isotope of boron, ^{10}B. To test structural materials which contain both hydrogen to reduce the energy of neutrons and boron to absorb neutrons of reduced energy, samples of two polyimides were made which contained varying amounts of either amorphous boron powder or boron carbide whiskers. The polymers used were a thermoset, PETI-5 from Imitec, and a thermoplastic, K3B from Fiberite. Both materials were made in pure form and with up to 20% by weight of the boron additives. The addition of boron in either form did not change the thermal properties of these materials significantly. However, the compressive yield strength and the tensile strength were both affected by the addition of the boron materials. A neutron absorption test using a PuBe thermal neutron source showed that a 0.5 cm thick sample of K3B containing 15% amorphous boron powder absorbed over 90% of the incident neutrons.

INTRODUCTION

With the advent of space stations, long duration space flights, and high-altitude flight, new emphasis has been placed on durable and lightweight materials with radiation-shielding capabilities. High-performance polymers have been the focus of much research for this specific purpose. Increased exposure to solar ($<1GeV$) and galactic (up to $10^{10}GeV$) cosmic rays in high-altitude aerospace and space environments can provide potential health hazards to humans [1]. Fortunately, these sometimes large, high-energy nuclei prefer interactions with smaller, lighter nuclei such as hydrogen. Hydrogen containing polymers thus provide good protection against these high-energy particles. Interactions between hydrogen and the larger nuclei produce lower energy thermal neutrons that can interact with the human body to produce harmful radioactive products. Therefore, an isotope with a relatively high abundance that can provide a high thermal neutron capture cross-section and stable, non-radioactive products is needed. With a natural abundance of 19.9%, a neutron capture cross-section of 3838 barns, and non-radioactive product nuclei, ^7Li and ^4He, ^{10}B can fulfill this role [2,3].

However, the prudent question involved in boron loading polymers is the effect it has on the thermal and mechanical properties of the neat resin. Two sources of ^{10}B have been loaded into high-performance polyimides: amorphous boron powder and 300 micron boron carbide whiskers. The boron carbide whiskers are of particular interest because of the possibility of improved mechanical properties in the resulting micro-composite formulated with thermosetting resins such as PETI-5 [4]. The goal of this paper is to analyze the effectiveness of ^{10}B as a neutron shield and the effects it has on the mechanical properties of high-performance polyimides.

EXPERIMENT

Radiation Shielding Tests

Neutron shielding studies were performed on a three Curie PuBe source at Virginia State University for samples of both pure K3B and 15% amorphous boron powder in K3B. Two samples 0.75 x 0.75 x 0.15 in were clamped together with a slightly smaller piece of indium foil sandwiched between. The specimen was then inserted into the neutron source for 1.5 hr in order to produce ^{116}In which is a β^- emitter with a half-life of 54 min [3]. Counts for the background radiation of the room were taken during the activation process. Counts of the activated material were taken for one min intervals every five min for at least one hr for each of the samples on an end-window Geiger counter.

The amount of neutron shielding can be seen with a plot of ln activity (counts/min) vs. time (min). The background radiation in counts/min was subtracted from each of the data points and the resulting data for each sample were plotted. A linear regression of the data was performed to extrapolate the line to the activity at time = 0. These two initial activities determine the amount of shielding.

Mechanical Testing

Three types of mechanical tests were performed on the 5 sets of specimens for both PETI-5 and K3B. The 5 sets of specimens for each polymer includes the following: pure, 10% boron carbide whiskers, 20% boron carbide whiskers, 10% amorphous boron, and 20% amorphous boron. All of these are weight percentages of the additive to the respective polymer.

Compression tests were performed solely on the five sets of PETI-5 specimens. The test method was ASTM D695 [5]. Six samples were machined to $1.0 \times 0.5 \times 0.3$ in. Each of the samples was fitted with a 2-directional (one axial, one transverse) strain gauge on each side. However, the strain exceeded the range of the gauges so an extensometer was employed to determine the stress-strain curves. The samples were tested on a Baldwin-Tate-Emery 120-kip (120,000 lb. full scale) universal test plant. This was coupled with a SATEC systems controller and an Analogic ANDS 5400 data acquisition system to record the data from the extensometer and strain gauges.

Prior to testing, the width and thickness of each sample were measured at several points along the sample to determine its cross-sectional area. Each of the samples was then loaded into the test stand with the help of an alignment tool and the top plate of the compression fixture was lowered to hold the sample. Once the extensometer was attached, the test was started with a speed of 0.05 in/min. Real time stress vs. strain curves were generated from the extensometer readout.

The test results were used in determining the ultimate compressive strength and the compressive chord modulus of elasticity on each set of PETI-5 specimen. The ultimate strength (or stress, σ) was calculated from:

$$\sigma = F / A_0 \qquad (1)$$

where F is the load and A_0 is the cross-sectional area. The compressive chord modulus of elasticity, which measures the stiffness of the material, was calculated using the slope of the extrapolated linear portion of the stress/strain curve. The strain of the material, ε, was calculated by the change in length (Δl) as a result of the applied load divided by the initial length.

$$\varepsilon = \Delta l / l_0 \qquad (2)$$

A total of ten 3 × 6 in plates, five each for PETI-5 and K3B, were molded for the tensile tests. Six dogbones specimens were machined from each plate for a total of 60 samples which were 6 in long, 0.14 in thick, and 0.375 in wide at the ends. The gauge area of each sample was 0.375 in long and 0.125 in wide, the dimensions prescribed by ASTM D638 [6]. The gauge area for each sample was then measured before testing on an Instron test stand. The elongation of the material was measured using an attached MTS extensometer. The tests were run using a load rate of 0.05 in/min for samples P1-P4 and 0.02 in/min for specimens P5 & K1-K7 with points taken at 4 Hz. Two channels were set up for the data where channel 1 reported load and channel 2 reported extension. Testing was discontinued after complete failure of the material.

Calculations were made in accordance with ASTM D638 for the ultimate tensile stress (equation 1) and the modulus of elasticity. The latter was found in a way similar to that of the compressive chord modulus by using the slope of the stress/strain curve.

Eleven 3 × 6 in plates were used to machine flexural tests specimens for the five different sets of PETI-5 and K3B samples. An extra specimen set, K3B neat resin molded by Fiberite was also tested. The molded plates were machined to dimensions of 3 in long, 0.5 in wide, and 0.125 in thick as prescribed by ASTM D790 [7]. The width and thickness of each sample were measured before testing on a MTS 50 kip test stand with an attached three-point bend fixture. Displacement in the middle of the sample was measured using a ±0.05 in DCDT. Three channels were set up accordingly for the acquisition of the load, the displacement from the DCDT, and the crosshead displacement. The load rate was set at 0.0533 in/min as prescribed by the standard.

The maximum fiber stress and the tangent modulus of elasticity were determined in accord with ASTM D790. The maximum fiber stress (S) represents the ultimate strain in the outer fibers and was calculated by the equation:

$$S = 3PL / 2bd^2 \qquad (3)$$

where P is the load, L is the support span, b is the width of the sample, and d is the depth of the sample. The maximum strain (r) in the outer fibers was calculated using equation 4 below

$$r = 6Dd / L^2 \qquad (4)$$

where D is the maximum deflection of the center of the sample. Finally, the tangent modulus of elasticity (E_B) was calculated using equation 5

$$E_B = L^3 m / 4bd^3 \qquad (5)$$

where m represents the tangent line of the slope of the stress/strain curve.

RESULTS

Radiation Shielding Tests

The results of thermal neutron irradiation of pure K3B and 15% amorphous boron in K3B are shown in Figure 1 below. For pure K3B, the linear regression gives an initial activity in ln [counts/min] of 8.4 while 15% amorphous boron gives an initial value of 6.1. These values thus represent a relative shielding value of 90.6% for the 15% amorphous boron in K3B vs. pure material.

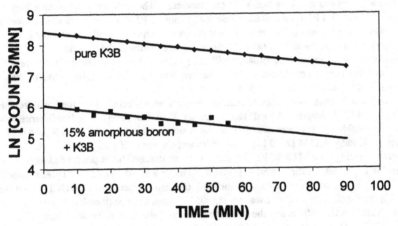

Figure 1. Decay of ^{116}In activated between samples of pure K3B or K3B/15% amorphous B.

Mechanical Testing

The ultimate strength and modulus values for each of the three mechanical tests are reported in Tables I and II. For the compression tests, the 20% boron carbide whiskers showed the highest strength of the five sets of PETI-5 specimens tested with a value of 25.35 ksi. Similarly, the 20% boron carbide displayed the highest modulus, or stiffness, of the tested specimens at 2978.0 ksi. The 20% boron carbide additive thus adds a 9% increase in strength over the pure material and almost a two-fold increase in stiffness.

Table I. Compression and Tensile Test Data

Sample	Compression Test		Tensile Test	
	Ultimate Compressive Strength (ksi)	Compressive Chord Modulus (ksi)	Ultimate Tensile Strength (ksi)	Modulus of Elasticity (ksi)
PETI-5				

Pure	23.20	1695.0	18.878	487
10% Boron Carbide	24.61	2412.8	17.121	324
20% Boron Carbide	25.35	2978.0	12.547	426
10% Amorphous Boron	18.24	1566.6	15.795	596
20% Amorphous Boron	16.51	1839.6	7.605	503
K3B				
Pure	X	X	7.019	496
10% Boron Carbide	X	X	10.943	586
20% Boron Carbide	X	X	9.057	534
10% Amorphous Boron	X	X	6.847	479
20% Amorphous Boron	X	X	5.449	414

Table II. Flexural Data.

Sample	Flexural Test	
	Ultimate Fiber	Tangent Modulus
PETI-5	Strength (ksi)	of Elasticity (ksi)
Pure	28.172	546.962
10% Boron Carbide	28.269	684.115
20% Boron Carbide	25.042	799.445
10% Amorphous Boron	18.912	486.291
20% Amorphous Boron	14.180	454.526
K3B		
Pure	9.263	407.358
10% Boron Carbide	14.692	536.639
20% Boron Carbide	13.233	618.616
10% Amorphous Boron	10.47	500.693
20% Amorphous Boron	6.71	373.225

The tensile tests showed slightly different results from those of the compression tests. In PETI-5, the 10% boron carbide exhibited the closest strength values to the pure material with a 9% decrease. However, the 10% amorphous boron displayed the highest stiffness value at 596 ksi; an 18% increase over the pure material. This result could possibly be the result of the whiskers interrupting the cross-linking of PETI-5 which particularly affects its tensile properties. However, the results for K3B are remarkably different from its thermosetting counterpart. The 10% boron carbide additive increases the tensile strength by 56% with a value of 10.9 ksi vs. 7.0 ksi for the neat resin. Since K3B is a thermoplastic and does not exhibit cross-linking capabilities, the 10% boron carbide increases the stiffness of the thermoplastic by 18%.

Finally, the flexural tests demonstrated the effectiveness of micro-composite behavior of the boron carbide whiskers. In PETI-5, the 10% boron carbide actually enhanced the flexural strength of the polymer by 100 psi. Similar to a composite, the stiffness of the polymer increased with higher percentages of boron carbide. The 20% boron carbide increased the modulus of the polymer by 46%, while the 10% boron carbide improved the stiffness by 25%. The results for the K3B flexural tests are identical to those of PETI-5. The 10% boron carbide exhibited the best flexural strength properties with a value of 14.7 ksi, 59% higher than the pure polymer. Furthermore, the 20% boron carbide displayed the highest modulus value at 619 ksi with 10% boron carbide following at 537 ksi. These two values represent a 52% and 32% respective increase in stiffness values over the pure thermoplastic.

CONCLUSION

Inclusion of boron into polymers provides neutron shielding with little or no degradation of mechanical properties. Due to the high thermal neutron capture cross-section of boron, small amounts provide high shielding capabilities, which minimizes the potential negative effects on the mechanical properties of the polymer.

Several conclusions can be drawn. First, the 10% boron carbide whiskers emerges as the best additive of the four tested. In PETI-5, the 10% boron carbide actually enhances the strength and stiffness of the polymer in the compression and flexural tests. In the tensile tests, the 10% boron carbide only decreases the strength of the polymer by a marginal amount. In K3B, the 10% boron carbide offers the highest strength and stiffness of all sets of materials tested. Second, the thermoset PETI-5 clearly possesses the superior mechanical properties over the thermoplastic K3B. The cross-linking in PETI-5 adds strength and stiffness over that of K3B. Coupled with the neutron shielding abilities of the boron carbide whiskers, PETI-5 provides a lightweight, durable material for aerospace applications.

ACKNOWLEDGEMENTS

This research was made possible under a cooperative agreement with the National Aeronautic and Space Administration, Langley Research Center in Hampton, VA.

REFERENCES

1. L.W. Townsend, J E. Nealy, J.W. Wilson, and L C. Simonsen, **NASA TM-4167**, p.1, 1990.

2. G. Friedlander, J.W. Kennedy, E.S. Macias, and J.M. Miller, *Nuclear and Radiochemistry*, 3rd ed., John Wiley, New York, 1981, p. 601.

3. W.D. Ehmann, and D.E. Vance, *Radiochemistry and Nuclear Methods of Analysis*, John Wiley, New York, 1991, p. 487.

4. J. Milewski, Polymer Composites, **13**, p. 224 (1992).

5 American Standard Test Method (D695).

6 American Standard Test Method (D638).

7. American Standard Test Method (D790).

Vapor Grown Carbon Fiber Reinforced Aluminum Matrix Composites for Enhanced Thermal Conductivity

J.-M. TING*, C. TANG**, AND P. LAKE**

*National Cheng Kung University, Tainan, Taiwan, 701, jting@mail.ncku.edu.tw

**Applied Sciences, Incorporated, Cedarville, OH, 45314, USA

Abstract

Aluminum matrix composites reinforced with high thermal conductivity vapor grown carbon fiber (VGCF) were developed for improved thermal efficiencies in electronic devices. The carbon fiber was heat treated to increase its thermal conductivity. Various aluminum matrix composites were fabricated by the densification of fiber preforms using a pressure casting technique. Uniformity of the density was examined using optical microscopy. A scanning electron microscope equipped with a microprobe was utilized to examine the mechanical integrity of the composite. Mechanical properties, including tension, compression and flexural properties, were measured. While the results of the mechanical property measurements indicate moderate values, the composite exhibited remarkable thermal conductivity that reached 642 W/m·K, three times that of aluminum, at a fiber volume fraction of 36.5%, following closely the rule of mixture.

Introduction

Electronic devices with greater power output levels are required for advanced systems such as spacecraft. Current and future chip designs indicate heat flux levels on the order of 100 W/cm^2 at the chip level, and 25 – 50 W/cm^2 at the module level.[1,2,3,4,5,6] Innovative methods are therefore being explored to improve thermal efficiency through either external or internal enhancement. Improved materials for heat sinks, an external enhancement, and substrates, an internal enhancement, represent critical technology for ensuring the operation of future generations of electronic devices.[7] The aerospace and electronics industry has, therefore, identified a need for significant research and development in advanced packaging technologies, including associated materials and cooling strategies. This paper presents a new class of composite materials for use as heat sinks in multi-chip modules (MCM).

Experiments

Vapor grown carbon fibers (VGCF), in the form of mat, were prepared using a technique described in reference eight.[8] Preforms of VGCF mat with fiber volume fractions ranging from

15% to 50% were fabricated. Both one-dimensional (1D) and two-dimensional (2D) preforms were made. Aluminum matrix composites were fabricated by infiltrating the preforms with molten aluminum using a pressure casting technique.[9] Composite specimens were then prepared for measurement of thermal conductivity, tensile properties, compression properties, and flexural properties.

Results and Discussion

Aluminum matrix composites with eight different fiber volume fractions were obtained. The aluminum infiltration was found to be homogeneous, and the resulting composites exhibited little porosity as shown in Figure 1. The dark areas are cross sectional images of VGCF, which vary in diameter over a range from 2 – 8 μm. Considering the low fiber volume in the Al composite, the fiber distribution achieved relatively high uniformity. The choice of highly graphitic VGCF and the aluminum alloy prevented the formation of a carbide interface between the fiber and matrix material, which resulted in higher thermal conductivity of the material.

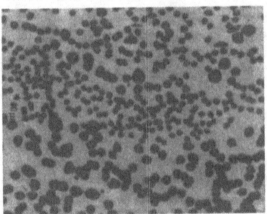

Figure 1: Microstructure of Al matrix composite.

Room temperature thermal conductivity of several 1D VGCF enforced aluminum composites were determined as listed in Table 1, which also includes the fiber volume fraction (FV), the measured density, and the specific thermal conductivity. It is clear that all the composites exhibited excellent thermal conductivity – higher than that of aluminum (~180 W/m·K). With only 15% fiber loading, the thermal conductivity of aluminum is doubled. The composite thermal conductivity is increased linearly with increasing fiber loading, as shown in Table 2. The highest composite thermal conductivity and specific conductivity obtained are 642 W/m·K and 263 (W/m·K)/(g/cc), respectively. The slopes of the fitted lines in Figure 2 are almost identical to estimated values based on rule of mixture, provided that the fiber thermal conductivity is taken to be 1500 W/m-K.[10]

ID	FV (%)	Conductivity (W/m·K)	Density (g/cc)	Specific Thermal Conductivity (W/m·K)/(g/cc)
1	15	381	2.57	148
2	22.1	406	2.53	160
3	23.2	413	2.51	165
4	26.1	474	2.50	190
5	27.9	534	2.50	214
6	36.5	642	2.44	263

Table 1: Thermal conductivity of Al composites.

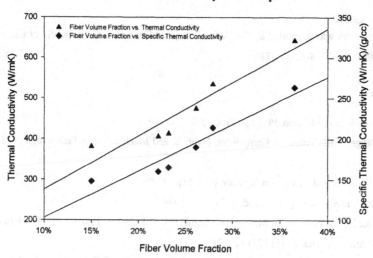

Figure 2: Relationship between thermal conductivity and fiber volume fraction.

Architecture	FV (%)	UTS	Compressive Strength	Flexural Strength	Flexural Modulus
2D	50	96.6 Mpa	75.5 MPa	119.2 MPa	-
1D	27	-	-	158.5 MPa	1.58 GPa
1D	38	-	-	168.3 MPa	1.62 GPa
1D	43	-	-	250.5 MPa	1.64 GPa
1D	50	121.5 Mpa	66.8 MPa	118.4 MPa	-

Table 2: Mechanical properties of Al composites.

Although the composite thermal conductivity is impressive, the composite mechanical properties are only moderate. The mechanical test results are given in Table 2. It appears that the flexural strength is proportional to the fiber volume fraction. However, at the highest fiber volume, i.e., 50%, the strength becomes lower. This could be due to fiber damage. More analysis would have to be conducted in order to explain the variations of mechanical strength in these composites.

Conclusion

Aluminum composites were fabricated using vapor grown carbon fiber as the reinforcement. Due to the excellent thermal conductivity of VGCF, the resulting composites exhibited very high thermal conductivity. However, further improvement of the composite mechanical properties is needed.

Acknowledgement

This work was supported by the National Science Council of the Republic of China under Contract #NSC87-2218-E006-071.

References

[1]. H.W. Markstein, Electron. Pkg. and Prod., p.40, Oct. 1991.

[2]. J.K. Hagge, IEEE Trans. on Component, Hybrids, and Manufacturing Technologies, **12** [2] 170 (1989).

[3]. R.D. Rossi, Hybrid Circuit Technology, p.35, Sep., 1989.

[4]. J.J. Licari, Electron. Pkg. and Prod., p.58, Sep., 1989.

[5]. M. Kokado, M. Yoshida, N. Miyoshi, K. Suzuki, M. Takaika, N. Tsuzuki, and H. Harada, IEEE J. of Solid State Circuits, **24** [5] 1271 (1989).

[6]. A. Goyal, R.C. Jaeger, S.H. Bhavnani, C.D. Ellis, N.K. Phadke, M. Azimi-Rashti, and J.S. Gooding, p.25, Proc. Semi-Therm 1992, Austin, TX, February, 1992.

[7]. M.A. Kuhlman and H. Sehitoglu, pp.110-8 in Proc. of Semi-Therm 1992, Austin, TX, February 1992.

[8] J.-M. Ting, " Processing-Microstucture-Tensile Property of Vapor Grown Carbon Fiber Reinforced Carbon Composite," p. 316, Proc. MRS Spring Meeting, San Francisco, CA (1995).

[9] A. Mortensen, L.J. Masur, J.A. Cornie, and M.C. Flemings, Metall. Trans. 20A, 2549 (1989).

[10] J.-M. Ting and M.L. Lake, Carbon, 33 [5] 663 (1995).

ARTIFICIAL AGING EFFECTS ON CRYOGENIC FRACTURE TOUGHNESS OF THE MAIN STRUCTURAL ALLOY FOR THE SUPER LIGHT WEIGHT TANK

P.S. Chen *, W.P. Stanton**
*IIT Research Institute, Metallurgy Research Facilities, Building 4628, Marshall Space Flight Center, AL 35812, poshou.chen@msfc.nasa.gov
**National Aeronautics and Space Administration, Materials and Processes Laboratory, Metals Processing Branch, Marshall Space Flight Center, AL 35812, william.stanton@msfc.nasa.gov

ABSTRACT

At Marshall Space Flight Center (MSFC), a new aging technique has been developed that can significantly enhance the cryogenic fracture toughness (CFT) of Alloy 2195, as well as reducing the statistical spread of fracture toughness values. This technique improves cryogenic properties by promoting T_1 nucleation and growth in the matrix, so that T_1 nucleation and growth can be suppressed in the subgrain boundaries. It also minimizes mechanical property scatter due to unavoidable variations in thermomechanical processing parameters and alloy chemistry.

INTRODUCTION

NASA has selected Al-Li alloy 2195 as the main structural alloy for the Super Light Weight Tank (SLWT) of the Space Shuttle. Cryogenic strength and toughness are critical to this application, since the SLWT will house liquid oxygen and hydrogen. The alloy must have higher strength and toughness at cryogenic temperatures than at ambient temperature, in order to avoid expensive cryogenic proof testing. To ensure proper quality control, NASA has imposed lot acceptance testing on Alloy 2195 plate before it can be used in the SLWT program.

Some commercial 2195 plates were rejected for the SLWT program, primarily due to low CFT that was found to be related to the density, size, and location of a precipitate labeled T_1. As T_1 precipitates increase in density at the subgrain boundaries, CFT decreases considerably. Therefore, attempts to improve fracture toughness focused on reducing density of T_1 precipitates at subgrain boundaries and enhancing the nucleation of T_1 precipitates in the matrix [1,2].

As a result, a new artificial aging treatment has been developed that can greatly improve CFT by controlling the location and size of T_1 precipitate. This treatment uses multi-step aging to prevent T_1 from precipitating preferentially at the subgrain boundaries, thus improving fracture toughness without sacrificing yield and tensile strength. In addition, this treatment minimizes mechanical property scatter due to unavoidable variations in thermomechanical processing parameters and alloy chemistry. This paper discusses the new aging treatment in detail and correlates it to the observed improvement in CFT to alloy microstructure.

TECHNICAL APPROACH

This study used an alloy with yield strength (YS) >73 ksi, which displayed a cryogenic-to-ambient fracture toughness ratio (FTR) <1 after isothermal aging [1,2]. In Al-Cu-Li alloys, FTR correlates well with the size and density of T_1 precipitates in the subgrain boundaries [2]. (See Figure 1.) High CFT can be achieved by suppressing T_1 precipitation at subgrain boundaries and enhancing T_1 nucleation in the matrix, eliminating premature fractures along precipitate-rich subgrain boundaries. Based on this finding, a series of step-aging treatments [1,2] were attempted to promote T_1 nucleation and growth in the matrix. (See Table I.)

285

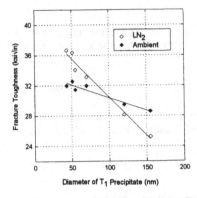

Figure 1. Fracture toughness versus maximum size of T_1 at subgrain boundaries. Fracture toughness decreases as T_1 size increases [1,2].

Table I. Matrix for Three- and Two-Step Aging Treatments

Aging Treatment	265 °F	270 °F	275 °F	280 °F	285 °F	290 °F
17		20 hr		20 hr		15 hr
18		20 hr		20 hr		10 hr
19		20 hr		20 hr		5 hr
20		20 hr		15 hr		10 hr
21		15 hr		20 hr		10 hr
22		15 hr		15 hr		10 hr
23	20 hr		20 hr		20 hr	
24		20 hr		40 hr		
25			20 hr		40 hr	

This effort started with a series of step-aging treatments that began with initial holding at low temperature (with high undercooling) to enhance formation of T_1 nuclei in the matrix. Then the furnace temperature was raised to permit each precipitate nucleus to grow and become a stable nucleus. These nuclei continued to grow during aging, with negligible dissolution into solid solution. Long-term aging at low temperatures also allowed T_1 precipitates to grow in the matrix before they could nucleate and grow at the subgrain boundaries. The most promising two-step aging treatment was selected for tensile and cryogenic properties evaluation.

EXPERIMENTAL PROCEDURES

Alloy 2195 (Al-4.0Cu-1.0Li-0.52Mg-0.42Ag-0.12Zr) was received as 1.7-inch-thick rolled plates, which were solutionized and stretched 3 percent at ambient temperature.

Tensile tests were carried out at ambient temperature, using flat tensile specimens to evaluate microstructural variation effects through the plate thickness. Uniaxial tensile properties were evaluated in the L, LT, and ST orientations, with at least two tests performed in each orientation. Fracture toughness tests were performed at ambient temperature and –320 °F. Plates were evaluated in the T-L orientation (notch parallel to the rolling direction) and the T-S orientation (notch parallel to the rolling direction) per ASTM E740. Specimens were fatigue precracked at 20 Hz, then tensile-tested to failure at a crosshead speed of 0.05 inch/min. Precrack length and maximum load to failure were factored into the standard equation. Simulated service tests were performed at -320 °F.

Microstructural characterization was performed using a JEOL 2000F transmission electron microscope (TEM) operated at 200 kV. Samples were jet-polished in an electrolyte (70% methanol-30% nitric acid) at -4 °F with an applied potential of 12 V.

RESULTS

Hardness and Microstructure

Table II shows hardness variation as a result of step-aging. The alloy was underaged at 270 °F to enhance precipitate nucleation in the matrix, then heated to higher temperatures to obtain a peak or near-peak aged condition while preventing preferential nucleation and growth of T_1 at subgrain boundaries. For Alloy 2195, a hardness of 90 to 91 HR$_B$ is roughly equivalent to 73 ksi YS in the L and LT orientations, which is the minimum strength requirement. Promising results were obtained for three-step aging treatment No. 1 (91.7 HR$_B$ after 55 hr) and two-step aging treatment No. 8 (more than 90 HR$_B$ after 270 °F/20 hr followed by 280 °F/40 hr).

Table II. Three- and Two-Step Aging Hardness Values for Rejected Lot of Alloy 2195 (950M029B)

Aging Treatment	265 °F	HR$_B$	270 °F	HR$_B$	275 °F	HR$_B$	280 °F	HR$_B$	285 °F	HR$_B$	290 °F	HR$_B$
1			20 hr	76.2			20 hr	86.2			15 hr	91.7
2			20 hr				20 hr				10 hr	88.9
3			20 hr				20 hr				5 hr	87.1
4			20 hr				15 hr	84.1			10 hr	88
5			15 hr				20 hr	84.4			10 hr	88.5
6			15 hr				15 hr				10 hr	87.6
7	20 hr	75.8			20 hr	83.2			20 hr	89.2		
8			20 hr	76.2			40 hr	90.7				
9					20 hr	79.1			40 hr	89.8		

TEM was used to examine the step-aged microstructures. After aging at 270 °F/20 hr, the matrix consisted of numerous fine θ'', with scattered T_1 precipitates. (See Figure 2a.) The large number of very fine θ'' indicated that an early stage of nucleation had taken place at 270 °F. Hardness increased rapidly while aging at 280 °F/40 hr, indicating near-peak precipitation of the strengthening phases. T_1 was the majority phase in the matrix and was also present at subgrain boundaries in sizes no coarser than the matrix T_1. (See Figure 2b.) Similar microstructural evolution was observed in other three- and two-step aging treatments.

Substantial microstructural differences were found between conventionally aged and step-aged materials. In conventionally aged alloy, T_1 was coarser and had a much higher density in the subgrain boundaries than in the matrix [1,2]. In step-aged alloy, the subgrain boundaries were almost devoid of T_1. (See Figure 3.) T_1 occasionally existed at subgrain boundaries, but it was not as dense as in the matrix. In the matrix, conventionally aged alloy contained more T_1 than θ' and θ'' while TS aging produced more θ' and θ'' than T_1. This finding clearly indicated that TS aging could achieve the same strength levels as conventional aging, by precipitating more θ' and θ'' in the matrix while preventing preferential T_1 precipitation at subgrain boundaries.

Mechanical Properties

Several step-aging treatments (Nos. 1, 8, and 9) were selected for tensile strength and fracture toughness evaluation. Tensile data indicated that TS aging could achieve the same YS and ductility levels as those produced by conventional aging [1,2]. (See Table III.)

The most noticeable improvement was CFT, for which the minimum requirement is 30 ksi √inch. (See Table IV.) TS aging improved CFT by more than 30 percent (from as low as 25.4 ksi√inch to about 34 ksi√inch) for a bad lot (Lot 950M029B) and by approximately 10 percent for a good lot (Lot 950M020F) of conventionally aged material. (See Figure 4.)

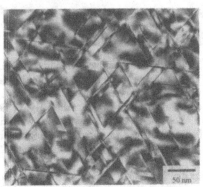

(a) TEM photograph showing early stage of precipitate nucleation and growth in the matrix

(b) TEM photograph showing final stage of precipitate growth in the matrix

Figure 2. Precipitate morphology after aging treatment No. 8 at (a) 270 °F/20 hr, and (b) 270 °F/20 hr + 280 °F/40 hr.

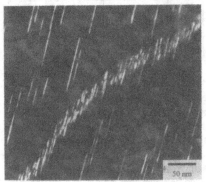

(a) Conventional aging (290 °F/30 hr)

(b) Two-step aging No. 8 (270 °F/20 hr + 280 °F/40 hr)

Figure 3. Dark field TEM micrographs showing subgrain boundary microstructures.

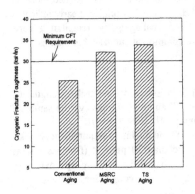

Figure 4. CFT data for alloy with TS aging No. 8 and conventional aging. TS aging greatly improved the CFT to meet the minimum requirement (30 ksi√in).

288

Table III. Tensile Properties for Aged Alloy 2195

Lot No.	Aging No.	Orientation	YS (ksi)	UTS (ksi)	%El
950M029B	1	L	77.4	80.9	8.5
		ST	70.3	83.4	5.5
950M020F	1	L	72.2	76.8	9.3
		ST	69.9	82.2	4.7
950M029B	8	L	79.4	84.5	9.5
		LT	76.8	84.6	9.9
		ST	69.6	82.2	4.5
950M020F	8	L	79.1	85.7	10.6
		LT	71.8	81.2	10.1
		ST	67.2	80.5	7.6
950M029B	9	L	77.4	81.1	9.0
		ST	71.8	84.1	4.8
950M020F	9	L	73.8	77.8	9.3
		ST	71.9	83.9	4.7

Table IV. Mechanical Properties for Aged Alloy 2195

Lot No. (3% stretch)	Aging	YS (ksi)	UTS (ksi)	%El	K_{IC} at a/2 (LN$_2$) (ksi$\sqrt{in}$)	K_{IC} at a/2 (ambient) (ksi$\sqrt{in}$)
950M029B	TS Aging #24	76.8	84.6	9.9	33.7	30.4
	Conventional	74.0	83.1	7.0	25.4	30.0
950M020F	TS Aging #24	71.8	81.3	10.0	37.6	34.8
	Conventional	76.1	83.4	8.0	34.9	32.9

Discussion

A special aging treatment was designed to obtain a desirable microstructure for Alloy 2195, resulting in much improved CFT. The major factors that affect CFT are matrix and subgrain boundary precipitates, especially T_1 at subgrain boundaries. In Al-Li alloys, improved CFT has been correlated to such factors as solidification of low-melting point impurities [3], reduced strain localization in closer and more widely spaced slip bands [4], increased homogeneity of plastic deformation from increased strain-hardening capacity [5], and delaminating toughening on fracture surfaces [6,7]. Yet these mechanisms do not account for the improvements observed here, because alloy chemistry, thermomechanical processing, grain size, and YS do not exhibit distinguishable differences among specimens made from the same lot.

If preferential T_1 precipitation can be prevented at subgrain boundaries, this study strongly suggests that Alloy 2195 has inherently higher fracture toughness at cryogenic than ambient temperature. Subgrain boundary T_1 precipitation is probably the most important factor influencing CFT for Alloy 2195. Since mechanical properties could be considerably impacted by any change in subgrain boundary microstructure, excessive subgrain boundary precipitation should be avoided in alloys intended for use in cryogenic environments.

Compared to conventionally aged specimens, TS-aged specimens have a significantly different microstructure and yet possess much higher CFT with nearly the same YS. Similar YS levels were observed, which can be qualitatively correlated to microstructural characteristics (e.g., type, size, distribution, and density of strengthening phases T_1 and θ''). Initial holding at low temperature (270 °F) increases the number of precipitate embryos. Subsequent aging at 280 °F enables precipitate particles to coarsen slowly without dissolving, increasing the total number of precipitates. Additional strengthening is provided by the much higher number of θ' and θ'' precipitates in TS-aged materials, making their YS comparable to that of conventionally aged materials. As aging continues, T_1 will eventually nucleate at subgrain boundaries and start to grow. However, this treatment allows matrix T_1 to precipitate and grow before the subgrain boundary T_1 does. Early coarsening of matrix T_1 greatly reduces the concentration of matrix Cu and Li, hindering the growth of subgrain boundary T_1 in a diluted Al-Cu-Li solid solution.

CONCLUSIONS

(1) A two-step (TS) aging treatment (270 °F/20 hr + 280 °F/ 40 hr) was developed which can achieve the same YS levels as those produced by conventional aging, while providing much improved CFT.

(2) Cryogenic properties were improved by controlling the size and location of T_1 precipitate. However, TS aging reduced the length of time that the materials were exposed to high temperatures, thus constraining T_1 nucleation and growth at subgrain boundaries and permitting the material to achieve much improved CFT.

(3) During TS aging, high tensile YS was achieved by promoting T_1 and θ'' nucleation in the matrix. The total density of T_1 and θ'' was higher than in conventionally aged materials.

REFERENCES

1. P.S. Chen and W.P. Stanton, NASA Technical Memorandum 108524 (1996).

2. P.S. Chen, A.K. Kuruvilla, T.W. Malone, and W.P. Stanton, J. Mater. Eng. & Per. 7, 682 (1998).

3. D. Webster, Metall. Trans. 18a, 2181 (1987).

4. K.V. Jata and E.A. Starke, Jr., Script. Metall. 22, 1553 (1988).

5. J. Glazer, S.L. Verzasconi, R.R. Sawtell, and J.W. Morris, Jr., Metall. Trans. 18a, 1695 (1987).

6. Y.B. Xu, L. Wang, Y. Zhang, Z.Z. Wang, and Q.Z. Hu, Metall. Trans. 22a, 723 (1991).

7. K.T.V. Rao, W. Yu, and R.O. Ritchie, Metall. Trans. 20a, 485 (1989).

DEVELOPMENT OF METALLIC CLOSED CELLULAR MATERIALS CONTAINING POLYMERS

S. KISHIMOTO, N. SHINYA
5th Research Group, National Research Institute for Metals
Tsukuba, Ibaraki, 305-0047 Japan, kishimot@nrim.go.jp

ABSTRACT

A new material for structures in space, which have a high energy absorbability has been developed using a powder particle assembling technique. Powder particles of polystyrene coated with nickel-phosphorus alloy layers using electroless plating were sintered at high temperature. A metallic closed cellular material containing polystyrene was then constructed.

The mechanical and ultrasonic properties of this material were measured at both room and high temperatures. The compressive tests of this material show a low Young's modulus and high energy absorption. Ultrasonic measurement shows that the attenuation coefficient of this cellular material is very large and would change due to increasing temperature. These results indicate that this metallic closed cellular material can be used for the space applications.

INTRODUCTION

The materials used in the structures in space are required to have many functions, which include light weight, high strength, high stiffness etc. Particularly, passive and active damping functions are becoming increasingly important in terms of vibration control and energy absorbability that has been required to protect structures from the impact of space dust.

Recently, cellular materials are receiving renewed attention as structural and functional materials. Cellular materials have unique thermal, acoustic and energy absorbing properties that can be combined with their structural efficiency[1]. Therefore, many kinds of cellular materials have been tested as damping and energy absorbing materials. However, most of the cellular solids are open cellular polymers, metal foams and woods. Though the closed cellular materials are thought to have many properties and applications for use in extraterrestrial space, there is a lack of technique to produce such fine closed cellular materials except for gas forming[2,3] or two dimensional honeycomb structures[1].

In this study, a new method to produce a metallic closed cellular material containing polystyrene has been developed. In addition, the mechanical and ultrasonic properties of this material have been measured, and the utilities of this material for the structures in space are discussed.

CONCEPTUAL PROCESS

The schematic process flow diagram of the metallic closed cellular material production is shown in Fig. 1. The process is as follows: 1) Powdered polymer particles are coated with a metal layer using electroless plating. 2) The powder particles are pressed into pellet compacts by cold isostatic pressing. 3) After sintering at high temperature in a vacuum, the closed cellular material is produced.

Mat. Res. Soc. Symp. Proc. Vol. 551 ©1999 Materials Research Society

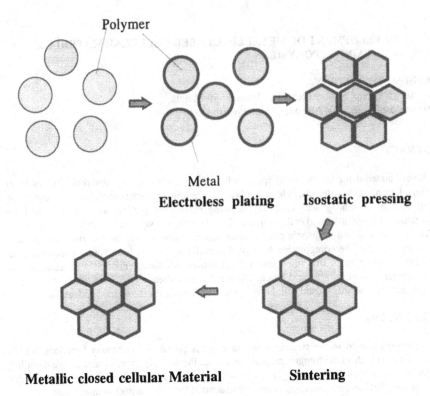

Polymer

Metal

Electroless plating   **Isostatic pressing**

Metallic closed cellular Material   **Sintering**

Figure 1. Flow diagram of metallic closed cellular material fabricating process.

EXPERIMENTS

Preparing the metallic closed cellular material

A thermal plastic polymer, polystyrene particles of 10 μm diameter (Japan Synthetic Rubber Co., Ltd.) were selected for this study. These polystyrene particles were coated with a 0.46 μm thick nickel-phosphorus alloy layer using electroless plating. These powdered polystyrene particles were pressed into pellets (green compacts) of about 8 mm and 16mm diameters and about 8 mm long by isostatic pressing. After this, these green compacts were sintered for 1 h at 800°C in vacuum.

Characterization

The microstructure of the green compacts before sintering and the cross-sections after sintering were observed using scanning electron microscope (SEM). To observe the cross-section of this material, the specimen was cut and the cross-section surface was polished using

292

emery paper (#600) and then 0.05μm Al_2O_3 powders. To measure the mechanical properties, compressive tests were performed at room temperature. In addition, ultrasonic measurements were carried out to estimate the attenuation coefficient of this material. The measurement was carried out with a 6.4 mm diameter probe generating a longitudinal wave of 10MHz at temperatures from room temperature to 100 °C.

RESULTS

Microstructure observation

Figure 2 shows an SEM image of the green compact after cold isostatic pressing. The surface of the polystyrene particles coated with the nickel-phosphorus alloy exhibits facets. An SEM image of the cross-section of this material after sintering at 800°C is shown in Fig. 3. In this figure, the cell walls of nickel-phosphorus alloy are observed as bright parts and the material inside the cell walls is observed as the darker parts. From the polishing method which is mentioned above, it was determined that the material on the inside cell walls is polystyrene.

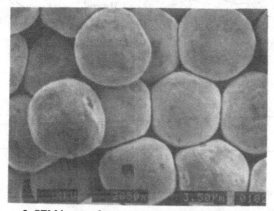

Figure 2. SEM image of green compact after isostatic pressing.

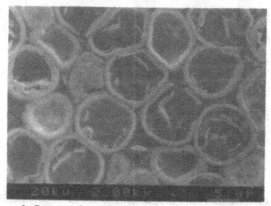

Figure 3. Cross-section of a metallic closed cellular material.

Compressive test

Compressive tests were carried out at room temperature. A typical example of the compressive test results is shown in Fig. 4. The stress-strain curve shows a linear elastic region, a long plateau where the stress gradually increases and a wavy region where the stress suddenly decreases and increases. Young's modulus was measured using the linear elastic region for each specimen, and the relationship between the Young's modulus and density of this material is shown in Fig. 5. The change in Young's modulus depended on the temperature.

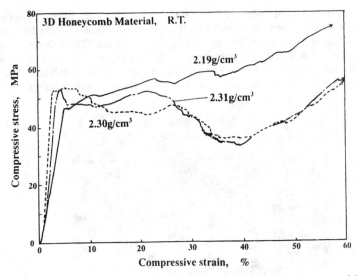

Figure 4. Compressive stress-strain curve for metallic closed cellular material.

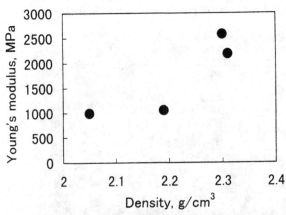

Figure 5. The relationship between Young's modulus and density of metallic closed cellular material

Ultrasonic measurement

Figure 6 shows the ultrasonic attenuation measurement results of this closed cellular material at temperatures from room temperature to 100°C. The attenuation coefficient (about 3.8 - 4.8 dB/cm) is larger than that of metallic materials, but smaller than that of polystyrene (15.2dB/cm). In addition, as the temperature increases, the ultrasonic coefficient of this material gradually increases.

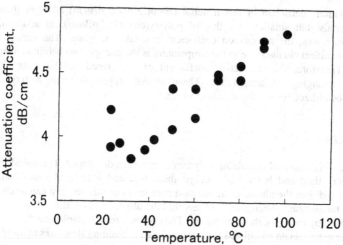

Figure 6. Relationship between ultrasonic attenuation coefficient of metallic closed cellular material and temperature.

DISCUSSION

Metallic closed cellular material

A metallic closed cellular material has been developed in this study. The density of this material is from 2.05 to 2.31g/cm³ which is fewer than that of an aluminum alloy. As Fig. 3 (a) shows, cell walls of a nickel-phosphorus alloy are observed as bright parts and the material inside the cell walls is observed as darker parts. The amount of emitted secondary electrons per a primary electron of the polystyrene is smaller than that of metals. During polishing, only 0.05μm Al_2O_3 powders were used. Therefore, the material of the inside cell walls is thought to be polystyrene. These results indicate that this metallic closed cellular material including the polymers can be produced by using this technique.

Energy absorption

As shown in Fig. 4, the stress-strain curve has a linear elastic region, a long plateau region and a wavy region. After the linear elastic region, cracks occur in the direction parallel to the stress axis.

It is postulated that the fracture initiates from a defect in this material. Therefore, if the metallic closed cellular material has only a few defects, the plateau area of the stress-strain curve will continue during the compressive test. As the presence of the plateau in the compressive stress-strain curve is responsible for the high energy absorption, this metallic closed cellular material seems to have a high energy absorbability.

Attenuation coefficient

The attenuation coefficient of this material (about 3.8-4.8 dB/cm) is larger than that of metallic materials, but smaller than that of polystyrene (15.2dB/cm). In addition, as the temperature increases, the attenuation coefficient gradually increases. The reason why the attenuation coefficient changes due to the temperature is the energy absorbability change of the polystyrene. Therefore, the attenuation coefficient of the closed cellular material can be controlled by changing the temperature. These results suggest that this material has the possibility to be utilized for passive damping materials.

CONCLUSION

A closed cellular material containing polystyrene has been developed. This metallic cellular material is very light and has a high energy absorption and a large ultrasonic attenuation coefficient. In addition, the ultrasonic attenuation coefficient of this material can be changed by changing the temperature. Therefore, it seems that the attenuation coefficient of this material can be controlled by changing the temperature. The obtained results emphasize that the metallic closed cellular material can be utilized for materials for the building structures in outer space.

REFERENCE

1. L.J. Gibson and M.F. Ashby, *Cellular solids, structure and properties*, Pergamon press, Oxford, 1988, pp.1-41.

2. J.T. Beals and M.S. Thompson, J. Mater. Sci., 19, p. 871 (1984).

3. N. Chan and K.E. Evans, J. Mater. Sci., 32, p. 5725 (1997).

MULTILAYER COATINGS FOR FOCUSSING HARD X-RAY TELESCOPES

A. Ivan [1,2], R. Bruni[1], K. Byun[3], J. Everett[1], P. Gorenstein[1], S. Romaine[1]
[1]Harvard-Smithsonian Center for Astrophysics, Cambridge, MA
[2]MIT, Dept. of Nuclear Engr., Cambridge, MA
[3]Harvard University, Cambridge, MA

ABSTRACT

Several multilayer test coatings for hard X-ray telescopes were fabricated using DC magnetron sputtering. The process parameters were selected from a series of trials of single layer depositions. The samples were characterized using X-ray specular reflectivity scans, AFM, and cross-sectional TEM. Additional measurements (stylus profilometry, RBS, and Auger analysis) were used in the optimization of the deposition rate and of the thin film properties (density, composition, surface/interface microroughness). The X-ray reflectivity scans showed that the combinations of reflector and spacer materials tested so far (W/Si and W/C) are suited for graded d-spacing multilayer coatings that present a constant reflectivity bandpass up to 70 keV.

INTRODUCTION

Scientific objectives for hard X-ray astronomy require the development of new, focusing telescope optics for the energy band above 10 keV. The proposed next X-ray mission, Constellation X, plans to use a hard X-ray telescope system with a sensitivity of 20-100 times over that of current non-focusing instruments. The Multilayer Facility at the Harvard-Smithsonian Center for Astrophysics is currently engaged in a feasibility study based on a new design for the coating of the X-ray optics. Our objective is the production of graded d-spacing multilayer coatings on the inner surface of figured replicated substrates. The technology for coating the inside of the optic shells has been proven in the fabrication of the mirror system for the AXAF observatory, but the new requirement for a reflectivity bandpass above 10 keV is increasing the complexity of the process. As shown in Figure 1, a single layer coating of Ir (300 Å thick, similar to AXAF) fails to reflect above 40 keV for a typical grazing incidence angle of 0.15°. A constant d-spacing multilayer structure of W/Si has a peak reflectivity at some fixed energy values where the Bragg condition is met, but has also very deep minima, which preclude its use as a constant bandwidth system. The third curve shows the calculated reflectivity vs. energy for a graded d-spacing W/Si multilayer. The simulation was based on the structure data obtained from the fit of the specular reflectivity scan at 8.05 keV of sample WSi42 (described later). The conclusion is that only the graded d-spacing multilayer design [1] can offer a useful reflectivity over a continuous range in the hard X-ray region.

Two deposition chambers have been built at our facility for testing runs and depositions of flight grade prototype mirrors. Both chambers use DC magnetron sputtering in argon atmosphere as a coating process, but they serve different purposes and have different geometrical configurations. One chamber, dedicated for test coatings, was used to produce the samples presented in this paper and will be described in the next section. The other chamber has two 26-in. long linear cathodes mounted back-to-back along a vertical axis. The cylindrical substrate shells can be rotated around this axis while the cathodes side-sputter onto the inside of the shells. This chamber will be used for the coating of protoype mirror shells and the work in progress is presented in a separate paper [2]. Because in the first stage of the feasibility study it is essential to identify candidate materials and to determine optimal deposition conditions, a large number of test coatings have to be performed first on inexpensive, flat substrates. Once good quality multilayers are produced in the test chamber, the process parameters are adapted for the runs in the larger cylindrical coating chamber. The following will describe the work completed in the test chamber for the W/C and W/Si systems.

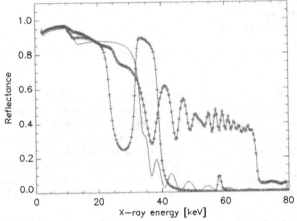

Fig.1: Reflectivity vs. energy at a grazing incidence angle θ = 0.15° for three types of coatings: single-layer (continuous line, no symbol), constant d-spacing multilayer (+), and graded d-spacing multilayer (◊). The single layer is a 300-Å thick Ir coating (similar to AXAF). The constant d-spacing multilayer is a W/Si coating with N=200, and a spacing d=90 Å. The graded d-spacing multilayer is the sample WSi42 described in the next section.

EXPERIMENTAL PROCEDURE

The deposition chamber for the test coatings is a stainless steel high-vacuum chamber (14 in. high and 24 in. diameter) with a typical base pressure of $2x10^{-7}$ T. The chamber is equipped with two 3-inch diameter magnetron sputtering cathodes and a central shaft-motor assembly that can rotate a platen. The configuration is shown in Figure 2. The distance from the cathode targets to the platen can be varied between 1 and 18 cm, but typically is 7 cm. The flat substrates are supported by the platen and can be moved in an oscillatory sweeping motion under the cathodes. It was determined that a sweeping motion with 2 RPM and 8° angular amplitude improved the thin film thickness nonuniformity and surface microroughness. The test substrates are 2-inch <111> silicon wafers and 1-inch superpolished fused silica substrates.

The first step of the program was to determine deposition rates and surface microroughness for single layer coatings. A series of test depositions on Si were used to vary the process parameters, such as: argon backpressure, cathode current, cathode gap (target to ground shield), and target to substrate distance. Based on profilometry measurements of step heights (samples thickness range: 300-800 Å) and on surface microroughness measurements (AFM in tapping mode), optimal process parameters were selected for W, Si, and C, and are presented in Table I.

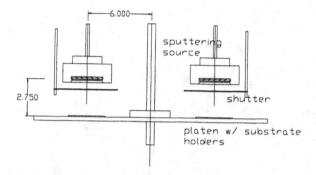

Fig.2: Side view of the sputtering cathodes setup in the test deposition chamber. The dimensions shown are in inches. The motion of the substrates and shutters is computer controlled.

Table I: Optimal deposition parameters in the test chamber for single layers of W, Si, and C. The argon pressure was 2.5 mT and the target to substrate distance was 7 cm. The cathode power supplies were run in current regulated mode. The cathode gap was 2.3 mm for all targets.

Target material	Cathode curent [mA]	Deposition rate [Å/s]	Surface roughness from AFM σ_{rms} [Å]
W	30	0.39 (±0.03)	1.6
C	90	0.15 (±0.01)	1.8
Si	75	1.00 (±0.06)	1.9

The deposition rates are low, but compared to higher attainable rates, they present some advantage for the quality of the thin films: low microroughness, stability of the deposition process over a long period, and little surface contamination with particulates.

After selecting the process parameters, the next step is to design and fabricate the multilayers and test their X-ray reflectivity at low grazing angles. The design made use of the IMD software package [3] and was based on a power law formula [4] for the thickness of each layer in the graded d-spacing multilayer stack:

$$z(i) = \frac{a}{(b+i)^{0.25}} \qquad (1)$$

where a and b are parameters to be determined, and i is the layer index number (i=1 for the topmost layer, N for the bottom layer). Based on the known deposition rates, the necessary sputtering time could be calculated for each layer.

As an intermediate step to check the quality of the deposition, we ran a series of constant d-spacing coating tests and checked the calibration and reproducibility of the sample structure by performing specular reflectivity scans at 8.05 keV. As an example, Figure 3 shows the reflectivity scans for samples WSi34 and WSi35 that were deposited in two different runs using the same set of process parameters. Table II compares the design and fit data from the two scans.

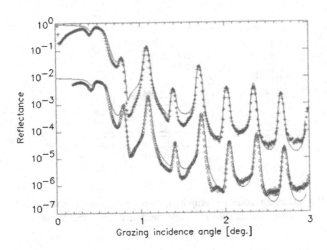

Fig.3: Measured specular reflectivity scans for two samples with identical deposition parameters: WSi34 (◊) and WSi35 (+). The reflectivity curve for WSi34 was shifted down by two decades to allow for comparison. Shown in continuous line are the reflectivity curves corresponding to the best fit (see Table II). A standard specular reflectivity scan setup was used: a collimated beam of 8.048 keV (Cu K_α) X-rays incident at a grazing angle θ on the sample was detected by a collimated NaI(Tl) detector positioned at 2θ relative to the beam direction.

Table II: Design and fit structure data for samples WSi34 and WSi35. Both samples had N = 20 bilayers of W/Si on Si wafer. The first two columns show the thickness for the W and Si layers, respectively. The last two columns show the effective interface roughness/diffuseness for the two types of interfaces: W/Si and Si/W. The multilayer spacing d = d_{Si} + d_{Si} was constant for these samples.

	d_W [Å]	d_{Si} [Å]	$\sigma_{W/Si}$ [Å]	$\sigma_{Si/W}$ [Å]
Design	67.5	67.5	-	-
WSi34 (fit)	69	63	1.1	4.9
WSi35 (fit)	71	63	2.0	5.5

The relative error from one sample to another is better than 3% (which includes the undetermined experimental errors due to sample curvature and misalignment of the sample with the beam), indicating good reproducibility. Only one sample (WSi34) was measured for surface microroughness by AFM, with an average $\sigma_{Surface}$= 2 Å in agreement to the reflectivity fit value (1.5 Å).

After fabricating a series of constant d-spacing multilayers for both W/C and W/Si systems, we proceeded with graded-d designs based on eq. (1). The number of bilayers used ranged from 40 to 200. The substrates are typically full 2-in. Si wafers, but a small number of runs were repeated on 1-in. diameter superpolished fused silica substrates. Table III presents some significant results from the fit of the reflectivity scans performed on W/Si and W/C multilayers. Shown in Figure 4 is the plot of measured reflectivity vs. grazing angle, together with the fit curve, for sample WSi42 (N=200).

Table III: Design and fit values for the d-spacing ($d_W + d_{Si\ or\ C}$) and the γ ratio (d_W/d) for a series of graded d-spacing multilayers. The values for d and γ vary with the depth in the multilayer due to the grading formula in eq. (1). One exception is the design for CW30 and CW31 that had a constant γ value. All the experimental data were from specular scans at 8.048 keV.

Sample	N	d_{top}-d_{bottom} design	d_{top}-d_{bottom} fit	γ_{top}-γ_{bottom} design	γ_{top}-γ_{bottom} fit
CW30	40	161-36 Å	160-36 Å	0.444	0.368
CW31 (repeat of CW30)	40	161-36 Å	180-36 Å	0.444	0.366
CW32	50	137-30 Å	126-28 Å	0.617-0.500	0.340-0.439
WSi39	100	141-30 Å	139-26 Å	0.568-0.500	0.594-0.523
WSi42	200	218-28 Å	203-25 Å	0.546-0.536	0.555-0.640

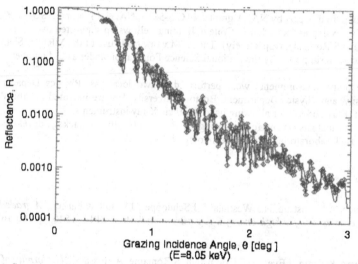

Fig.4: Specular reflectivity scan for a graded d-spacing multilayer sample (WSi42) with N=200 bilayers. Details of the fit results are in Table III.

RESULTS AND DISCUSSION

The fit results presented in Table III show a satisfactory agreement between the design and the fit values for the d spacing and the γ ratio (d_W/d). However, as illustrated in Fig.4, it is difficult for the fit to reproduce all the features of the experimental reflectivity curve. This is due to the fact that the 2N interfaces have a distribution of roughness/diffuseness parameters (and thickness non-uniformities) that cannot be totally replaced by a couple of effective roughness values. The complexity of the structure increases when N and/or the gradient of the d-spacing distribution increase. Therefore, graded-d samples with N>50 are not expected to be fitted very precisely using only a small number of fit parameters.

The reflectivity scans, peformed only at Cu K_α energy, are nevertheless very useful because they provide quick and direct evaluation of the technological goal to obtain high reflectivity at grazing incidence angles above the critical value. The fit data can be used to simulate the reflectivity at higher energies and indicate whether higher energy scans should be performed or not on a particular sample.

CONCLUSIONS

The W/Si and W/C graded-d systems studied offer a reference for a study of similar coatings in the large chamber. In parallel with these depositions, future work in the test chamber will continue with Ni/C, Pt/C, and other target combinations. Further research will include residual stress studies in the multilayers produced.

ACKNOWLEDGEMENTS

This work was supported in part by NASA grants NAG 5-5095, NAG8-1194 and NSG-5138. A Smithsonian Fellowship and a Radcliffe College Bunting Fellowship supported two of the authors (A.Ivan and S.Romaine, respectively). The AFM work made use of the MRSEC Shared Experimental Facilities supported by the National Science Foundation under award number DMR94-00334.
The X-ray reflectivity measurements were performed at two locations: Physics Department, Harvard University, and Physics Department, Boston University. We are indebted to Prof. Peter Pershan and Prof. Karl Ludwig for allowing us to use their X-ray instruments.
The reflectivity data analysis was performed using the IMD v.4.0 software package developed by David L. Windt, Bell Laboratories, Lucent Technologies.

REFERENCES

[1] F.E.Christensen, A.Hornstrup, N.J.Westgaard, J.Schnopper, J.Wood, K.Parker, "*A graded d-spacing multilayer telescope for high-energy x-ray astronomy*", Proc. SPIE, 1546, pp.160-167, 1992

[2] R.Bruni, A.Ivan, K.Byun, J.Everett, P.Gorenstein, S.Romaine, A.Hussain, "*Uniformity of coatings for cylindrical substrates for hard x-ray telescopes*", these proceedings

[3] D.L.Windt, "IMD version 4.0", http:// www.bell-labs.com/user/windt/idl/imd/index.html

[4] K.D.Joensen, F.E.Christensen, H.W. Schnopper, P.Gorenstein, J.Susini, P.Høghøj, R.Hustache, J.Wood, K.Parker, "Medium-sized grazing incidence high-energy x-ray telescopes employing continously graded multilayers", Proc. SPIE, 1736, pp.239-248, 1993

Carbonization and/or Nitridation of Titanium and Zirconium for Increasing Melting Temperature

M.Nunogaki*,Y. Susuki**, K.Kitahama*, Y. Nakata*, F. Hori***, R. Oshima*** and S. Emura*
*ISIR, Osaka University, Ibaraki,Osaka 567-0047, Japan, nunogaki@sanken.osaka-u.ac.jp
**Department of Pysics, Osaka Kyoiku University, Kashiwara, Osaka 582-26, Japan.
***RIAST, Osaka Prefecture University, Sakai, Osaka 599-8570, Japan

ABSTRACT

Samples of Ti and Zr metals have been carbonized or nitrided at high temperature by means of reactive plasma processing. Thickness and hardness of modified layers increased with the processing temperature and time in an experimental range below 1500°C for less 5h. The maximum hardness of TiC-layer modified at 1300°C for 3h was about 5000Hv(kg/mm^2) and the maximum thickness of ZrC-layer formed at 1400°C for 5h reached about 100μm. A TiC-layer was analyzed with EPMA, XRD, EXAFS and PAS(Positron Annihilation Spectroscopy). It was confirmed that the sub-surface layer of Ti metal was the mixture of TiC ceramics and Ti metal, and that the thickness of TiC with NaCl type structure was very large.

INTRODUCTION

Bulk ceramics such as TiC, NbC, ZrC, TiC, VC, ZrN, TiN, AlN, ZrO$_2$, etc. have been manufactured by sintering. These ceramics are superior materials in hardness, erosion and heat resistance. Ceramics formed by sintering are too hard to finish mechanically with correct dimensions, and ceramics produced by coating are liable to peal off due to thermal and mechanical stress at the interface between ceramics and a base material. We have demonstrated a new concept, by means of plasma processing, to transform surfaces of some metals to ceramics having the properties of functionally gradient material on abrasion resistance at the interface and of toughness at the modified layer.

The new concept is put into practice by the following two processes; "plasma processing "as the main processing and "ion implantation " as the preparative processing. The plasma processing is performed to induce one or two reactions of the followings; carbonizing, nitriding and oxidizing. The ion implantation and occasionally, a relativistic electron beam irradiation are used when it is effective on assisting the main processing. For instance, the pre-implantation of Mo- or Ti-ion to pure Al oxidized by air enabled nitriding and lowered the activation energy on nitrogen diffusion [1]. As another example, in the case of nitriding iron, the variation of fluence of Ni-ion pre-implantation changed controllably the thicknesses of ε - and γ'-phase nitrogen compound layers in the modified surface [2]. The ceramics described in this paper were produced simply by reactive plasma processing at high processing temperature without any particle implantation.

EXPERIMENT

Apparatus

Generating a reactive plasma of homogeneous density around at samples and achieving samples at high temperatures were regarded as of major importance on constructing the apparatus. The apparatus was composed of two chambers; one was the plasma chamber for a PIG (Penning Ion Gage) discharge under the gas pressure of about $1\sim4\times10^{-2}$Pa, and another was the processing chamber to be filled with plasma diffused from PIG plasma as shown in Fig. 1. The processing temperature of the sample placed on the

303

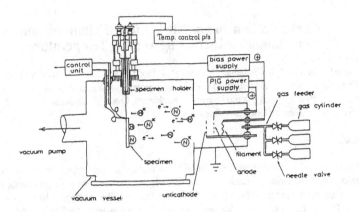

Fig.1. Conceptual drawing of an apparatus for reactive plasma processing at high temperature.

sample holder was controllably kept at high temperature above about 900°C. The dc bias of 1 kV was applied between the diffused plasma and the sample (as the cathode) to accelerate ions in diffused plasma towards the sample. The apparatus is different from the usual ion nitriding devices in the following three points: (1) Sample temperature can be elevated over 1000°C, independently on plasma production. (2) A powerful MW (micro wave) plasma is applicable for production of reactive plasma for scaling up. (3) A homogeneity of modified layer is obtainable even if the geometrical shape of the workpiece is complicated, as the ion sheath is extremely thin. In carbonizing treatment with this apparatus, sooting of samples which had often taken place in the usual ion nitriding devices, was not observed.

Procedure

Test samples of about $5 \times 8\,mm^2$ in area and 1 mm in thickness were polished mechanically with emery paper. The polished samples were annealed for 3h under the residual gas pressure of $\sim 10^{-3}Pa$ at 350°C. Samples of Ti and Zr were carbonized or nitrided in a temperature range of 700°C to 1500°C for 2 $\sim$ 5 h. Characterization of the modified layer was mainly made on ZrN with Vicker's hardness tester and PAS, and on TiC with EPMA, XRD and EXAFS.

Thickness and Hardness

Figure 2 shows the cross-sectional surfaces of TiC-, ZrN- and ZrC-mixed modified layers formed in the base metals of Ti and Zr, respectively.   The mean thickness / hardness of each modified layer; (1) TiC, (2) ZrN and (3) ZrC were about $10\,\mu m/5000kg/mm^2$ for 3h at 1300°C, $60\,\mu m/3000kg/mm^2$ for 3h and 100 $\mu m/1300kg/mm^2$ for 5h at 1400°C, in order. A glowth of TiC layer seen in the figure (1) from the cross-sectional surface of the base metal of pure Ti was brought about due to the difference of hardness on polishing the cutting surface.

Figure 3 shows the variation of compositions; C and Ti at the cross-sectional surface of TiC- mixed layer shown in Fig. 2(1). Carbon content obtained with EPMA decreased gradually in a microscopic scale around at the interface, whose distance from the surface was about $10\,\mu m$ with close agreement to one

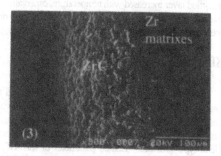

(1)
TiC ceramics transformed from Ti metal by plasma processing at 1300°C. Thickness; 10 μm, Hardness: 5000kg/mm², Processing time; 3h.

(2)
ZrN ceramics transformed from Zr metal by plasma processing at 1400°C. Thickness; 60 μm, Hardness: 3000kg/mm², Processing time; 3h.

(3)
ZrC ceramics transformed from Zr metal by plasma processing at 1400°C. Thickness; 100 μm, Hardness:1300kg/mm², Processing time; 5h.

Fig.2. SEM pictures of cutting surfaces of TiC-, ZrN- and ZrC-mixed layers.

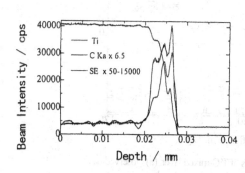

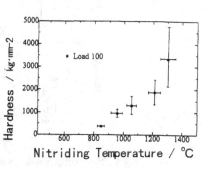

Fig.3. Component analysis with EPMA of cross-sectional TiC-mixed layers formed at a sub-surface of Ti.

Fig.4　Vicker's hardness of ZrN surface vs processing temperatures

measured with SEM. This modified layer with the cermics graded layer can be recognized to be a kind of functionally gradient material on the function of thermal and mechanical stress.

Figure 4 shows Vicker's hardness of ZrN-layer against the processing temperature. Hardness of modified layer increased with temperature, especially, over about 900°C, which was about half of the melting point of pure Zr. It was ascertained that the diffusion coefficients of C and N in Ti and Zr metals were extraordinarily large over at about 900°C as suggested in reference [3].

XRD

The TiC-mixed layer modified at 1300°C was analyzed with XRD as shown in Fig.5(a). Rietveld analysis of X-ray powder diffraction data was made on a program RIETAN-94 [4]. Simulations of TiC with space group Fm3m and metallic Ti with space group P6₃/mmc are shown in Figs.5(b) & (c), respectively. From these data, it was noted that majority of the TiC-mixed layer was TiC compound with NaCl type structure. Lattice parameter of this TiC was a =4.289(1) Å, which value was about 0.9% smaller than reported value on bulk a=4.3274 Å [5]. This would be explained by size effect of the TiC grains. When the grain size was small, surface tension was estimated to increase. That is, lattice size reduced so much.

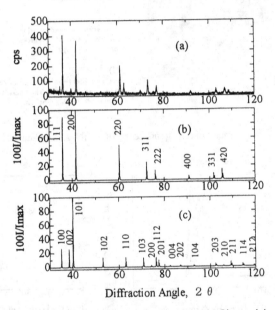

Fig.5　X-ray diffraction diagrams of TiC-mixed layer. (a) Observed data. (b) TiC; simulated. (c) Ti-substrate; simulated.

From intensity of TiC relative to that of metallic Ti, it was noted that thickness of the TiC-mixed layer was considerably large. Diffraction intensity of 110 Mirror index was extraordinarily larger than those of other indices, which indicated that the grains of metallic Ti were preferably oriented along the <110> direction. Not assignable peak was observed at around 2θ =35°, which needed further investigation.

EXAFS

The EXAFS spectrum of TiC-mixed layer in sub-surface layer of Ti metal formed by the reactive plasma processing at 1300°C was observed at room temperature at Beamline BL01B1 in Spring-8 (Harima, Hyogo Prefecture) with a fluorescence mode. In Fig.6, a spectra (a) shows the excitation spectra around Ti K-edge of TiC, and the EXAFS oscillation is shown in the spectra (b) which was extracted from the spectra (a), and the diagram (c) gives the radial distributions around the Ti atom, which is obtained by Fourier transformation of the EXAFS oscillation. Here, its phase correction was not made. A peak around 1.7Å corresponded the distance of Ti-C, and a strong peak at 2.6Å gives the distance of Ti-Ti . Another peaks were assigned as presented in Fig.6 (c). The existences of strong EXAFS oscillation observed over $k = 16$Å^{-1} suggested that this artificial TiC had large hardness. It would be originated from a large surface tension due to the smaller lattice constant than one of bulk TiC as pointed out by XRD analysis.

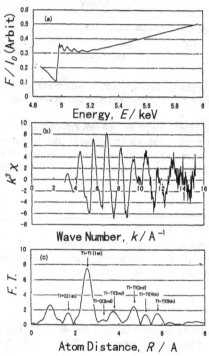

Fig. 6 EXAFS spectrum of TiC-mixed layer. (a) Excitation spectrum of TiC-mixed layer around titanium K--edge. (b) EXAFS oscillation extracted from spectrum (a). (c)Radial distributions around Ti atom, which was obtained by Fourier transformation of (b).

PAS

By means of PAS with a ^{22}NaCl isotope sandwiched by thin Kapton films, the doppler broadening spectrum of annihilation γ-ray from ZrN-mixed layer and Zr metal was measurement at room temperature. The resolutional γ- energy of Ge detector was about 581eV at 122 keV of ^{57}Co. The spectrum was drawn with a total count of 2.5×10^7.

The S-parameter is defined as the ratio of the central area of gamma peak to the total area of the spectrum, the difference of peak hight corresponds directly to that of electron density. In this case, there

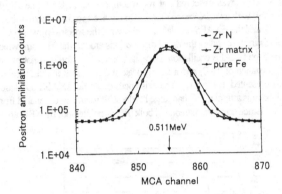

Fig. 7. Positron annihilation doppler broadening spectrum. S-parameter of ZrN-mixed layer was larger than that of Zr metal. Free space in ZrN-mixed layer was created by processing.

exist the differences in the doppler broadening profiles between ZrN-mixed layer and Zr metal. The result shows that positron annihilation occurred with smaller amount of core electrons in ZrN- layer than in Zr. metal.   This means that free volumes such as incoherent boundary between ZrN-layer and Zr metal has created, or that the density of valence electron in the modified layer has changed due to the processing.

CONCLUSION

A kind of new type ceramics mixed with TiC, ZrN or ZrC and matrixes of respective native metal; Ti and Zr, respectively, has been formed in a sub-surface of each native metal by means of the reactive plasma processing. Further investigation will be continued for establishing the concept on other metals and alloys.

ACKNOWLEDGMENTS

This work was supported by the Scientific Grant of the Ministry of Education in Japan. EXAFS spectrum was detected as an attempt under the Proposal 1998A0219-CX in Spring-8.

REFERENCES

1) M.Nunogaki, H.Suezawa and K.Miyazaki, Vacuum, **39**, 2-4, p.281(1989).
2) M.Nunogaki, H.Suezawa, S.Nishijima and T. Okada in *Intn'l Conf. on Evolution in Beam Applications*, (JAERI & IAEA,1991),pp.108-111.
3) B. Edenhofer, *Heat Treatment of Metals*, *1*(1974),pp.24.
4) F.Izumi, *The Rietveld Method*, (R.A. Young. Oxford University Press,. Oxford,1993),Cap.13.
5) *Powder Diffraction file #32-1383*,(International Center for Diffraction Data, 12, Campus Blvd. Newton. Sq. Pen. 19073-3273 USA).

AUTHOR INDEX

SUBJECT INDEX

Printed in the United States
By Bookmasters